ARCHAEOLINGUA

Central European Series 2

Woodland and Forests in Medieval Hungary

Péter Szabó

BAR International Series 1348

2005

Published in 2016 by
BAR Publishing, Oxford

BAR International Series 1348

Archaeolingua Central European Series 2
Woodland and Forests in Medieval Hungary

ISBN 978 1 84171 694 7

COVER IMAGE *by Helena Lukášová*

Word processing, desktop editing and layout: András Kardos
Copy editor: Alice M. Choyke

BAR Publishing is the trading name of British Archaeological Reports (Oxford) Ltd. British Archaeological Reports was first incorporated in 1974 to publish the BAR Series, International and British. In 1992 Hadrian Books Ltd became part of the BAR group. This volume was originally published by Archaeopress in conjunction with British Archaeological Reports (Oxford) Ltd / Hadrian Books Ltd, the Series principal publisher, in 2005. This present volume is published by BAR Publishing, 2016.

Printed in England

BAR
PUBLISHING

BAR titles are available from:

BAR Publishing
122 Banbury Rd, Oxford, OX2 7BP, UK
EMAIL info@barpublishing.com
PHONE +44 (0)1865 310431
FAX +44 (0)1865 316916
www.barpublishing.com

Editorial

This is the second title in the Archaeolingua Central European Series of BAR, a joint publication of BAR and Archaeolingua. The Series started in 2002 with the two-volume proceedings of the 31st International Symposium on Archaeometry, held in Budapest in 1998. Although this work focused on a specialised field of study and therefore targeted a relatively small group of archaeologists, we wish to emphasize that the aim of the series has been to create a multidisciplinary forum. The present book represents our efforts to fulfil this aim.

It is also our intention to publish the works of young researchers, and, in accordance with the title of the series, to support research on topics that are in the forefront of Central European archaeological studies. We are pleased to acknowledge that this work, which was originally prepared as a doctoral dissertation, does not simply form part of a current trend in research, but also covers a period of high importance in the environmental history of the Carpathian Basin. Moreover, it is also a significant contribution to European environmental history in general, which integrates the BAR Archaeolingua Central European Series into a wider academic context.

This book is part of a long-standing co-operation between Archaeolingua and the Department of Medieval Studies at the Central European University, Budapest, where the author's dissertation was defended. We have already jointly published a collection of essays on environmental history, entitled *People and Nature in Historical Perspective*. As our plan is to publish two volumes every year in the Series, we hope that our co-operation with CEU will continue to be successful.

We are proud to present this book to the readers, and wish them an interesting journey through *Woodland and Forests in Medieval Hungary*.

Elisabeth Jerem

Contents

PART I: WOODLAND

PART II: FORESTS

Foreword

by Oliver Rackham

Twenty years ago, when investigating the history of trees and woods in England, I speculated on how the study of cultural landscapes and historical ecology would develop in other countries and other continents. I would not have put Hungary high on the list of those countries. It has had a succession of violent takeovers by outside invaders – by the Magyar themselves, Mongols, Turks, Hapsburgs, and Soviets – each of which destroyed much of the archives and might have erased most of the evidence in the landscape itself. I would have thought it impossible to reconstruct the infrastructure or functioning of the Hungarian countryside more than a century or two back.

In this book Péter Szabó, an ingenious and resourceful young Hungarian scholar, proves me wrong. Hungary has been far more rewarding than I would have expected. The written evidence may be relatively meagre (much of it consisting of boundary perambulations in charters, as in Anglo-Saxon England) but he makes the most of it, to portray convincingly what the pre-Turkish Hungarian landscape looked like and how medieval woodland functioned and was used. In combining this with evidence still visible on the ground, he does not (as many scholars do) treat trees as mere environment: for him trees are wildlife, and he takes due account of how the different species react to people's actions and neglect.

It turns out that there are many similarities between the landscape histories of Hungary and England, despite their very different cultural and political histories and the much greater proportion of woodland in Hungary. There begins to emerge a common tradition of cultural landscapes in north and central Europe, involving such things as coppicing, royal Forests, common or private woodland, and pollarding. This book is an important and unexpected link in establishing that tradition.

Acknowledgements

This book was originally written as a PhD dissertation at the Department of Medieval Studies, Central European University, Budapest. Many people helped me prepare it. First and foremost, I would like to thank my dissertation supervisor, József Laszlovszky, whose continuous support was one of the driving forces behind my work, and without whose encouragement I could not have started in the first place. Richard C. Hoffmann read several drafts of this work and had countless useful observations. He also helped me to establish many contacts all over the world through his network of environmental historians. Pál Sümegi guided me in palaeoenvironmental matters. Zsolt Molnár and János Bölöni offered their kind assistance in the complicated world of modern ecology. Faculty and doctoral students of the CEU Medieval Studies Department also deserve my gratitude. János Bak discussed issues of social history with me, Katalin Szende was a most helpful reader of the first draft, and Béla Zsolt Szakács had the solutions to all my art-historical questions. Edit Sárosi had many valuable comments on the text, and Andrea Kiss was always ready to share her knowledge with me. I would also like to thank the departmental staff (Csilla Dobos, Dorottya Domanovszky, and Annabella Pál).

I cannot be grateful enough to my friend Péter Banyó, who taught me palaeography. Petra Mutlová was kind enough to check my Latin translations. If the English of this work is enjoyable, that is largely the result of the efforts of Alice M. Choyke. I would also like to mention the late Pál Engel, whom I had the fortune to know personally, and who has influenced the way I see medieval Hungarian history more than anybody else. Last, but not least, I would like to thank Oliver Rackham, who taught all of us how to look at trees and woodland in a meaningful way, and who encouraged me to do it my way.

Financially, the completion of the original dissertation was possible due to the doctoral scholarship I was granted at the CEU.

Finally, something personal. I am usually not the type of person who would seek a justification for his subject: one topic is as good as any other. In fact, I find it somewhat annoying when people say: "Yes, but what is this good for?" As an historical ecologist – that is, doing something unusual – I hear this question more often than I would wish to. Recently, however, I happened upon a sentence in G. F. Peterken's *Woodland Conservation*, which is perhaps my answer to the question. Here it is: "Any wild creature going about its business can give the human observer a balancing perspective on his own existence."

List of Illustrations

List of Abbreviations

ÁMTF – Györffy, György. 1963–1998. *Az Árpád-kori Magyarország történeti földrajza* (A historical geography of Hungary in the Árpádian period). 4 vols. Budapest: Akadémiai Kiadó.

AnjOkl – Kristó, Gyula, ed. 1990–. *Anjou-kori Oklevéltár* (Charters from the Angevin period). Several volumes not numbered continuously. Budapest and Szeged: n.p.

AnjOkm – Nagy, Imre and Gyula Tasnádi Nagy, ed. 1878–1920. *Anjou-kori okmánytár* (Charters from the Angevin period). 7 vols. Budapest: Magyar Tudományos Akadémia.

ÁÚO – Wenzel, Gusztáv, ed. 1860–1874. *Árpádkori Új Okmánytár / Codex Diplomaticus Arpadianus Continuatus*. 12 vols. Pest and Budapest: n.p.

BTOE – Kumorovitz, Lajos Bernát, ed. 1987. *Budapest történetének okleveles emlékei* (Charters of the history of Budapest). Vol. 3. Budapest: Budapesti Történeti Múzeum.

CD – Fejér, György, ed. 1829–1844. *Codex Diplomaticus Hungariae Ecclesiasticus ac Civilis*. 11 vols. Buda: n.p.

CDES – Marsina, Richard, ed. 1971–1987. *Codex Diplomaticus et Epistolaris Slovaciae*. 2 vols. Bratislava: Academia Scientiarum Slovaca.

Csánki – Csánki, Dezső and Antal Fekete Nagy, ed. 1890–1941. *Magyarország történelmi földrajza a Hunyadiak korában* (A historical geography of Hungary in the age of the Hunyadi). 5 vols. Budapest: Magyar Tudományos Akadémia.

DAP – Tóth, Melinda, ed. 1975–1978. *Documenta Artis Paulinorum*. 3 vols. Budapest: MTA Művészettörténeti Kutatócsoport.

Df – Magyar Országos Levéltár Diplomatikai Fényképgyűjtemény (Hungarian National Archives Diplomatic Photo Collection).

Dl – Magyar Országos Levéltár Diplomatikai Levéltár (Hungarian National Archives Diplomatic Collection).

DHA – Györffy, György, ed. 1992. *Diplomata Hungariae Antiquissima*. Budapest: Akadémiai Kiadó.

DRMH – Bak, János M., György Bónis, Leslie S. Domonkos, and James Ross Sweeney, ed. 1989–1996. *Decreta Regni Mediaevalis Hungariae*. 3 vols. Idyllwild, California: Charles Schlacks, Jr. Publisher.

FNESz – Kiss, Lajos, ed. 1988. *Földrajzi nevek etimológiai szótára* (An etymological dictionary of geographical names). 2 vols. Budapest: Akadémiai Kiadó.

HO – Nagy, Imre, ed. 1865–1891. *Hazai Okmánytár / Codex Diplomaticus Patrius*. 8 vols. Győr and Budapest: n. p.

HOkl – Nagy, Imre, Farkas Deák, and Gyula Nagy, ed. 1879. *Hazai Oklevéltár 1234–1536* (Hungarian charters 1234–1536). Budapest: Magyar Történelmi Társulat.

Károlyi – Géresi, Kálmán, ed. 1882–1897. *A nagy-károlyi gróf Károlyi család oklevéltára* (Charters of the Károlyi family of Nagy-károly). 5 vols. Budapest: n. p.

KMTL – Kristó, Gyula, ed. 1994. *Korai magyar történeti lexikon (9–14. század)* (Early Hungarian historical lexicon (ninth-fourteenth centuries)). Budapest: Akadémiai Kiadó.

Kolozsmonostor – Jakó, Zsigmond, ed. 1990. *A kolozsmonostori konvent jegyzőkönyvei* (Protocols of the convent of Kolozsmonostor). 2 vols. Budapest: Akadémiai Kiadó.

Lexicon Latinitatis – *Lexicon Latinitatis Medii Aevi Hungariae*. 1992–1999. 5 vols. Budapest: Argumentum.

MEO – Tagányi, Károly, ed. 1896. *Magyar Erdészeti Oklevéltár* (Charters of Hungarian forestry). 3 vols. Budapest: Országos Erdészeti Egyesület.

MES – *Monumenta Ecclesiae Strigoniensis*. 1874–1999. 4 vols. Esztergom: n. p.

MNyTESz – *A magyar nyelv történeti-etimológiai szótára* (A historical-etymological dictionary of the Hungarian language). 1967–1984. 4 vols. Budapest: Akadémiai Kiadó.

MRT – *Magyarország Régészeti Topográfiája* (Archaeological Topography of Hungary). 1966–1998. 10 vols. Budapest: Akadémiai Kiadó.

Oklevélszótár – Szamota, István and Gyula Zolnai, ed. 1902–1906. *Magyar oklevél-szótár* (Hungarian charter dictionary). Budapest: Hornyánszky Viktor.

Podmaniczky – Lukinich, Imre, ed. 1937–1943. *A podmanini Podmaniczky-család oklevéltára* (Charters of the Podmaniczky family of Podmanin). 5 vols. Budapest: Magyar Tudományos Akadémia.

PRT – Erdélyi, László and Pongrác Sörös, ed. 1902–1916. *A pannonhalmi Szent-Benedek-rend története* (A history of the Benedictine order of Pannonhalma). 12 vols. Budapest: Pannonhalmi Szent-Benedek-rend.

RA – Szentpétery, Imre and Iván Borsa, ed. 1923–1987. *Regesta Regum Stirpis Arpadianae Critico-Diplomatica*. 2 vols. Budapest: Akadémiai Kiadó.

RDES – Sedlák, Vincent, ed. 1980–1987. *Regesta diplomatica nec non epistolaria Slovaciae*. 2 vols. Bratislava: Academia Scientiarum Slovaca.

Sopron – Nagy, Imre, ed. 1889–1891. *Sopron vármegye története. Oklevéltár* (A history of Co. Sopron. Charters). 2 vols. Sopron: Sopron vármegye közönsége.

SRH – Szentpétery, Imre, ed. 1937–1938. *Scriptores Rerum Hungaricarum*. 2 vols. Budapest: MTA.

Sztáray – Nagy, Gyula, ed. 1887–1889. *A nagymihályi és sztárai gróf Sztáray család oklevéltára* (Charters of the Sztáray family of Nagymihály and Sztára). 2 vols. Budapest: n.p.

Tripartitum – Werbőczy, István. 1897. *Tripartitum opus juris consuetudinarii inclyti regni Hungariae per magistrum Stephanum de Werbewcz personalis praesentiae regiae majestatis locum tenentem accuratissime editum*. Budapest: Franklin Társulat. Reprint 1989.

VeszpReg – Kumorovitz, Lajos Bernát, ed. 1953. *Veszprémi regeszták* (Regesta from Veszprém). Budapest: Akadémiai Kiadó.

Zala – Nagy, Imre, Dezső Véghely, and Gyula Nagy, ed. 1886–1890. *Zala vármegye története. Oklevéltár* (A history of Co. Zala. Charters). 2 vols. Budapest: Franklin.

Zichy – Nagy, Imre, Iván Nagy, Dezső Véghely, Ernő Kammerer, and Pál Lukcsis, ed. 1871–1931. *A zichi és vásonkeöi gróf Zichy-család idősb ágának okmánytára* (Charters of the Zichy family of Zich and Vásonkeö). 12 vols. Budapest: n. p.

ZsOkl – Mályusz, Elemér and Iván Borsa, ed. 1951–2001. *Zsigmondkori Oklevéltár* (Charters from the Sigismund period). 7 vols. Budapest: Akadémiai Kiadó and Magyar Országos Levéltár.

Notes to the Text

Measures used in this book

1 modern acre = 4000 square metres
1 hectare = 10,000 square metres
1 royal *iugerum* (in the fifteenth century) = 8442 square metres
1 royal *falcastrum* – unit of measurement for meadows (in the fifteenth century) = 8442 square metres
1 royal plough (in the fifteenth century) = 150 *iugera* = 126.63 hectares

Before the fifteenth century, the *iugerum* was probably smaller, and one plough consisted of 120 *iugera*. The actual size of the *iugerum* could differ in each region. These measures lived on as the *usualis* (as opposed to royal) *iugerum* and plough, and are very difficult to define.

Periods in the Carpathian Basin

Palaeolithic – to ca. 10,000 BC
Mesolithic – ca. 10,000 – 6000 BC
Neolithic – ca. 6000 – 4500/4400 BC
Copper Age – ca. 4500/4400 – 2700/2500 BC
Bronze Age – ca. 2700/2500 – 900/800 BC
Iron Age – ca. 900/800 BC – end of first century BC
Roman – ca. beginning of AD first century – AD 400
Migration – ca. AD 400 – 895
Conquest – AD 895 – 1000
Árpádian – AD 1000 – 1301
Late Middle Ages – AD 1301 – 1526
Early Modern – AD 1526 – nineteenth century

Prehistoric dates are given in actual calendar years BC, and are based on Visy 2003a.

Place-names

If a settlement falls outside present-day Hungary, I use its present name. To make identification easier, I name the country in which it is now. Country codes: Slovakia (Sk), Ukraine (Ua), Romania (Ro), Serbia and Montenegro (Sr), Slovenia (Slo), Austria (Au), Croatia (Cr). In the Appendix, H also appears, referring to Hungary. Deserted settlements I call by their medieval name in whichever tongue they appear in the original document. The counties I refer to are the historic counties of the Hungarian Kingdom at the end of the fifteenth century, as found in Engel 2001c. For a gazetteer of place-names, see the same work.

PART I

WOODLAND

Chapter 1

Introduction

The scope of any study on woods has to be defined in relation to the two determining factors in woodland history: people and nature. Most woodland studies have so far concentrated on the former. However, what people think of woods and the way they perceive them do not necessarily have anything to do with what is actually happening in real woods. Forests are a favourite literary subject, especially in medieval literature. If, to quote a well-known example, we read stories of Robin Hood hiding in Sherwood Forest, that does not mean that the territory of this Forest was covered with trees in the Middle Ages or that Robin Hood really lived there.[1] These stories tell us much about the way their educated medieval writers imagined and contextualized woodland, but little about the actual English woods of the time. On the whole, literary works belong to the field of anthropology rather than ecology. One step closer to the physical reality of woods, which is my primary concern, is what people have said about them in non-literary contexts. In Hungary, many scholarly works have been dedicated to "forest-legislation," mostly management guidelines, which began to be published from the sixteenth century. It has occurred to few, nonetheless, to examine whether these laws had any effect on what was going on in woods or whether they really signified the beginnings of woodland management in this country.

Nature, on the other hand, is almost completely neglected in historical works. Historians tend to forget that trees are not artefacts but have their own properties and behaviour. Moreover, woods are much more than many trees together. The community of plants and animals that make up a wood is a most complicated subject to understand and has its own share in the making of the history of woodland. Certainly the best example to illustrate this is the age-old misconception that trees disappear when people cut them. Many documents of such cuttings are listed in historical books as evidence of woodland clearance. However, anyone who has ever been to a wood in the first spring after cutting will know that most trees, other than conifers,[2] sprout new growth again after felling. This property of trees, in fact, was the basis for all woodland management before the appearance of modern forestry, and will be constantly referred to throughout this work.

Studies of woodland, thus, have run along two lines. Those dealing with the historical aspects have confined themselves to the general or literary written record. Present-day woods, on the other hand, have been examined by botanists and ecologists who tend to disregard the human impact.[3]

Historical ecology, which is the methodology of this work, lies in-between these two approaches. It is history in the sense that it uses sources that were not produced to answer its questions. Researchers do not carry out experiments as in ecology, but gather pre-existing and often well-hidden data. It is ecology in the sense that nature is accepted as having an active role in the events. Human impact and natural processes are interpreted together and with respect to each another. Historical ecology concerns actual woods and what people did in those woods.

The general concept behind this book has been motivated by English landscape history and historical ecology. Ever since William Hoskins' famous words from his *The Making of the English Landscape*: "For those who know how to read it aright, the English landscape is the richest historical record we possess,"[4] researchers have tried to interpret landscape as an independent source that substitutes or complements information from written evidence. Objects that had attracted little attention before (ditches, banks, remains of ploughing, fields, sunken lanes) have become important.[5] While landscape historians have always focused on man-made elements, the study of historical ecology sprang from the recognition that "nature" can also be read as an historical source in the form of tree-rings, vegetation patterns, hedge-composition, etc.[6] Although the reasons are often unclear, we know that time and change are in as close a relationship in the history of nature as they are in that of human society. Thus, observing and recording seemingly "natural" features can be a dating tool in itself. Hooper's hedge-dating rule will serve here as an illustration.[7] By comparing the number of tree and shrub species in over two hundred hedges with the age known from written documents, Hooper concluded that the number of tree and shrub species multiplied by one hundred approximately equalled the age of the hedge. As a result, today one can examine a hedge with no written documentation, and, if it contains five tree and shrub species,[8] one can argue that

[1] In fact, the contrary is true. Documents show that Robin lived in Barnsdale and already by 1086, Sherwood Forest was, at most, thirty percent wooded. Holt 1982; Rackham 1986, 293–294. There are countless scholarly works on Robin Hood. For a bibliographical list, see the Robin Hood Project at the University of Rochester: www.lib.rochester.edu/camelot/rh/rhhome.stm.

[2] There are exceptions, such as yew or cypress.

[3] In an otherwise good book (Mátyás 1996, 265), I was amazed to read that "human activity has affected woods for some 10,000 years . . . Until recently, however, this effect was merely clearance to gain pasture and arable. Woodland, as a source of energy, has only played a significant role since the beginning of modern times." (Translation mine. Henceforth all translations are mine except when indicated otherwise.) The truth is quite the opposite. Woodland, as a source of energy, played a significant role *until* the beginning of modern times.

[4] Hoskins 1955, 14.

[5] There is a vast amount of literature on landscape history nowadays. It would be pointless to provide a list here; useful starting points for orientation are Aston 1985 and Bowden 1999. The two journals dedicated to the subject are *Landscape History* and *Landscapes*.

[6] Sheail 1980; Rackham 1978.

[7] Pollard, Hooper and Moore 1974.

[8] Counting is done in a standardised way.

it is likely to be half a millennium old.[9] The above example partly explains why I chose to follow the English ecological methodology. There are other schools in woodland historical studies – above all the German and French traditions[10] – but these, besides being far too "historical," require much and high quality written evidence, which, as we shall see in the chapter on sources, was not available to me. In other words, the parallel use of written and non-written sources in studying woods and Forests in Hungary is not only a commendable method, as it would be in any case, but it is also inevitable in such a poorly documented period as the Middle Ages. I have adopted some material from books written by archaeologists, but my own fieldwork has been an integral part of the research. The archaeology of woodland has been very little studied in Hungary, so that the fieldwork presented here will, in many cases, be experimental.

This work is not intended as only a simple comparative study between England and Hungary. The English example will serve as the "working hypothesis" itself. That is, assuming that what was happening in medieval English woods and Forests was not specific to the place but rather to the period, I will try to test for similar features in medieval Hungary. How much can such an approach be justified? First and foremost, one always has to be aware of regional differences. We should not push theories too far and suppose that medieval English and Hungarian woodland histories coincide any more than the social and political histories of the two kingdoms. Furthermore, environmental conditions were even more different than social ones. Yet, continuously accumulating evidence attests that woodland management was essentially the same throughout medieval Europe.[11] My studies, then, form part of a general European trend, in which researchers follow a similar line of investigation and arrive at similar results with, of course, the local colour.[12]

1. 1. Traditional Woodland Management

This book is about medieval *woods* and *Forests*. For the modern reader – and the writers of contemporary dictionaries – these two words mean the same. We know, however, that in England (and some more countries of Western Europe) in historical terms this is not at all true.[13] In historical terminology, a place where there are trees is called a *wood* or *woodland*. This is in contrast to *Forest* (with capital F to distinguish it from the modern word "forest") which is a legal term. As this work will first examine woods and then Forests, I shall describe my working hypotheses in this order.

Owing to the works of Oliver Rackham, and later also many others, the basic principles of traditional English woodland management – the "art of woodmanship" – are clear.[14] There are three ways (other than forestry) to handle trees.[15] The first is woodland, where woodmen take advantage of the fact that most trees grow again when they are felled. The second is wood-pasture, where trees and grazing animals share the same habitat. Grazing destroys young trees, thus there always has to be some solution to allow for trees and animals at the same time. The third way is individual trees in hedges and fields.[16]

As mentioned above, the most important principle in traditional woodland management is that non-coniferous trees grow again when they are cut. The stump, the base of the tree left behind after cutting, becomes a *stool*, from which numerous *shoots* grow. Stools are a permanent feature of ancient woods. Shoots are cut at short intervals (approximately ten years). This is called *coppicing*.[17] Stools get larger with every coppicing; it is common to find ones that are several metres across. They are often hollow inside and have an interrupted perimeter. How long a stool is able to produce new shoots is not known but the oldest ones are about a thousand years old and still alive. What kills them is too much grazing or overshadowing by taller trees. Ancient woods comprise many stools, together called *underwood*, and many *maiden* trees, also known as *standards*. Maidens are those trees that originated from seed and have only one, clearly discernible trunk that branches only at a considerable height. An often-used name to denote underwood with a scatter of maiden trees is coppice-with-standards. Modern foresters, naturally, are aware of coppicing, but they consider it a primitive form of forestry, and also think that the stools "wear out" after a few cycles of coppicing, in other words they will not grow new shoots any more. As many ancient stools attest, this is not true.

Maiden trees and coppice shoots are used for different purposes. Shoots would hardly make a good gatepost, but a twenty-metre-tall maiden can only be turned into firewood by wasting much energy, which, ironically enough, is exactly

[9] The matter, of course, is not this straightforward, and there are many exceptions.

[10] See for example Hilf 1938; Bertsch 1951; Semmler 1991. A good summary of the German research is Schenk 2000. Rubner 1965; Devèze 1961; Devèze 1966a; Devèze 1966b.

[11] Wales: Linnard 1982. Scotland: Smout 1997; Smout 2003. Ireland: Jones 1986; Pilcher and Mac an tSaoir 1995. Germany: Warde 2000; Brandl and Schmidt 1998; Radkau 1996; Schenk 1995. France: the giant coppice stools in *Archéologie de la France rurale de la préhistoire aux temps modernes* 1986, 139; the woodbank with a hornbeam stud in Guilaine 1991, plate XI; Ostermann and Reif 2000; Beck, Braunstein and Philippe 1998; Zadora-Rio 1986; Boissière 1986. Italy: Salvestrini 2000; Moreno, Piussi and Rackham 1982; Moreno 1986; Redon 1996.

[12] For example François Duceppe-Lamarre wrote about Flandre, Hainaut, and Artois: ". . . il n'existe pratiquement aucune étude d'archéologie forestière dans cette région." This statement is followed by a description of "l'impulsion des Britanniques," similar to what I have presented. Duceppe-Lamarre 1998. See also Beck and Delort 1993, 11, with basically the same statement.

[13] There is a whole monograph dedicated to a single Forest: Rackham 1989. This book in its entirety tries to explain what makes Hatfield (Essex) a Forest and not a wood. A short explanation of the phenomenon can be found in any of the author's books on the British landscape.

[14] Rackham 1975; Rackham 1976; Rackham 1980. This now has a revised edition published by Castlepoint Press in 2003. Rackham 1986; Rackham 1989; Rackham 1996. (The second and last entries are not the same. The first version of the book was, in theory, revised in 1990, but actually much of it was rewritten, and thus can be considered a separate piece. I work from the first, 1996 paperback edition of this new text.) Peterken 1981; Peterken 1996.

[15] Trees in orchards and towns are outside the scope of this book.

[16] There is probably a fourth tradition, unknown in the British Isles (and Hungary): the combination of trees and meadow. This land-use is best known in Sweden and Finland (Swedish *löväng*, translated into English as "pollard meadow" or "coppice meadow"), but is present in Norway, Spain, Italy, or Bulgaria. Hæggström 1983; Hæggström 1998.

[17] From French "couper" meaning "to cut."

what happens nowadays in many weekend-houses. Even Neolithic farmers knew that shoots are much more useful for firewood than the original standard tree.[18] Almost all medieval depictions of firewood are of *faggots*: bunches of coppice shoots tied together (*Fig. 1*)[19].

The process of coppicing itself was also pictured many times.[20] The clearest representation known to me is in a fifteenth-century Flemish book of hours, in the Hungarian National Library.[21] Here, under March, we can see a man with a special knife in his hand. On his left are rather young shoots, on his right, the already coppiced stools (*Fig. 2*)[22].

It may appear that standards are a "natural" form of reproduction, whereas stools are merely "artificial," but the matter is not that simple. There are, for example, elm trees that do not use seeds but clone themselves with the help of their root system.[23] A well-known example is hazel, which, when overshadowed by taller trees, does not flower, but sends up shoots.

In wood-pasture, coppice stools cannot be used to produce shoots, because the territory is accessible to grazing animals. Young shoots are the finest possible part in the diet of cows or goats. If the shoots are eaten in July, they will reappear next spring, however, after a few occasions the stool will cease to produce shoots and rot away. The solution, then, is to cut the tree not close to the ground, but at a height of about two or three metres, so that animals cannot reach the new growth. As far as the shoots are concerned – for the tree, if you like – the height at which shoots grow is of no significance. This is quite the opposite for the person who has to work them: it is not easy to use a saw or an axe while balancing on top of a ladder. In the end, we are facing a method born out of an uncomfortable need. It combines the height of maidens with the firewood production of stools: all this is called *pollarding*.[24] The permanent part of a pollard tree is called a *bolling*. Hedgerow and free-standing trees can be standards, stools, or pollards (*Fig. 3*)[25].

The first part of my work will centre around these principles, and will try to establish the history of Hungarian woodland in relation to them. No such work has ever been carried out in Hungary, although the history of forestry is a well-developed subject. It is important, however, to differentiate between woodmanship and forestry.

Fig. 1. King Nebuchadnezzar orders the burning of a young Jew on a fifteenth-century miniature from the Hungarian King Matthias' Graduale. The servant on the right brings more firewood: coppice shoots tied up in a faggot.

Fig. 2. Coppicing in a Flemish book of hours. Where this picture was made (or in the tradition it was produced) March was thought to be the right time to coppice.

1. 2. Forestry and Its History

Modern forestry deals with trees in ways different from the ones described above. Chiefly, it is engaged in the creation and maintenance of plantations. These, although they are called "forests," resemble cornfields more than woodlands. Before they can be established, the ground is cleared of all vegetation (if there were trees there, this means removing the stumps with

[18] Rackham 1979.

[19] Reproduced from Soltész 1982, 136–137. In the original biblical story (Dan. 3: 13–25) there were three men, and they did not burn. Also note the way the servant is carrying the faggot, and compare to a mid-thirteenth century charter, in which some people were obliged to "bring wood on their back to the house . . . of the Abbot" of Pannonhalma (Co. Győr). *ÁÚO*, vol. 2, 8. "portare ligna super dorsum ad domum . . . Abbatis"

[20] For some examples, see Hausen 1984, 103–104.

[21] Cod. Lat. 396. Facsimile edition with an essay is Soltész 1983. This book of hours is often, even in scholarly works, referred to as "Hungarian," which suggests that it was made in Hungary. This is far from the truth: it might have been ordered in Hungary, but even that is more than questionable and has no relevance to its style, whatsoever.

[22] Reproduced from Soltész 1983, fol. 3v.

[23] This is not called coppicing but *suckering*. It produces individual trees of the same genetic make-up.

[24] Rackham 1991.

[25] Reproduced from Rackham 1996, 9.

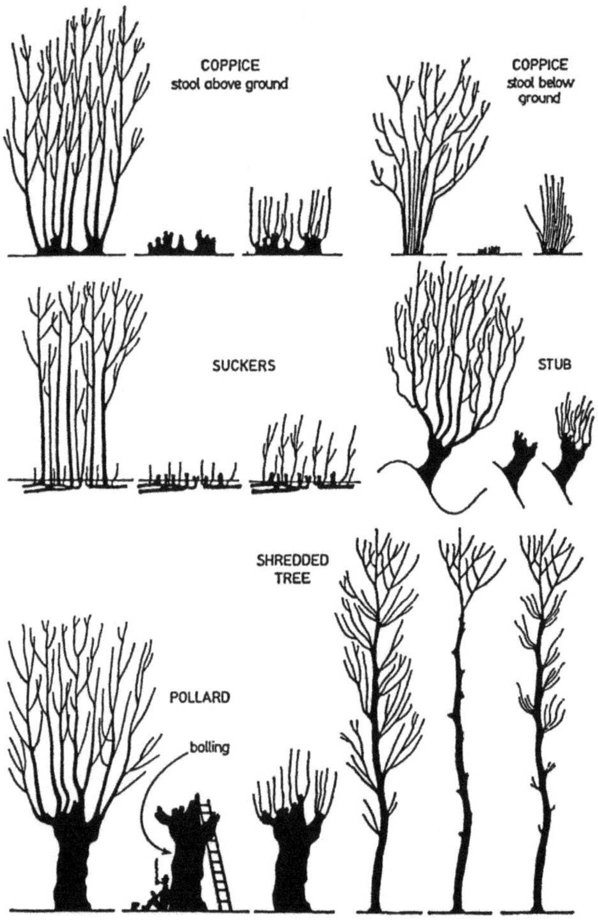

Fig. 3. "Ways of managing wood-producing trees. For each method the tree, or group of trees, is shown just before cutting, just after cutting, and one year after cutting."

heavy machinery) and is ploughed – hence many plantations display miniature versions of ridge-and-furrow, trees being always on the ridge.[26] Then, either seeds are sown or young trees (all the same species) nursed in special gardens are planted (rather the latter). They are protected all their lives: competing plants are cut or poisoned by pesticides. All this is not enough to keep them alive: every third planted tree in Hungary dies young and has to be replaced.[27] Afterwards, when the trees are considered mature (usually less than one hundred years) they are harvested, and the whole process starts all over again.[28] Foresters count plantations – be they composed of foreign species or native ones – the same as woods. This is based on the fact that the appearance of both woods and forestry plantations is dominated by trees; however, this is as far as the similarities go. Equating plantations with woods ignores the most important feature of woods: their capacity for self-renewal. The Hungarian word for wood – "erdő" – originates from the verb "ered," which means "to grow by itself."[29] In sharp contrast to this, when plantations are left alone, they collapse: larger trees fail to reproduce,[30] and the whole area falls back to an earlier phase of succession.[31] The reason for this is ultimately simple. A wood, as I have said before, is much more than many trees together. Large trees appear at the end of a long process, following many, though undoubtedly less spectacular, plants and also animals. Trees cannot survive alone; if people stick young seedlings in the ground, they do not create the whole community that deserves to be called a wood. Plantations, therefore, are not my concern here.

As for present-day woods, they are also all managed by forestry methods. This management leaves a very characteristic mark on them: one sees mostly even-aged trees, preferably of one species only. This is achieved by a rather simple method: foresters regularly cut all trees and shrubs that do not fit their ideal pattern. Woods are carefully mapped, and areas of varying size are cut in a not very flexible system. Modern forestry, nonetheless, is not this simple. Its overall aim is to produce *maiden* trees, that is, trees growing from seed and having the least possible number of branches, which would reduce their "quality," that is, their price. The most widely used method is "gradual dark-felling," developed in this country by Antal Majer.[32] The basic principle of this method is that some trees are left standing while all others around them are cut so that seeds from the ones that were left behind can develop into seedlings, although planting trees in woods is also common. Recently, however, there have been debates about alternative, more "natural" kinds of woodland regeneration, and most importantly, grubbing out woods and replacing them with plantations is now mostly discouraged.[33]

Histories of woodland written in Hungary so far have really been histories of forestry. Their aim was to find the methodology of forestry in the past, and trace the continuous development from the primitive methods of our ancestors to the highly advanced techniques of our days. This has been stretched back to the Middle Ages. When Pál Csőre, the author of the only Hungarian monograph dedicated to medieval woods, concluded that "the Middle Ages in general were characterised by unmethodical forestry,"[34] he was criticising medieval people for something they did not have the chance to carry out, much in the same way as if he had blamed them for not having chainsaws instead of axes.

The birth of modern forestry in Hungary is well documented. Especially after the foundation of the Academy of Forestry (originally that of Mining) in Banská Štiavnica (Sk; Co. Hont), a theoretical interest arose in the origins of forestry.[35] The greatest achievement of this was undoubtedly

[26] This may be a somewhat older custom in countries where drainage is a problem. In the Netherlands, in certain ancient woods there are old coppice stools on mini-ridges. Kindly told me by Joep Dirkx.

[27] Halász 1994, 57–58.

[28] For the methodology applied in this country, see Danszky 1973, vol. 1, 655–756.

[29] *MNyTESz*, s.v. "erdő."

[30] This does not happen to black locust plantations. Black locust (*Robinia pseudoacacia*) acts more like a weed, and once planted is extremely difficult to get rid of. Even when the stumps are removed from the ground with heavy machinery, they live on happily wherever they are placed. Hence the strange banks with black locust hedges on them, often seen between plantations.

[31] For succession, see Pickett, Collins and Armesto 1987.

[32] Majer 1973.

[33] For example, see Solymos 2000; Bartha 2001.

[34] Csőre 1980, 227.

[35] From the abundant literature, see for example Hiller 1993. An early summary is Vadas 1896. For the history of forestry in the mining regions, see Magyar 1983.

the three-volume collection of Károly Tagányi: *Charters of Hungarian Forestry*, published in 1896.[36] Tagányi wrote a long preface to the first volume, an essay on the history of forestry, which in many respects is still the best summary we have. The collection devotes some attention to the Middle Ages, albeit discouragingly little. Up to the year 1526, it contains only 108 documents. Tagányi himself apologised for this, and claimed that he had only collected those charters with immediate connections to methods of woodland management. Unfortunately, many modern scholars have overlooked these few sentences and have taken it for granted that there was no more medieval material than this.[37] Tagányi's book is most useful for the eighteenth and nineteenth centuries, although criticism has evolved about its one-sidedness.[38] After the researches of Tagányi and many others,[39] it is clear that modern forestry was initiated in Hungary mainly from German speaking areas: it was the Habsburg dynasty, ruling after the battle of Mohács (Co. Baranya) in 1526, who started to foster the business employing German speaking specialists. The changes originally focused on the northern mining cities, such as Banská Bystrica (Sk; Co. Zólyom), Kremnica (Sk; Co. Bars), or the above-mentioned Banská Štiavnica, which had a German population. If German forestry has had a deeper impact on Hungary than on many other countries, these are certainly among the reasons.

In sum, the first part of this book will be an attempt to examine medieval and earlier sources *not in the forestry but in the woodmanship tradition*. This difference is probably the single most important aspect of my work. I shall attempt to establish how wooded medieval Hungary was, and how the woods were grubbed out or managed. This part will involve some necessary consideration of pre-medieval times as well, because the Hungarians certainly did not arrive in a pristine and untouched landscape. Also, woodmanship did not end in the sixteenth century. In interpreting today's woods, the Early Modern Period is at least as important as the Middle Ages, however, covering it thoroughly would require a separate book.

I shall write about the whole of the medieval Hungarian kingdom. Why, one may ask, have I not restricted myself to a smaller region? As relevant as this question may be, the answer is rather simple. Hungary is not rich enough in medieval documents to permit such a work, even though one of the wooded Forests to be discussed in the second part of the book, Bakony, is exceptionally rich in written sources. A concrete example will illustrate my point more clearly. The 1454 perambulation of the village of Bere (Co. Veszprém in Bakony) talks about a certain "ancient *troncus* among stones."[40] Although other perambulations in the region contain some more references to *tronci*, based on sources from Bakony only, I could not identify the meaning of *troncus*. If I find, as I actually did, a more verbose allusion to the same phenomenon three hundred kilometres to the north in a document from a village in present-day Slovakia, I have to use that as well. Moreover, I have to be satisfied that I managed to find anything at all.[41] This methodology is, of course, nothing new. Most Hungarian medievalists work this way, and what is more important, it has been proved useful in the history of agriculture as well. Márta Belényesy discussed the history of Hungarian field systems and medieval agricultural development using precisely the same methodological framework.[42]

1. 3. Medieval Forests

In England, as we have seen, a wood was not the same as a Forest. The word "Forest" most probably comes from Latin *foris est*, ie. something that is outside of something.[43] While some linguists may argue for another origin, the people of medieval England definitely perceived their Forests as outside of something, namely outside of common law. After Hastings, in England a Forest "was an area of roughland on which the king or some other magnate had the right to keep deer and to kill and eat them."[44] In practice this meant that at some point in time the king announced that this or that piece of land was afforested (this had nothing to do with trees!) where henceforth special laws protected red, fallow, and roe deer, and wild swine. Whatever their original purpose, Forests were producers of large amounts of venison for the royal table, and sources of income paid for offences against Forest Law. The letters of Henry III, filed in the Close Rolls, carefully record how many deer the king wanted from his Forests: numbers were often very high. At Christmas, 1251, the king had on his tables more than 800 deer, 1300 hares, and lots of other smaller animals.[45] The other aspect is aptly demonstrated by Henry II's campaign against Scotland, for which in the year 1176 the biggest revenue was the money coming from Forest offences.[46] Using Forests as sources of income was also known elsewhere in Europe. In France, for example, in the mid-thirteenth century, around a quarter of the royal revenues came from Forests.[47] Forest Law, an almost unsolvable puzzle of eyres and assizes, is a research subject in its own right.[48] It must be understood that a Forest was where Forest Law

[36] *MEO*.

[37] Furthermore, although Tagányi was an excellent scholar, he inevitably made some mistakes, which have been repeated endlessly. For example, in the *regesta* of the foundation charter of the Pauline abbey of Porva in Bakony (1450) he wrote that the monks had the right to sell wood from Bakony Forest. *MEO*, vol. 1, 25. In contrast, the original says "nullatenus tamen concedimus eis de ipsa silva alienis vendere seu distribuere quovismodo." Dl 14 424.

[38] Magyar 1993, 146.

[39] For general bibliographies, see Gerlai 1936; Kolossváry 1966. An important collection is Kolossváry 1975.

[40] Df 201 343. "antiquum troncum inter lapides"

[41] Elemér Mályusz, the most prominent medievalist of the first half of the twentieth century once wrote that having few written sources could, in fact, be an assett. In Hungary, so he argued, one could aim at analysing *all* sources, which would be unrealistic in France or Germany. He was otherwise not famous for such optimism. Mályusz 1967.

[42] Belényesy 1958a; Belényesy 1958b; Belényesy 1960; Belényesy 1964.

[43] Literally outside the door.

[44] Rackham 1989, 2.

[45] Rackham 1989, 52.

[46] Poole 1958, 339.

[47] This amount declined rapidly from the thirteenth century on, and arrived at a minimum at the end of the fifteenth century. Devèze 1961, vol. 1, 65–72.

[48] Young 1979; Holdsworth 1938.

applied. There were Forests, Dart*moor* Forest for example, where there were no trees at all. Many Welsh Forests were also treeless. Most Forests included some woodland, yet all in all, they were not any more wooded than England in general. English Forests were, however, "the end of the chain of developments,"[49] with many earlier phases that also provide relevant analogies for Hungary. I shall return to this question with a more detailed treatment in Chapter 9.

The central question in the second part of my work is whether the distinction between woodland and Forest was present in medieval Hungary as well. A preliminary search of the available material suggested that although the difference was never as clear-cut in Hungary as in England, similarities were strong enough to justify the separate usage of "wood" and "Forest," and to carry out research about the latter, whatever specific meaning(s) it may finally acquire. Contradictions in terminology should not be underestimated: I am using a language where Forest and wood have their well-defined historical meanings, but I am writing about a country where this is not exactly the case. Nonetheless, I did not want to create further confusion in a situation which is confusing enough as it is, therefore I chose to stretch the meanings of the original English words to cover their Hungarian counterparts rather than to create new English vocabulary.

I shall examine two Forests, Pilis and Bakony. These two regions were royal property and operated in a form comparable to medieval English Forests, at least in the thirteenth century. They were both wooded Forests, and ancient woodland is still preserved in them, although largely transformed by modern forestry. I shall also study these ancient woods considering the results of the first part of this work. In a sense, the woods of these Forests will serve as case studies to test how much can be achieved in historical woodland studies on a regional basis.

1. 4. Geography and History

The geographical scope of this study is the Carpathian Basin (*Fig. 4*). This large unit comprises some 300,000 square kilometres, and is very variable. On the north, the east, and the south-east it is bordered by the Carpathian Mountains, which give their name to the whole region. The southern boundary is marked by the Danube and the Sava rivers, while in the west we find the easternmost fringes of the Alps in Austria. This geographical unit mostly coincides with the medieval kingdom of Hungary. Croatia and Dalmatia, although in a personal union with Hungary from the twelfth century, are ecologically different and fall outside the scope of the present study, except for Slavonia, the region between the Drava and the Sava rivers. Bosnia and other parts of the former Yugoslavia, which only at certain points belonged to the Kingdom, are not my concern, either. Present-day Hungary, formed after the Trianon peace-treaty of 1920, is much smaller than the medieval kingdom. In this book, "Hungary" means the kingdom, which included parts of today's Slovakia, Ukraine, Romania, Serbia and Montenegro, Croatia, Slovenia, and Austria. What is called what in which historical period is a very problematic issue, which may be difficult to understand for someone who is not from East-Central Europe. However, I wish to avoid any misunderstandings: "Hungarian" in this book has nothing to do with nationality, and perhaps even less with the people who call themselves Hungarians today. It simply means someone or something that has to do with the Kingdom of Hungary. Hungarian speaking people were only one, although dominant, group in the multinational population of this kingdom.

At the first sight, the Carpathian Basin has three basic components. All around there are high mountains: peaks of two thousand metres or more, pines, sheep, and a mountainous way of life. West of the Danube, which bisects the Basin, there is the rolling countryside of smallish hills and numerous woods that we call Transdanubia. East of the Danube is a whole different world. This flat region, the Great Hungarian Plain (Alföld) is what tourists, and nineteenth-century poets, mostly identify with Hungary. The Plain is very much agricultural, and has been so for millennia. Today, woods are almost entirely missing. There are outliers to each region. The Eastern Carpathians extend themselves to the west, enclosing a triangle with lowland inside, called Transylvania.[50] This lowland itself is much more hilly than the Alföld, thus resembling the Transdanubian region. Because of its special climate, environment, and history, Transylvania is usually considered a separate unit. There is, finally, a smaller version of the Great Plain in the northern part of Transdanubia, aptly called the Little Plain. In this work, I shall mostly focus on the Transdanubian region with some attention directed towards the Great Plain. The high mountain regions are very much different in woodland ecological terms, because they are dominated by conifers.

A closer look reveals further variations. Three large climatic zones meet here: the Atlantic in the west, submediterranean in the south, and continental in the east. In addition, the mountains create their own subcarpathian zone.[51] Soils are also assorted, both on a macro and micro level. Based on bedrock types, four main types can be differentiated: loessy, sandy, compact bedrock, and alluvial soils. These can be divided into dozens of smaller units.[52] Consequently, vegetation is also diverse.[53] But it is not only climate and altitude that formed the natural vegetation in prehistory. Rivers are a most important factor in the history of the Carpathian Basin. Besides the two main rivers (the Danube and the Tisza), there are countless smaller rivers that kept, so some argue, as much as one third of the whole Plain periodically flooded.[54] These "green corridors" heavily influenced their environment and provided migration routes for prehistoric plants, animals, and peoples alike.[55]

---

[49] Wickham 1994, 161.

[50] This is not the same as what is called Transylvania today. The latter is bigger and includes the easternmost parts of the Great Plain as well.

[51] Bacsó 1960. More recent, woodland specific studies are to be found in Kircsi 2000.

[52] Stefanovits 1981.

[53] "Consequently" is a dubious term. The relationship between soil and vegetation, namely which changes which, is a matter of debate. Willis, Braun, Sümegi and Tóth 1997. That is partly why the lack of woodland soils should not be automatically taken as an indication that the natural vegetation of a given piece of land is not woodland, as in Németh 1998.

[54] Somogyi 1987, 61–64; Sümegi and Kertész 1998.

[55] Sümegi and Kertész 1998, 150–153.

Probably the most important feature of the whole Basin is that many different smaller habitats make up the whole in the form of a mosaic.[56]

As far as the landscape is concerned, the three basic regions are markedly different. The Carpathians appear largely as coniferous woodland with the settlements in clearings. The Transdanubian region still has much woodland, mostly on the mountain range that stretches from north-east to south-west and is known as the Hungarian Middle Mountain Range (*Magyar Középhegység*).[57] Both Pilis and Bakony form parts of this. On the hilly lowland areas, woods are less numerous and are scattered and discontinuous. Settlements take the form of both hamlets and nucleated villages, roads are numerous and winding. As opposed to this, the Alföld is a landscape dominated by large villages, straight roads and very little ancient woodland. All this is similar to the English *ancient countryside* versus *planned countryside*, or the French *bocage* versus *champagne*. There is, however, a fundamental difference in the historical development of these countries. While, for example in England, ancient and planned countryside are already present in the first available written documents, the present outlook of the Alföld is a much more recent phenomenon. In the Árpádian period (eleventh-thirteenth centuries) the Plain had many more and much smaller settlements. These disappeared in the Late Middle Ages and large nucleated settlements were formed.[58] On the natural side, the Alföld suffered the greatest damage in the nineteenth century, when drainage works reduced flood-plains to a fraction of their original territory, completely destroying the previous ecological systems. The workforce employed to achieve this was something comparable to the creation of the Great Wall of China. The reason why the Plain lacks native woodland can be partly explained by this, but the matter is more complicated. I shall return to the problem in Chapter 5.

The history of the landscape must be viewed against the history of the people that live in it (*Fig. 5*). Although many other peoples inhabited the Carpathian Basin before the Hungarians, the first useful documents were written after 1000. The Magyars conquered the Basin around 900 (the "Conquest") and the first king, St. Stephen was crowned in 1000.[59] St. Stephen managed to resolve the mighty task of organising the whole Basin into one well-functioning unit. He set up the outlines of both lay and church administration.[60] His successors (the Árpád dynasty)[61] ruled the country until 1301. They were followed by the Hungarian branch of the Angevins: Charles Robert (1301–1342)[62] and his son, Louis the Great (1342–1382). From 1387 to 1437 Sigismund of Luxembourg (also Holy Roman Emperor) ruled as king, then after a brief period, the Hunyadi family gained control. First, John became governor in a country which was for all intents and purposes kingless, then his son, Matthias, was properly crowned (1458–1490). Thirty-six years after the death of this great Renaissance king, the Ottoman army defeated the Hungarians at Mohács in 1526, where King Louis II also died. This is usually considered as the end of the Middle Ages in Hungary, because the kingdom fell apart.[63] The northern parts fell under Habsburg control (Hungary proper), in the east Transylvania became independent (although vassal to the Ottomans), while the south was part of the Ottoman Empire until the end of the seventeenth century. Traditional periodisation of Hungarian medieval history, thus, is different from the Western European model. Until 1301 we speak about the *Árpádian period*, from 1301 to 1526 the *Late Middle Ages*. These are not merely dynastic categories: the Árpádian period was an age marked by an intricate network of social positions, whereas in the Late Middle Ages the whole society was organised into a dichotomy of noblemen versus peasants (called *iobagiones* in Latin, from the Hungarian *jobbágy*). Economic conditions also differed in the two main periods. In the first two hundred and fifty years, larger estates controlled the agriculture; in the Late Middle Ages peasants worked their individual plots and allodial lands were at a minimum.[64] The age of transition was the thirteenth century, aggravated by the Mongol invasion of 1241–1242.[65]

[56] Sümegi, Hertelendi, Magyar and Molnár 1998, 183–185.

[57] Other nations would probably call these "mountains" hills. The highest peak in the whole range is 1016 metres.

[58] There was another great age for hamlets. The origins of these were long thought to be medieval, but recently they have been shown to be early modern. The hamlets themselves mostly disappeared in the twentieth century.

[59] Until recently, there has been no general book covering the history of medieval Hungary in an international language and on an acceptable scholarly level. This has changed with the publication of Engel 2001. This comprehensive, very high quality work can be recommended to all those interested in the subject. It also includes an exhaustive bibliography that lists all major publications in Western languages.

[60] Györffy 1983.

[61] Members of the Árpád dynasty were not Stephen's direct descendants, but those of his cousin, Vazul. Árpád himself was more of a mythical forefather. Nonetheless, for medieval kings, the Árpád dynasty was equated with St. Stephen. When talking about legitimacy, they continuously stressed that they were descendants of Stephen.

[62] Charles Robert was first crowned in 1301, but for a long time he was far from ruling the kingdom. After years of fighting and dealing with a competing king (the Czech Wenceslas, son of Wenceslas II, King of Bohemia), he was crowned again by the papal legate in 1309. The ceremony, however, had to be repeated the following year, this time with the "holy crown," thus reassuring legitimacy.

[63] This really only happened after the fall of Buda in 1541.

[64] István Szabó 1975.

[65] Szűcs 1993b.

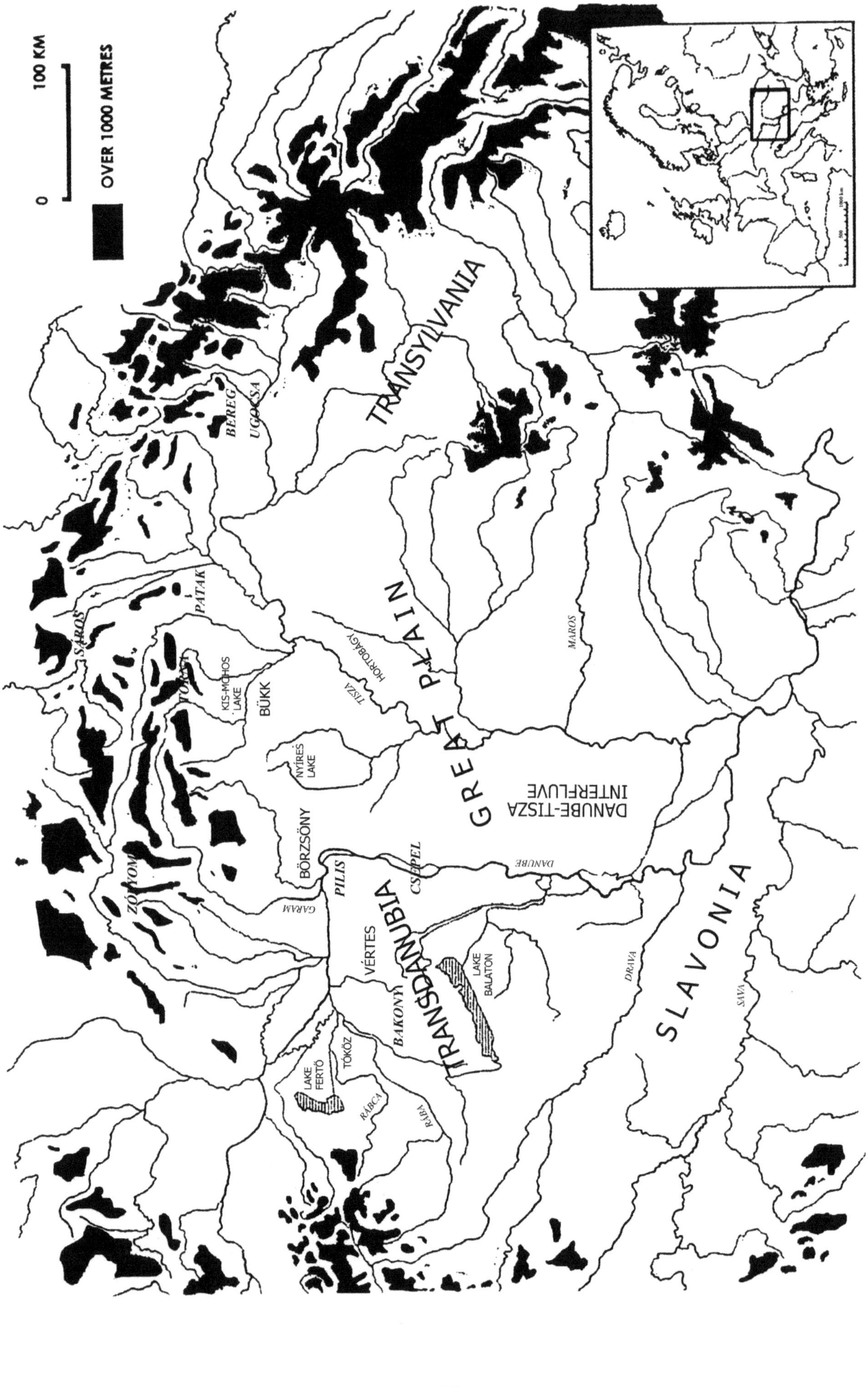

Fig. 4. The Carpathian Basin with major rivers, mountains, and regions mentioned in the text. Areas with Royal Forests appear in bold italic.

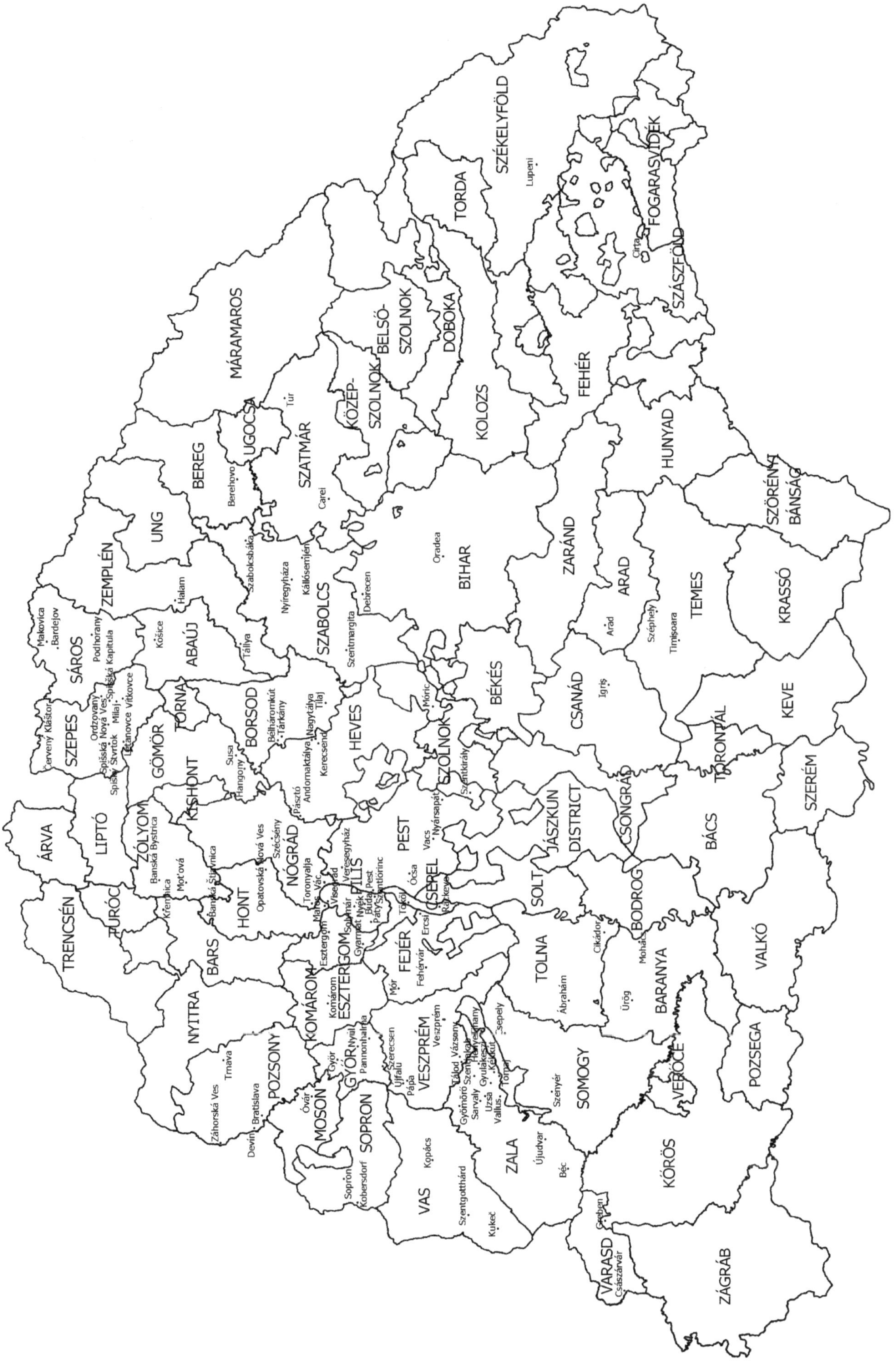

Fig. 5.The Carpathian Basin with the historic counties of the Late Middle Ages and all the settlements mentioned in the text.

Chapter 2

Sources and Methods

2. 1. Written sources

What we do not have

The Hungarian source material is very much one-sided. Narrative sources are few, and theoretical writings on agriculture are entirely missing. Some account books survive from the fourteenth century, there is a fair number from the fifteenth, but sadly little is to be found in them concerning woodland incomes. What the researcher is left with are legal documents, mainly records of lawsuits and those of rights over landed property. This does not imply that I shall analyse only court-cases, but almost any written data will be extracted from a piece that has its origins in the application of medieval Hungarian law. This distortion cannot be eliminated, but one must not forget that it exists.

The reasons for this lack of balance are manifold. First, little has survived of the archival material of the medieval Kingdom of Hungary.[1] The greatest single loss is the disappearance of the Royal Archives, which shared a similar fate with many other, smaller collections: they were lost while the owners (in this case the whole royal court) fled before the Ottomans. In a highly – to Western-European eyes unbelievably – centralised country, such as the Hungary of that time, the Royal Archives would have been of immeasurable importance in most affairs, but especially so in the case of royal lands, such as Bakony and Pilis Forests. Nonetheless, one should be aware that fragments at least of all possible source-types most probably survived, thus mourning over what has been lost only partly relieves the historian of having to account for the situation as a whole. Erik Fügedi, for example, pointed out that Hungarian genealogies are difficult to compile not, as is usually stated, because most family archives were damaged, but, as demonstrated by the examination of intact surviving material, because nobles were not interested in the maternal side of the family.[2]

This leads us to the second, and perhaps more important factor behind the uneven nature of the source-material. Certain types of sources are not present until the end of the Middle Ages, and when they are finally available, they are of little value to the woodland historian. General conscriptions of land were not compiled in this country. Why and how these should appear is difficult to explain. In England for example, the Domesday Book is perhaps the result of William the Conqueror's curiosity to learn what he possessed, but then followed a two-hundred-year gap until the mid-thirteenth century, when "the practice of record keeping suddenly advanced" and "there is a continuous stream of documentation."[3] In Hungary, surveys do not appear until the fifteenth century. There had, by that time, been a constant preoccupation with land in the written record for some five hundred years. Charters keep talking about estates, but no one seemed interested in putting onto paper what the particular piece of land in question consisted of, and what each part was worth. We shall see later on that the means to find this out were available from early on, yet some other impetus was needed to put those means to work.

Then, around 1500, a number of great surveys called *urbaria*, were compiled,[4] which clearly illustrate another problem, deeply rooted in late-medieval Hungarian society. From the fourteenth century on, the Hungarian agrarian population was organised around small personal plots, and for two centuries greater manorial lands were non-existent.[5] The result was that the income of the landlord derived from the individual peasant households. Therefore conscriptions of estates did not record the quantity and quality of the land, but simply the amount of money each household owed for using that land. A typical *urbarium* is a list of villages belonging to an estate with a list of the inhabitants along with the sizes of their plots and the amount of money to be paid.[6]

What we still have[7]

There is a rather unlucky tradition in Hungary to call every piece of non-literary medieval writing a *charter*.[8] As a consequence, the term "charter" in this book will be used in the widest possible sense. I shall use four main types of these charters: perambulations, terriers, estimations, and general charters. In the following, I shall briefly describe all four types. More detailed quantitative and qualitative analyses will be carried out in later chapters.

2. 1. 1. Perambulations

"The first boundary starts on the southern side of that same land in a *virgultum* next to the big wood of the village called

---

[1] For a general description of what is left, see Hunyadi 1997–1998. See also Mályusz 1968.

[2] Fügedi 1970, 7.

[3] Rackham 1980, 9.

[4] For *urbaria* in general, see Maksay 1957. For a full list of the published ones, see Kredics and Solymosi 1993, 14. Since then, the *urbarium* of the Greben (Co. Kőrös) estate has appeared in print: E. Kovács 1998. Taxation surveys existed from the first half of the fifteenth century. Only four survive, all from the same year, 1427. However, these are, by their very nature, not concerned with land. Engel 1989.

[5] István Szabó 1975, 39.

[6] Sometimes services to be rendered to the lord are recorded and in them bits and pieces of information about woods may survive. For example, in the *urbarium* of the bishopric of Veszrém: "Item ad festum nativitatis Domini tenentur duo plaustra lignorum ad castrum portare." Kredics and Solymosi 1993, 82.

[7] A more detailed account of this is Szabó 2003.

[8] The medieval collection of the National Archives in Budapest is called the Diplomatic Archives and Diplomatic Photo-collection, although many types of sources are kept there from privilegal charters through account books to manorial records.

Verestow that belongs to the monastery of Almad, beside the public road that leads from Bilige in Durguche, at the place that is called Echefaya, where there are two boundaries. Thence proceeding right next to that great wood, in a way that one boundary always exists on a large tree from that wood, whereas another one in the *virgultum*, by continuous double boundaries it goes straight towards the north. Leaving the surroundings of the aforementioned wood, it goes to a small arable land that exists in the *virgultum*, where there are two boundaries, one in a shrub that is commonly called *gerthan*, the other in earth on its western side. From here proceeding directly in that *virgultum*, passing two boundaries it comes to another ancient wood of the aforementioned monastery, in whose southern corner there are two boundaries, one in a large oak-tree, the other in a shrub. Thence it descends to an arable plot, and passing some boundaries it comes to an *angularis* boundary, which separates from the territory of the aforementioned village of Verustow and from the territory of the village of Chepel, where there are two boundaries. Then it turns to the east, and by continuous double boundaries turning right it comes to a certain hill, next to which a bit further away there are two boundaries, which separate from the territory of the aforementioned village of Chepel, and from the territory of Gatal, son of Peter of Balag. Thence turning back to the south, it comes directly to the wood of the sons of Peter, where there are three boundaries, from where turning along that wood it reaches a certain ancient tree that is commonly called *magal*, where there are three boundaries. From here, going right around that wood, it enters a certain *virgultum*, and in that *virgultum*, beside the aforementioned wood, there are three boundaries. Thence turning somewhat to the west, passing three boundaries, on a certain ancient road it leaves the *virgultum* and enters a meadow to three boundaries, of which one is in a shrub that is called *gymulchyn*. Thence it goes to a certain *nemus* of plums beside the houses of the son of the aforementioned Peter and those of the sons of Gothalum, where there are three boundaries. Thence turning somewhat towards the west, beside their village it comes to three boundaries, which directly face the church of the sons of the said Peter and Gothalum. From there, again leaving the surroundings of the son of Peter beside the territory of the sons of Gothalum somewhat turning towards the east, it comes to a *virgultum*, at whose entrance there are three boundaries, of which one is in a small pear. Thence proceeding towards the east it comes to the aforementioned public road, which goes from Bilege to Durguche, and by that road it returns to the place called Echefaya, where the first boundary started; and by these said borders and boundaries the aforementioned territory is distinct."

The boundary of Barnag (Co. Veszprém), 1284.[9]
(An *angularis* boundary is where three or more boundaries meet. Further special vocabulary will be dealt with later on.)

Perambulations are like ancient woods: each is similar, yet distinct.[10] A lot of information is written in them about the landscape, but some are more useful than others. Here, I have deliberately chosen one that describes a wooded part of the countryside, however, there are hundreds of medieval perambulation without woods or even trees.

As this example clearly demonstrates, a perambulation was a symbolic as well as a practical equivalent for modern maps.[11] Whenever need arose – there was a controversy, a new settlement was founded, etc. – neighbouring people gathered, and in a supposedly Indo-European ritual,[12] they walked around a certain piece of land and noted down its most conspicuous features, and they also made some themselves. The practice, which in Hungary survived way into the nineteenth century,[13] created what was probably the most common feature in the medieval Hungarian landscape: earthen boundaries. The above perambulation has around thirty boundaries that are simply called *meta*. In other documents, it is explicitly said that these were earthen boundaries (*meta terrea*). Many charters stress that they were made to purpose during the perambulation (*metam terream cumulassent*). Trees, when selected for boundary signs, were often also surrounded by a pile of earth (*meta terrea circumfusa*). One still sees those earthen boundaries, especially in wooded areas, where they have better chances of survival than in open farmland, but at present it is impossible to tell how old a particular boundary is, since there have been no attempts at locating those that can be shown to be ancient.

Perambulations come in various lengths and in varying purposes. The Barnag perambulation is of moderate size;

[9] *Zala*, vol. 1, 94–95. "Prima meta incipit a parte meridionali eiusdem terre in virgulto iuxta magnam silvam ville monasterii de Almad Verustow vocate, prope viam publicam per quam de villa Bilige itur in villam Durguche in loco qui Echefaya dicitur, ubi sunt due mete; et inde progrediendo iuxta eandem magnam silvam immediate, ita videlicet, quod una meta semper existit super magna arbore de predicta silva, alia vero in virgulto cum continuis binis metis vadit directe ad partem septemtrionalem; transeundo autem vicinitatem predicte silve, venit ad modicam terram arabilem in virgulto existentem, ubi sunt due mete, una in dumo vulgariter gerthan dicto, alia in terra iuxta eandem a parte occidentis; inde progrediendo directe in eodem virgulto, interpositis duabus metis, venit ad aliam antiquam silvam monasterii prenotati, in cuius silve angulo meridionali sunt due mete, una in magna arbore ilicis, altera in dumo; inde descendit ad terram arabilem et directe eundo interpositis metis venit ad metam angularem, que separat a terra predicte ville Verustow et a terra ville Chepel, ubi sunt due mete; adhuc flectitur ad orientem et cum continuis binis metis recte vergendo venit ad quendam collem iuxta quem parum remote sunt due mete, que separant a terra predicte ville Chepel et a terra Gatal filii Petri de Balag; abinde reflectitur ad partem meridionalem et directe pergendo venit prope silvam filii Petri ubi sunt tres mete; abinde flectendo prope eandem silvam attingit quandam arborem vulgariter magal dictam, ubi sunt tres mete; hinc circa ipsam silvam immediate eundo, intrat in quoddam virgultum et in ipso virgulto iuxta predictam silvam sunt tres mete; abinde flectendo paulisper ad occidentem mediantibus tribus metis in quadam veteri via exit in pratum ad tres metas, quarum una est in dumo gymulchyn dicto; dehinc vadit ad quoddam nemus pruni prope domos filii Petri prenotati et filiorum Gothalum, ubi sunt tres mete; deinde flectendo parumper ad occidentem, prope villam eorumdem venit ad tres metas, que ecclesiam filiorum Petri et Gothalum predictorum directe rescipiunt; abinde iterum exeundo a vicinitate filii Petri iuxta terram filiorum Gothalum paulisper declinando ad orientem, pervenit ad virgultum, in cuius introitu sunt tres mete, quarum una est in parva piro; deinde pergendo versus orientem cadit in viam publicam prenotatam, per quam itur de villa Bilege venitur ad villam Durguche, et in eadem via revertitur ad locum Euchefaya dictum, ubi prima meta inchoavit; et sic premissis terminis atque metis predicta terra est distincta."

[10] For more on perambulations, see Szabó 2002.

[11] Harvey 1980.

[12] Belényesy 1955b.

[13] There is a whole book devoted to early modern perambulations: Takács 1987.

laconic ones are merely a few lines long, whereas verbose ones go on for pages and pages. The boundary they describe can be a settlement's whole boundary, one part of it, a boundary within one settlement, two or more settlements with a common boundary, any combination of these, and more. The first Hungarian charter that came down to us in its original (from 1055)[14] contains several perambulations. It tells us that the practice already existed, but it also reveals that the sophisticated Latin terminology, which is later associated with it, was yet to be born. Early perambulations are a welcome relief from the often monotonous later ones. These documents increase in number in the thirteenth century, which was a time of great changes in land-ownership. From 1300, there is a constant flow of perambulations, although a statistical analysis would probably show increases and decreases.

Woods, woodland trees, and free-standing trees were often mentioned in perambulations. In the Barnag document we read about several types of woods. Where the original uses the Latin *silva*, I translated it as wood. However, *virgultum* and *nemus* were also mentioned. These should not simply be equated with woodland for two reasons. First, one should never rush to suppose that two different words have the same meaning simply because in our time the difference is invisible. Secondly, and this demonstrates the kind of information one can extract from perambulations, the Barnag charter has a place where a wood adjoins a *virgultum*: for the contemporaries these two terms meant different things.

Individual trees are less problematic to identify. Usually the document tells us what type of tree we are concerned with. Some complications might arise, of course, especially when post-Linné terminology does not fit earlier practices. Here, for example we read about "a certain ancient tree that is commonly called *magal*." Today, Hungarian *magyal* is the evergreen holly (*Ilex aquifolium*); however, that tree is not native to this part of the country, obviously due to our harsh winters. It took some research to find out that *magyal* in the Middle Ages meant an oak, in one interpretation the sessile oak (*Quercus petraea*).[15] To make matters more complicated, *ilex* in medieval documents should also be understood as referring to oak, perhaps the pedunculate oak (*Quercus robur*), although I am largely sceptical about expecting such botanical precision from the sources.[16] Such problems are not unique to the Hungarian situation. In Northern Italy, for example, early medieval documents did not use the word *quercus* at all. Instead, *rovere* referred to oaks, however, not to the tree that is now called *rovere* (*Quercus petraea*) but most probably to *Quercus robur*, whose modern name (*farnia*) is again missing from the sources.[17] Nonetheless, most cases are clear, and on a basic level perambulations are reliable sources for the identification of oaks, limes, hornbeams (*gyertyán* in Hungarian, also mentioned in the example), etc. How we interpret this information is more problematic. Trees in perambulations do not necessarily represent the trees of the given region in general, since most boundary-trees were not woodland trees. On the other hand, trees in woods could easily serve as boundary signs, providing some extra feature – an earthen boundary, a cross, a picture – was added to them.

2. 1. 2. Terriers

> "First and foremost one *iugerum* of land that belongs to the plot in the middle of that same village Zabady exists around the plot of Prencel on the north. Item, one *iugerum* exists at the place called Kuestelek, in manured land, beside the village of Zabady next to the lands of Ladislaus, son of Chepanus, on the north. Item, another *iugerum* there at the same place, at the end of the lands of the said Ladislaus on the east, and next to the lands of Peter, son of Bochk, on the west. Item, at the place called Urcuthfeu, three *iugera* between the lands of Pethe on the north and those of Georgius on the south. Item, at the same place beside the sign of a wooden cross, three *iugera* that lean towards Yspansag wood. Item, six *iugera* at the place called Rumlutheghaz, of which four are in the vicinity of the lands of Blasius on the south, and two are next to the lands of Paulus, son of Cosmas, on the east . . ."

Extract from the description of the 103.5 *iugera* of land that certain nobles from the village of Szabadi (Co. Veszprém) donated to the chapter of Veszprém (Co. Veszprém), 1314.[18]

These documents do not have their own medieval name. I call them terriers, which covers the main idea behind their making: a description of territory. Scholars tend to confuse them with perambulations, but there is no reason other than the fact that they both talk about land.[19] Although terriers are mostly concerned with arable, they sometimes include meadows, or even woods. The most usual reasons for their commissioning were donations, last wills, and divisions within a family. It also has to be noted that they required individual ownership of land, and not common fields with redistribution of the furlongs every so many years. The amount and location of common fields as opposed to private arable in medieval Hungary is little known,[20] yet it is clear that sociological conditions restrict the availability of this source-type perhaps more than any other.[21]

---

[14] *DHA*, 145–152. Original here means that the actual piece of parchment is extant.

[15] Reuter 1965.

[16] For more on this, see Chapter 7.

[17] Montanari 1979, 36–37.

[18] Df 200 789. "Primo et principale unum iugerum terre sessionalis in medio eiusdem ville Zabadi circa sessionem Prencel ab aquilo existit. Item, unum iugerum in loco Kuestelek vocato in terris fimatis prope villam Zabadi iuxta terras Ladislai filii Chepani a parte aquilonis existit. Item, aliud iugerum ibidem in eodem loco in fine terrarum dicti Ladislai ab oriente et iuxta terras Petri filii Bochk ab occidente. Item, in loco Urcuthfeu dicto tria iugera inter terras Pethe ab aquilone et Georgii a parte meridionali. Item, in eodem ambitu prope signum crucis ligneum tria iugera versus silvam Yspansag finem tendencia. Item, sex iugera in loco Rumlutheghaz vocato quorum quatuor in vicinitate terrarum Blasii a meridie et duo iuxta terras Pauli filii Cosme ab oriente existunt…"

[19] A typical example is Belényesy 1958b. The document in question concerns a division of estates with an extensive terrier, and has nothing to do with perambulations.

[20] Tagányi 1950. Originally published in 1894, Tagányi argued that common fields appeared rather in the south than in the north, and rather in lowlands than uplands. Nonetheless, he concluded that they must have been present throughout the whole country. Sadly enough, no one has touched the topic since then. For a summary of the question, see Laszlovszky 1999.

[21] For more on this, see Zatykó 2003.

In lucky cases, a terrier may cover a settlement in its entirety, which allows the researcher to reconstruct both the inner structure and the fields of the settlement, and then to compare that with what can be seen in the landscape today, something Csilla Zatykó managed at a village called Csepely (Co. Somogy).[22] Other than this, one has to gather information by bits and pieces. In the example above, several place-names appear alongside a wood that comes with its individual name (Yspansag).

2. 1. 3. Estimations

> "Item, first in the said Hangon [they found] the private plot of the same Blasius, one peasant plot and a stone church with a cemetery, one plough of arable land, three ploughs of acorn-bearing woods, twenty *falcastra* of meadows, all to the royal measure, and a river on which a mill can be built. Item, in Susa two ploughs of arable with *nemora*, eight *falcastra* of meadows, four and a half peasant plots, a wooden chapel without a cemetery, one plough of *permissionalis* woods, of which woods only one quarter belongs to the same Blasius."

Extract from the estimation of Hangony and Susa (both Co. Gömör), 1423.[23]
(*Nemus*, here, is a type of woodland. *Permissionalis* woods are most probably coppice woods.)

Estimations are a most controversial source-type. At the first sight they appear to be the perfect tool for the landscape historian, yet the more carefully one approaches them, the more problematic they become.

Most estimations, like the one quoted here, seem to be economic documents, but in fact that is hardly more than mere coincidence. Land here only represents something that has a fixed value, hence the name. Estimations are jurisdictional sources, and were commissioned when for some reason (division, filial quarter, distributing the possessions of someone that was arrested, etc.) authorities wanted to know how much money someone's possessions were worth. Possessions included plots of peasants, land, buildings, livestock, mills, etc. Estimations have an out-of-this-world feel to them: everything had a fixed "price." One *aratrum* of arable was always three marks, regardless of its fertility, or, in other words, its actual value.[24] There survive several contemporary lists of what should be included in an estimation, the most famous of these is certainly the one in Stephen Werbőczy's *Tripartitum*, the great early sixteenth-century compilation of Hungarian customary law.[25]

Problems with estimations are innumerable, mostly because there is hardly any secondary literature about them.[26] We know little of when they developed, how many of them have come down to us, how reliable they are, or how far we can push their interpretation. *Yet estimations are the only medieval sources that allow us to – at least partly – answer the crucial question of how wooded Hungary was in the Middle Ages*. In theory this could mean detecting changes in time as well, as it is possible in England to compare the Domesday Book with the Hundred Rolls of 1279,[27] but in reality estimations do not come in sufficient numbers and quality before the fifteenth century.

2. 1. 4. General Charters

> "Over that case there was no contradiction, and the arable on the northern side until the pinewood freely belongs to the said people. In which assertion there was a consent among the oft-mentioned Forest-guards, that is Benke centurio, Zunka, Weoda and Martinus, who were the executors to this case with all their people, both old and young who appeared in the case of this said land. And they will fully give up all prosecutions and actions of the whole case that has arisen over the said land under the condition that the people of the said church shall not in the future try to grub out any living tree. Furthermore, the said abbot obliges himself to pay to the said guards a sum of three or four marks, if with their other people, they perform a sincere oath over this matter . . ."

Extract from a peaceful agreement in a case between the abbot of Pannonhalma (Co. Győr) and the Forest-guards of the village of Kenyeri (Co. Veszprém), 1258.[28]

> "Geyza, once king of Hungarians . . . donated [to the hospitallers of Esztergom] the freedom to take five cartloads of wood every day from his very own wood called Ples for the daily uses of their house."

---

[22] Zatykó 1997. See also Major 1959.

[23] Ila and Borsa 1993, 140. "Item primo in dicta Hangon propriam sessionem eiusdem Blasii, unam sessionem iobagionalem et ecclesiam lapideam sepulturam habentem, terras arabiles ad unum aratrum, silvas glandinosas ad tria aratra, prata ad viginti falcastra regalis mensure se extendentia, fluvium, in quo potest fieri molendinum; item in Susa terras arabiles cum nemoribus ad duo aratra, prata ad octo falcastra, quintamdimidiam sessiones iobagionales, capellam ligneam sine sepultura, silvas permissionales ad unum aratrum, de quibus scilicet silvis solum quarta pars esset eiusdem Blasii."

[24] Fügedi 1998, 32–33. The price was not created out of the blue, but was fixed in the thirteenth century, when this was the average value. Hóman 1916, 506. See also Szűcs 1984, 366.

[25] *Tripartitum*, I, 133–134. How "customary" is the *Tripartitum* is, of course, debatable. Further estimation lists are Dl 32 012, Dl 48 204, Df 283 678, Df 283 386. A discussion and comparison of these may be found in Bónis 1972, 162–185. A list Bónis did not know about is Df 252 476. On the practical aspects of the process, see Bogdán 1978, 44–46, 53–55, 66–69.

[26] The development of the word itself has been studied in Rus 1999.

[27] Rackham 1996, 59.

[28] *ÁÚO*, vol. 2, 315. "Super quam causam contradiccio nulla habebatur, et terram arabilem a plaga aquilonari usque ad pinetum dictis populis libere pertinere; in quam assercionem consenserunt sepe memorati custodes silue, videlicet Benke centurio, Zunka, Weoda et Martinus, qui litis huius fuerant executores cum universis sociis suis senioribus ac iunioribus, qui omnes in facto dicte terre comparuerunt, et omnis litis prosecucioni et accioni, quam super terra memorata habere videbantur, omnino renunciarunt tali condicione, quod populi dicte Ecclesie imposterum nullam arborem viventem audeant extirpare; insuper eciam supra notatus Abbas obligauit se dictis custodibus silvarum expensas ad tres vel quatuor marcas refusurum, ita tamen, si cum ceteris sociis suis super hoc ipsum serio prestarent sacramentum."

Extract from a charter of Pope Urban III, where the donations of Géza II (King of Hungary 1141–1162) to the hospitallers of Esztergom (Co. Esztergom) were recited, 1187.[29]

"General charters" is evidently an umbrella term intended to include everything that does not fall within the three main types described above. I mention these two examples to illustrate that there is a fairly wide choice, although the two basic groups are court-cases and privilegal charters.

These types of sources are even less designed for use by the historical ecologist than the previous ones. Whether one finds out something useful or not is purely accidental. The charters I shall be using are mostly connected to the history and administration of Royal Forests. Evidence of woodland management, such as the one quoted above, are almost exclusively found in these sources. There are kinds of documents, for example court-cases describing illegal wood-cutting, that are available in larger numbers and can be utilised in a quantitative sense.

2. 2. Archaeological Sources and Methods

Anything we know about the days before the letter – in practice in Hungary any time before AD 1000 – we know from archaeologists. However, the usefulness of archaeological sources does not cease when written material appears; in the Middle Ages they provide an independent angle for the questions we ask. What is important and stressed in the Introduction, is that we are concerned with more than a simple integration of traditional archaeological data within an ecological discussion. Archaeology itself has developed new ways to discover and analyse new sources that are particularly useful for our purposes. On the one hand, entirely new methods were born (such as aerial photography) to study hitherto neglected landscape elements (such as ditches). On the other hand, more and more emphasis has been laid on the environment at many "traditional" archaeological sites. Remains of plants and animals are recorded and studied, providing valuable insights into past landscapes. Here, I shall outline the basic sources and methods that are relevant in woodland studies. I write *sources* and *methods*, but we have to keep in mind that in archaeology the two cannot be separated as clearly as in traditional history. With written sources, the "method" is straightforward: one reads them. All the rest is interpretation. Is, however, an ancient piece of wood a source? If so, there are a number of ways to read it, each beyond the abilities of all but specialists. These specialists publish studies in a form that is meant to be understandable. It is these studies that we have to approach critically, not the original material. We have to be aware of the advantages and disadvantages of each method, both in its study of the material remains and in its interpretations. Because archaeology is notorious for destroying its own study material, scientific studies are often the only "sources" we have.

2. 2. 1. Pollen Analysis and Related Methods

Pollen analysis studies the powder flowering plants produce to fertilise their matching partners. Pollen can be carried by insects or wind: different plants have chosen one or the other solution in their evolution, therefore some produce more pollen than others. If not affected by the air or deposited in basic soils, pollen is a very resistant material; in certain places, especially in acidic soils and peat bogs, pollen is preserved as it was deposited layer by layer, year by year, going back several millennia. It is then possible to take a core and count the number of pollens of different species (or sometimes only different genera) in the individual layers. Counting is to be understood literally, thousands of pollens in one layer, and there may be several layers.[30]

Once the amount of pollen from different species is available, the analyst can establish the proportion of each plant within the whole. What to do then is more problematic, because there is no direct relationship between the proportion of pollen by a plant and its proportion in the actual vegetation. Oak, for example, produces a lot of pollen, lime produces only a little and hazel pollen varies with the amount of light the plant gets. In the past few decades, several methods were developed to overcome this difficulty. First, pollen proportions are now modified from absolute proportions to relative ones, which may reflect vegetational patterns better. Secondly, independent methods are applied alongside with pollen analysis. These include *charcoal analysis* (to see if any burning occurred), *geo-chemical analysis* (different chemicals leach into the deposit under different conditions), and *malacology* (based on the fact that certain snail-like animals prefer particular types of habitat).[31] By using a combination of these, light is shed on what was going on in the woods of the Carpathian Basin before the coming of the Magyars.

2. 2. 2. Archaeobotany

Archaeobotany is the science of macro plant remains from excavations. Of all the natural scientific methods, this has probably had the longest continuous development in Hungary and has produced considerable results.[32] This discipline, however, is more suitable for studying what plants people cultivated and what they were eating than in discovering the landscape around them. Much information (e.g. about the appearance and significance of wheat, or the types of cultivated fruit trees) still helps the historical ecologist to learn more about past woodland in an indirect form. In fortunate cases, as we shall see with the excavations of the wells of the Royal Garden in Visegrád (Co. Pilis) in Chapter 11, immediate conclusions are also possible.

2. 2. 3. Dendrochronology

Dendrochronology researches the annual rings of trees (in Europe, mostly oak).[33] It uses both dead and living trees. Dead

[29] Knauz 1863, 131. "Geyza quondam rex Hungarorum . . . ad cotidianos (!) usus ipsius domus singulis diebus de propria silva sua, que vulgo Ples nuncupatur, quinque currus lignorum portandi liberam . . . contulit facultatem."

[30] For a general handbook, see Faegri, Kaland and Krzywinski 1989. The basic Hungarian reading is Zólyomi 1936.

[31] See for example Sümegi, Hertelendi, Magyari and Molnár 1998.

[32] Gyulai 2001. An English version of this book with an extensive catalogue of finds is forthcoming by Archaeolingua, Budapest.

[33] Basic readings are for example Stokes and Smiley 1968; Schweingruber 1983. A very instructive book is Baillie 1982.

trees survive in standing buildings (this rarely happens in present-day Hungary) or in archaeological excavations (this occurs more often). In Europe and in North-America, where there are cold and hot seasons, trees do not grow all year but spend some time at rest, which is visible in the cross-section of any tree. How wide one annual ring is is partly determined by external factors such rainfall, light, and temperature. In smaller or bigger geographical regions, trees of one species show a similar pattern in their rings. If several samples are available, it is possible to put together a long chronology with the help of *crossmatching*. Crossmatching is based on the fact that a pattern of about 25–30 years must be unique, thus sequences of 70–80 years or so, if they match over a smaller part, will take the chronology further back in time. In certain parts of the world, chronologies are several millennia long. When this *master chronology* is done, individual finds can be dated with amazing precision as opposed to radiocarbon dating, for example. Many disciplines, above all climatology, benefit from the work of dendrochronologists. Woodland studies, too, are aided by tree-rings. Pollard cycles, which did not leave much written evidence even in England, were established by superimposing the widths of annual rings on the master chronology in Hatfield Forest.[34] In France, what people did in medieval woods was researched with tree-rings.[35] The results of dendrochronology will be used in this work to a far lesser degree than what would be desirable. The reason for this is that systematic research is a recent development in Hungary: for a start, there is no master chronology available that would stretch back to the Middle Ages.[36]

2. 2. 4. Botany and Ecology

Woods themselves are sources. Their composition (what plants are found in them and in what proportion) is largely influenced by people. *Botany* is the science that tries to understand plants as individual living beings; *ecology* focuses on how plants form communities. In the same sense, *flora* is the list of all plants that exist in a place regardless of their abundance, while *vegetation* is the same plants "considered in terms of the abundance of each species and its relation to other species."[37] Most Hungarian woods have been carefully described and put into categories by phytocoenologists,[38] but these categories will not have much to say to the non-specialist. Besides over-complicated terminology, the other problem is that in Hungary the tradition is to concentrate on *potential vegetation*. This means that many botanists investigate what plants *should* grow in one place, not what is there in reality. Human influence on plant composition is usually restricted to modern forestry, and the possible impact of earlier practices is neglected. Nowadays a lot is written about how plants interact among themselves, but more often than not those works look like mathematics and not something that was written for people to read. The biggest difference between ecology and historical ecology is that the latter is "to a large extent, history and not pure science. . . Usually we have to make do with information which exists and has survived for reasons unconnected to its usefulness to us, and whose quality needs to be critically assessed."[39]

2. 2. 5. Woodland Archaeology

This type of landscape archaeology studies features that are specific to woodland.[40] In other words, it is archaeology *of* woodland rather than *in* woodland. Features that are unrelated to woodland and were only preserved in woods (such as medieval field remains or bronze age tombs) are excluded from it.[41] The most typical phenomena recorded through *fieldwork* and interpreted later are pollard trees, coppice stools, and woodbanks. In Hungary, living trees rarely survive from the Middle Ages. Nonetheless, one has to keep in mind that until the birth of modern forestry, woods were most probably managed the same way, thus trees more than three hundred years old will tell us about the old practice. The most important principle here is that ancient trees display in their outward appearance whatever people did to them in the past. Both pollards and coppice stools survive in large numbers in Carpathian Basin and await investigation. Woodbanks and ditches protected coppice woods in many parts of medieval Europe, and it is a highly important question whether this was the same in Hungary as well. This is not only a theoretical problem: woodbanks are such a powerful tool in deciphering the history of individual woods that their presence would add a large new dataset to the scarce source material. Very little is known about the archaeological features of woods in Hungary. This book will be the first attempt to address the question.

2. 2. 6. Topography and Toponyms

Human settlements do not usually form part of woodland and Forest studies. In my case, they need to be included for two reasons. Firstly, because in the Forest studies I am concerned with geographical regions rather than individual woods, where settlements are an integral element in the landscape; and secondly, because settlements represent another possible source-type, which, given the fact that I work in a poorly documented period and an even more poorly documented topic, must not be left out of the discussion.

I will include not only those settlements that have written records, but also those that archaeologists found during their *fieldwalks*. Fieldwalking is when scholars systematically survey a region – walk over it – and collect finds, usually potsherds, which are problematic to date, but less so

[34] Rackham 1989, 246–248.

[35] Lambert 1996.

[36] Grynaeus 1996; Grynaeus 1997. Some examples of the kind of historical ecological conclusions that can be drawn from the available material: Grynaeus 2003.

[37] Rackham 1980, 401.

[38] Soó 1964–1985. This is a "German" type of "handbook," it is barely possible to *lift* the individual volumes with one hand.

[39] Rackham 1980, 16.

[40] The classic handbook is Rackham 1996, originally published in the *Archaeology in the Field* series.

[41] Although they often provide valuable information. In the above cases, they prove that the woods in question must be secondary.

than settlement-names.[42] Methodological implications of fieldwalking will be discussed in Chapter 12.

Names of settlements (and those of their smaller units) ought to be useful to us, but they are difficult to interpret. In theory, they should give us a *terminus ante quem* of when one settlement was established on purely linguistic grounds. Although some scholars still think it is possible, others have come up with the frank yet discouraging suggestion that in Hungary place-names are of not much use in dating.[43] The biggest problem is that the largest block of medieval settlement-names are derived from personal names, and this mountain of information at present looks like dense rock: linguists have merely scratched the surface, but are not able to say what might be inside. Nonetheless, it has been argued that some toponyms contain important information for the woodland historian: the *–aszó* ending, for example, may refer to a clearing in a wood, something comparable to the well-known English *–ley*.[44] According to another theory, settlement names ending in *–falu* ("village") are a sign of internal colonisation, thus, woodland clearance.[45] It seems that meaningful results can be best achieved by analysing *groups* of similar settlement names.[46]

In Hungary, most ancient woods have their own name. These can also be analysed. As always, one should not draw straightforward conclusions – Somos (*som* in Hungarian is "cornel") in Bakony was probably not a cornel-wood but rather a wood with at least one conspicuous cornel bush. If this sounds obvious, why should a Tölgyes (*tölgy* is "oak") be interpreted as an oakwood? As in the case of settlement names, groups of similar woodland names should be analysed together and in the context of woodland history. Such a work, however, remains to be conducted.

Old maps and *aerial photographs* can also provide useful topographical information. Well-known early small-scale maps, such as the depiction of Hungary by Lazarus (1528), were not meant to be accurate in the modern sense and are meaningless for our purposes. In Hungary, the first good quality large-scale estate and cadastre maps were produced in the eighteenth century, and became widespread in the nineteenth. The first Ordinance Survey, covering the whole of Hungary at the scale of 1:28,800, was carried out in 1782–1784. It was accompanied by a questionnaire about the location, buildings, woods, waters, etc. of each settlement from a military point of view. These rich descriptions mostly remain unpublished to the present day. The sheets of the first OS, nonetheless, do not come near those of the second (1829–1866) and the third (1869–1884), which, by the way, were produced independently of the first.[47] These beautiful maps accurately depict every wood and attempt to show individual trees as well. They should be the basis for all historical woodland studies in Hungary (*Fig. 6*).

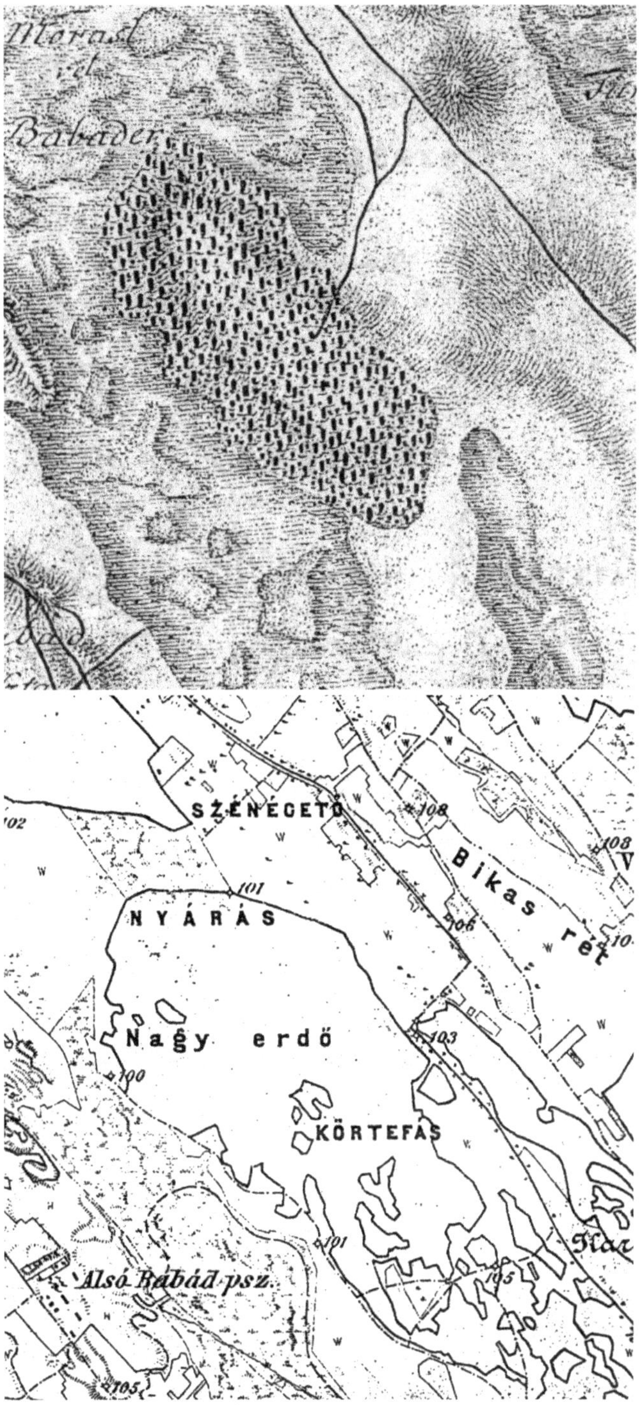

Fig. 6. Nagyerdő in Ócsa (Co. Pest) on the first (1783) and third (1886) Ordinance Survey. The shape of the wood has remained virtully the same, except for the southern part. The wood, with minor alterations, still looks like this today. Note the different styles of the two sheets, not only in the attention to detail but also in the representation of woodland.

Aerial photography is a relatively recent development in Hungarian archaeology.[48] The reason for this is sadly simple: between 1950 and 1989 flying over the country and taking pictures was a suspicious activity only allowed for military purposes. Suspicion, however, had a positive effect as well: the whole country was surveyed in the 1950s, and these by now historic pictures, which record the countryside before the radical changes of collectivization, are today available for study.

[42] An example of how powerful a tool fieldwalking can be when applied to its full potential is Astill and Davies 1997.
[43] Kristó 1976.
[44] Takács 1980, 135–145. See also Chapter 4.
[45] Kázmér 1970, 56–64, distribution map on page 63.
[46] A classic in settlement studies applied the same method with great success: Heckenast 1970.
[47] A good summary of Hungarian map-making is Csendes 1980.
[48] Visy 2003b.

Chapter 3

The Prehistory of Woodland

What does the historian of medieval woods have to do with prehistory? The answer lies in the continuous development of the landscape. Native woods, as we shall see, had greatly diminished in extent by AD 900 and many of those still in existence were managed. *Wildwood*, the untouched primaeval woodland, if any, lingered on only in outlying regions. Clearance probably started in the Mesolithic and was already a large-scale activity, as in other parts of Europe, by the Neolithic. The Middle Ages must, thus, be understood as part of a longer process.

3. 1. Wildwood

Wildwood was very much unlike the managed woods of today, but it is difficult to imagine what the wildwoods of the Carpathian Basin may have looked like. Most of the popular ideas about wildwood are derived from rainforests, which, being in an entirely different climate without seasons, are of little help to us. One solution is to look at the remaining wildwoods of the United States (a comparable climate with trees that look similar to the European ones though they are rarely the same species). The other option is to examine those woods in Central Europe that were set apart as "woodland reserves."[1] These are not wildwoods, but simply areas of woodland where human impact in the past had been relatively small and where all tree cutting and other management activities have been banned. The oldest reserves in this region are about one hundred and fifty years old. According to Csaba Mátyás, their characteristic features are the following: they consist of a multitude of tree species grouped on smaller territories in a mosaic-like fashion; large trees dominate the general view; there are many fallen and rotting trees; microclimatic conditions are variable; and there is no close relationship between the size and the age of the individual trees.[2] Our knowledge of the similarities and differences between European and American "wildwoods" was greatly advanced by the publication of George F. Peterken's *Woodland Conservation and Management* and his *Natural Woodland*.[3] Peterken observed that "even under the most favourable circumstances, natural forests never quite achieve the complete and intimate intermixing of species and ages of popular supposition."[4] Even-aged trees tend to form groups. Any wood where all ages are present in all places can only be artificial.[5] This is the result of a system where regeneration is not easy (there are always many old trees that allow little sunlight down on the ground, although some trees are more tolerant of this than others) and usually takes place in the gaps left by dying dominant trees.

Wildwood was not static. It changed through time along with changes in climate on the one hand, and its own dynamics on the other. "Whatever the theoretical ideal, natural forests have in fact had little chance of ever developing a completely stable state."[6] This also applies to tree composition, which means that the historical dominance of oak in the lower-lying regions of the Carpathian Basin may turn out to be a myth. In wildwood, which kind of tree had the upper hand for a generation in a given area very much depended on such seemingly minor issues as having plenty of seeds in the right year in the right place, faster rates of growth (genetically) than the neighbouring trees, having favourable climate in a year when there was space to grow, etc. All these factors have been in place for hundreds of millennia, but the time that concerns us directly (the Holocene) began after the withdrawal of ice from the Carpathian Basin about twelve thousand years ago.[7] Until recently, the prehistory of woodland could be described only in general terms. It was the works of Bálint Zólyomi that influenced our understanding of how wildwood changed.[8] Zólyomi worked, among other things, with cores in pollen deposits from Lake Balaton, the largest lake in East-Central Europe. With its huge catchment area, the Balaton is certainly very useful for drawing general conclusions about woodland history in the Carpathian Basin.[9] Zólyomi's research resulted in a four-phase periodisation: 1. *pine-birch age* (until 8000 BC). Mostly *Pinus silvestris*, with some birch but generally not closed woodland. Cold and dry climate. 2. *hazel age* (8000–5000 BC). Oak, ash, and limewoods. Abundant hazel pollen shows that either woods were open or there were extensive hazel woods. Warm and dry climate. 3. *oak age* (5000–2000 BC). Hazel withdrew, probably because the canopy closed leaving insufficient sunlight for hazel to flower. Warm but increasingly wet climate. 4a. *beech age* I (2000–800 BC). Beech and hornbeam gain ground. Less oak. Cooler and wet climate. 4b. *beech age* II (800 BC – present). Slight warming, more oak.

This method, however, had serious drawbacks. Firstly, radiocarbon was not used in dating the pollen, dates were

[1] Broekmeyer 1993; Broekmeyer 1994; Průsa 1985; Mayer 1989. For the Carpathian Basin, see Fekete 1906; Midriak 1994; Czájlik 1994. There are sixty-three woodland reserves in present-day Hungary. Research, however, has barely started in them, and has focused on the dynamics of tree-communities. Bartha and Esztó 2001.

[2] Mátyás 1996, 84.

[3] Peterken 1981; Peterken 1996. His work was largely influenced by Jones 1945.

[4] Peterken 1996, 163.

[5] Peterken 1981, 3–6.

[6] Peterken 1981, 6. For disturbance, see Peterken 1996, 86–116.

[7] For an overview, see Roberts 1989; more recent research is summarised in Dickinson 2000.

[8] Csőre 1980, 35–37.

[9] Zólyomi 1936; Zólyomi 1952.

rather achieved through apparently circular reasoning.[10] Even if available, the radiocarbon dates were not calibrated, which means that one can never be certain whether they are correct or not.[11] Moreover, although Zólyomi did use methods to get relative percentages from the absolute proportion of species in the pollen record, this was not always carried out and, in any case, the methodology used was questionable. Secondly, general statements leave apart the intricate regional variation so characteristic of the Carpathian Basin. If one discusses the history of a particular wood, what does "beech age" actually tell us about the vegetation of that wood? As Zólyomi himself pointed out,[12] it would have been necessary to analyse smaller lakes to get information on particular pieces of woodland. Thirdly, and for the present purpose most importantly, the impact of human activities was completely neglected. Hints were inserted here and there, but always in general terms. These studies suggest that wildwood predominated and remained untouched until the Iron Age or even afterwards. Fourthly, parallel methods (malacology, charcoal analysis, geochemical analysis, sedimentology, etc.) were not applied.

More recently, there have been studies that examined deposits from smaller lakes and peat bogs, and made full use of the parallel methods mentioned above. Most interestingly, some of them compared the data with the number of known archaeological sites from different periods in the vicinity of the lakes, thus inferring conclusions about human impact on the environment.[13] These works have drawn a complicated and stimulating picture. It is now unquestionable that the Neolithic brought fundamental changes to the wildwood almost everywhere in the Carpathian Basin. Moreover, there is evidence, in the form of microcharcoal, for burning of the wildwood as early as 7000 BC, in the Mesolithic.[14] These smaller charcoal peaks tend to coincide with peaks in the hazel pollen. As mentioned above, hazel does not flower when overshadowed by taller trees. The broad-leaved wildwood formed by this time around the deposits in question would not have burned quite as easily as the previous coniferous woodland. Therefore hazel, as an indicator of open spaces, plus charcoal, even if in small amounts, may reflect deliberate burning of the wildwood by Mesolithic people.

3. 2. Kis-Mohos Lake

Without further generalisations, I would like to present one example and illustrate the kind of lessons palaeoenvironmental studies can teach the woodland historian. In the deposit of Kis-Mohos Lake, which is in a hilly region in the north-east of present-day Hungary, several anthropogenic activities were shown to have occurred in the Holocene.[15] Since the region lies beyond the territory of the first wave of Neolithic culture, people mainly appeared there in the Middle Neolithic (ca. 5400–4800 BC). Before this, the wildwood (oak, hazel, lime, hornbeam, elm, and ash) remained undisturbed. The Middle Neolithic was a period of unequalled woodland clearance. The charcoal concentration had two peaks, the first of these – indicating when people basically appeared in the area – was the highest ever. Half of the woodland disappeared. In general, Neolithic settlers created an open landscape around the lake, but cereal cultivation was not yet present. The large-scale production of pottery may be connected with a second charcoal peak, in other words, woods may have already been managed for firewood. Intensive woodland management used to be thought of as beyond the agricultural abilities of such early peoples as the Neolithic settlers, but now we know that, in fact, the contrary is true. The analysis of the Sweet Track (Somerset, England) showed evidence of a coppicing structure rather more sophisticated than the medieval one.[16] Because the original coppice shoots were found, there can be little argument with the conclusions. Pollen analysis provides more indirect data, but after the Somerset excavations the fundamental question is no longer "Is it possible?" but rather "Are my conclusions correct?" An interesting example of this is the analysis of a hollow in Næsbyholm Storskov (Denmark). Here, for a few hundred years after 4000 BC, lime-pollen was abundant while all other trees virtually disappeared. This fact lead the author to suggest that lime had been shredded on a large scale.[17] This is an elegant solution, however, it is equally possible that, for example, there existed a mixed coppice wood with very short cycles (which would prevent trees from flowering) containing lime standards. One must always keep in mind that after the Mesolithic, the lack of pollen from a species does not necessarily mean that the tree itself was missing.

Returning to northern Hungary, the records of Kis-Mohos Lake show that woodland largely regenerated until around 4000 BC (*Fig. 7*)[18]. Then oak, elm, lime, and hazel drastically declined, whereas beech, birch, and hornbeam increased. This has little to do with climate, but is allegedly connected to copper mining and smelting. This was followed by the appearance of agriculture proper, as can be judged by substantial erosion for the first time between 3200–2300 BC. This continued until 1500 BC, with beech and hornbeam in decline, while birch increased. Settlers of the Middle Bronze age, thus, continued to manage the woods and used temporary fields, which were later invaded by birch. Erosion was severe in the early Iron Age, and the structure of the remaining woodland changed. From 800 to 300 BC, the territory was once again abandoned, and the woodland regenerated with beech and hornbeam. Then, after 300 BC, Celtic tribes appeared in the neighbourhood. Erosion was very intense and oak pollen replaced hornbeam and beech. For the first time, hemp was cultivated around the lake. This impact was to be on-going until the Hungarian Conquest. After the AD tenth century, cereal and

[10] Dating, when based on archaeological stratigraphy and then fit into a pre-existing pattern, which, in turn demonstrates that layers are indeed connected to certain periods, seems to me to be a kind of circular reasoning.

[11] On radiocarbon and calibration, see Pilcher 1983; Taylor, Long and Kra 1992. See also the *Radiocarbon* journal.

[12] Zólyomi 1952, 521–522.

[13] Sümegi, Magyari, Dániel, Hertelendi and Rudner 1999; Sümegi 1998a; Sümegi 1998b; Willis, Sümegi, Braun, Bennett and Tóth 1998; Sümegi and Bodor 2000. There are many similar examples from other European countries. See for example Part Two in Birks, Birks, Kaland and Moe 1988. This part has eighteen articles, almost all based on the same principles.

[14] Sümegi 1998b, 194; Willis, Braun, Sümegi and Tóth 1997, 745. For the "human" interpretations of pollen records, see Behre 1986.

[15] The following is based on Sümegi 1998a.

[16] A trackway in Somerset is a wooden structure that interconnects islands in the peat bog. Rackham 1979.

[17] Andersen 1988. On the same topic, see also Rasmussen 1990a. For a critical view, Rasmussen 1990b.

[18] Reproduced from Sümegi 1998a, 383.

hemp cultivation dominated, and tree pollens declined. Since the seventeenth century, there has been continuous secondary woodland formation as people withdrew from the area around Kis-Mohos Lake. Nowadays we do not think of this environment as particularly habitable. This is reflected in the high amounts of tree pollen and in the absence of erosion.

As this example demonstrates, wildwood started to disappear from around 4000 BC and by 3500 BC the structure of woodland, or at least as much as can be detected in the pollen record, seems to be the result of management practices. The only time one is able to detect primaeval woodland regeneration on a larger scale is the Early Iron Age, but it is difficult to tell whether beech and hornbeam increased because they would have been the "natural" vegetation of the place at this time or because they grew nearby as the result of management practices and thus had better chances to colonise.[19] The pollen diagrams would make very little sense without comparison to the number of archaeological sites.[20] It is also interesting to note that this deposit does not confirm the popular belief that large areas reverted to woodland (let alone wildwood) before the Hungarian Conquest. Sümegi convincingly argued that the natural balance of the area had been irreversibly changed in the four centuries following 300 BC. After this time, the vegetation could not have returned to its "natural" state, even if it had had the chance. What we have to imagine as the medieval landscape (without cereals) was also formed around this time and persisted until the Early Modern Period. Medieval Hungarians, it seems, brought nothing fundamentally new to the landscape, but only exaggerated already existing effects.

We must also remember that every site tells a different story. In Tököl (Co. Pest), for example, the present open landscape was created in the Roman period.[21] At Nyíres Lake (near the Ukrainian-Hungarian border), the Roman period marked the starting point of woodland regeneration and peat-bog formation.[22] The history of specific places is determined by their ecological limitations and capabilities and by contemporary ideas of where it is good to live. For example, people in the Late Bronze Age favoured places where AD sixth century peoples, just like us today, did not care to dwell. To date, it appears that general conclusions on the regional variations of human impact on woodland will only be possible when more case-studies are carried out. As far as I can see, the most problematic question is how human activity influences the composition of woodland. There are two main reasons for this. Firstly, little is known in general about the intricate relationship between prehistoric wildwood and managed ancient woodland. When did the transition take place? What were the proportions of wildwood and managed woodland in different periods? What happened to managed woods when they were abandoned? How far can management change the composition of woodland? In England, there is certainly some connection between regional variations in ancient woodland and those of wildwood, which would suggest that vegetation can be very persistent.[23] Secondly, one has to consider the above-mentioned problem of pollen production. As I have discussed, pollarding and coppicing can prevent flowering, thus making certain species disappear from the pollen record if not in fact.

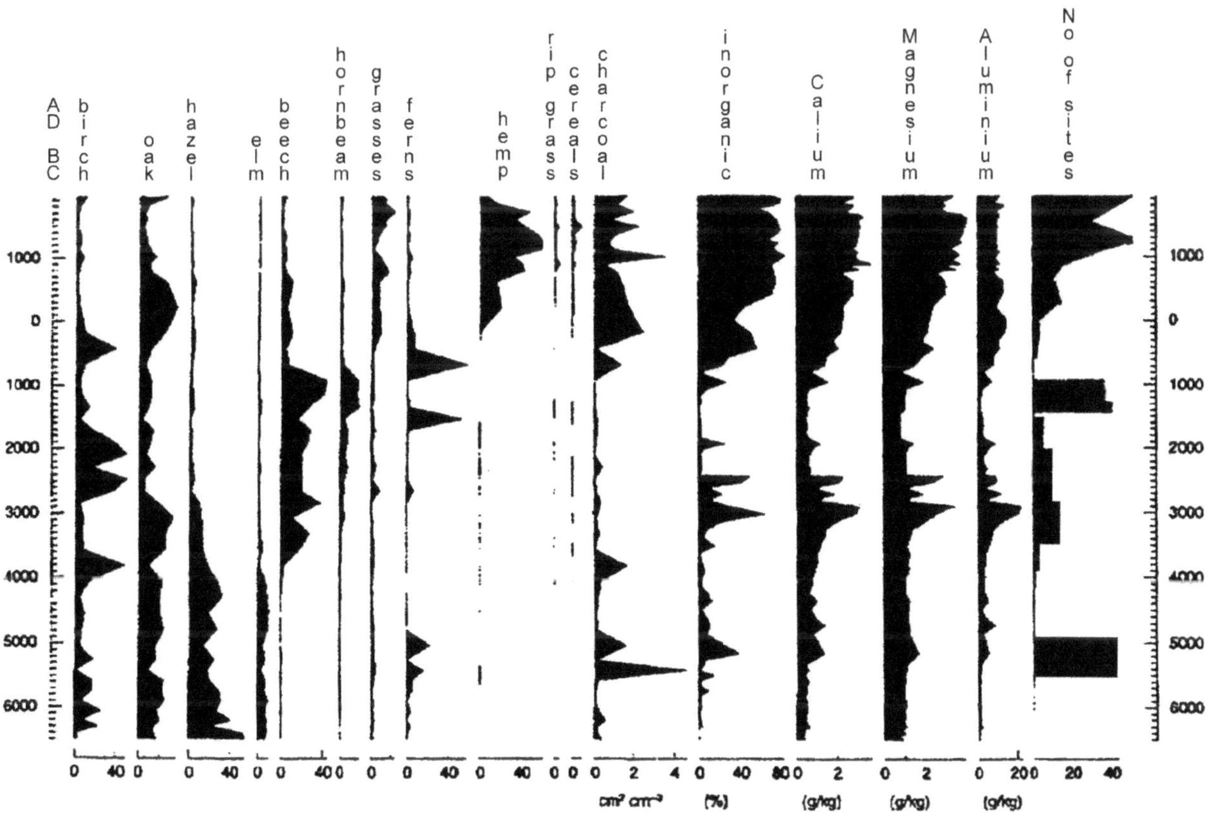

Fig. 7. Percentage pollen and spore diagram from Kis-Mohos Lake plotted against calibrated radiocarbon dates, together with charcoal concentrations.

[19] Although beech is a weak coloniser.

[20] It must be noted that Zólyomi 1952 was one of the first attempts anywhere to compare pollen diagrams with the number of archaeological sites.

[21] Sümegi and Bodor 2000, 89.

[22] Sümegi 1998b.

[23] Rackham 1996, 28–33.

Chapter 4

The Destruction of Woodland

Woods – and this cannot be repeated enough – are not destroyed by cutting. They are destroyed when the territory is transformed into land intended for some other purpose: arable, meadow, or pasture. The Anglo-Saxon poet very rightly described the ploughman as the "enemy of the wood."[1] Treeless lands, then, need to be maintained, otherwise in most places they revert to woodland. Thus, it is clear that population density and the abundance of woodland are closely related to each other.[2] Many woods in Hungary are secondary: at one point in history they were used for something other than wood-production. There are obvious signs of this: traces of former settlements, field remains,[3] or even ridge and furrow, and less conspicuous ones: certain vegetation patterns.

We have seen in the previous chapter that people had already cleared some land in the Mesolithic, but large-scale clearances started with the appearance of agriculture in the Neolithic. There are many theories about how Neolithic or later settlers destroyed wildwood, yet only some may be proved to be more than relevant ethnographic analogies. Experimental archaeology has demonstrated that it is possible to cut substantial trees with Neolithic axes,[4] but the problem is that one needs written evidence to be able to argue that certain things not only *could* happen but in fact *did* happen. The first thing to do in clearance was always getting rid of the trees, an immense task in itself, especially keeping in mind that wildwood was full of gigantic trees, both standing and fallen. Then the stumps needed to be removed. We know next to nothing about the method used for this.

In the following, I shall discuss different methods of woodland clearance in the Carpathian Basin. Parts of this chapter will already take us to the Middle Ages, because, as I have just mentioned, *written* evidence is very important in this matter. This also means that I shall not only be concerned with the destruction of wildwood, but also with the better known clearance of more or less managed woods, for which the proper term is *grubbing out*. This topic has a considerable scientific literature in Hungary, written mostly by ethnographers. Ethnography, as such, has a special place in the agricultural history research in Hungary. Because this country was industrialised considerably later than much of Western Europe, traditional life lingered on, especially in more remote areas, until recent times. Ethnographers, thus, could go to villages as late as the 1950s and study ancient ways of land management. For example, they could observe ridge-and-furrow in the making.[5] But they did more than that. Stimulated by field experience, they carried out high quality archival research, as well. I have already mentioned the works of Márta Belényesy, who was also an ethnographer. Another emblematic figure of this school was Iván Balassa, who investigated the history of plough and ploughing.[6] Woodland clearing techniques and implements of the eighteenth-nineteenth centuries were studied by Lajos Takács.[7] The most problematic aspect of his otherwise excellent books, and of all such ethnographic-historical studies, is whether it is really justifiable to project the results back to the Middle Ages, let alone earlier times. Firstly, ethnographers tend to generalise and talk about "pre-industrial society," however, historians know that such a thing never existed. Change, at times as rapid as nowadays, has always been present. Secondly, ethnographers were mostly concerned with the Early Modern Period, which, because of the Ottoman occupation, has no direct continuity with the Middle Ages in most of the country. Thousands of settlements were deserted, as we shall also see in Pilis and Bakony Forests, which were resettled in the eighteenth century. One of Takács' books is about such a resettling process, which, apparently "was very similar to the one which can be reconstructed from the written sources of the Middle Ages."[8] Nonetheless, how far similarities can actually go is not known.

4. 1. Burning

This technique is more suitable for some areas and for some tree communities than for others. Also, as Pyne put it, "there is not one fire but many. Each has its habitat, its traits, its behavior, its ecology."[9] Today's spectacular fires take place mostly in conifer plantations, where trees are set closer to one another than in woods. Conifers, in any case, are much more combustible than other trees. There is abundant palaeoenvironmental evidence for the spontaneous burning of conifer wildwood. It seems that non-plantation coniferous woods need to be burnt from time to time, it is part of their

[1] *Exeter Book Riddles* 1993, riddle no. 21. The solution is debated, it may be plough, but in our case that does not make a difference.

[2] Population density is not simply the number of people living in a region. It also depends on how many people a given amount of land can support.

[3] Nováki 1975–1977; Nováki 1984–1985; Nováki 1990; Torma 1981. These studies were all by-products of larger projects, such the *Archaeological Topography of Hungary*, or the excavations at Sarvaly (Co. Zala). Holl and Parádi 1982a. With Gyula Nováki retired, no one is around with an eye to discover field-remains. Recently, this situation appears to be changing. For a summary of the topic, see Laszlovszky 1999.

[4] Iversen 1956; Coles 1973; Steensberg 1957. It is difficult to imagine, though, how it would have been possible to cut the immense trees that are supposed to have dominated the primaeval woods.

[5] Dömötör 1954.

[6] Balassa 1973.

[7] Takács 1976; Takács 1980.

[8] Laszlovszky 1999, 435.

[9] Pyne 1997, 9.

normal life-cycle. Woodlands in the Carpathian Basin, however, are different. No one in their right mind would try to set fire on large flood-plain woods, but the climate in many places is arid, therefore flat, dry areas may have been apt for burning living trees.[10] Written evidence for this, contrary to all assertions, does not exist. Although one occasionally reads about a wood that burnt down,[11] there is no case where it was explicitly mentioned that this was done in order to destroy the wood, nor is there any hint that the land ceased to be woodland.

4. 2. Ring-Barking

This means that a "ring" is cut all around into the bark of the tree. Why this should kill the tree on the long run is explained by the way trees function. The trunk of a tree has five basic parts, four of which are clearly visible in its cross-section. The thickest and innermost part is the heart, which is dead and supports the tree. Then we see the sapwood, which is often lighter in colour and is living. The sapwood is surrounded by the bast, a soft layer, which is protected by the bark, a hard and dead part: the outside surface of the tree. The fifth, very thin component lies between the sapwood and the bast: it is called the cambium. It produces both sapwood and bast every year. The heart and the sapwood transport water from the roots; the bast carries solute organic material from the leaves. Cutting through the bark and the bast will stop the latter, vital process. It is important that the *whole* circumference of the tree is cut. If the bast remains intact at one place, the tree may stay alive. Ring-barking does not yield immediate results: trees might still be flowering for years or recover altogether. The ultimate goal, in sum, is to have a dry but standing tree, which is easy to burn.

It is assumed in all learned books on Hungarian woods that the destruction of wildwood took place with the help of ring-barking and burning.[12] This is certainly a plausible theory, but no data exists which positively proves it. Burning did occur, but whether it was aided by ring-barking or not, we shall never know. Why this belief is rooted so deeply in scientific consciousness is because ethnographic studies demonstrated its use in the Early Modern Period. From this it was extrapolated back some five thousand years to all "pre-modern societies."

4. 3. Aszó: Linguistic Evidence for Ring-Barking?

One of the most debated words in Hungarian landscape history is connected to ring-barking. *Aszó* occurs many times in toponyms (and exclusively there) from the earliest records on, and because it may be of crucial importance, it is worth taking a closer look at its linguistic context. Many scholars have cast their votes for one or another meaning and origin of this word.[13] These two aspects usually run parallel, however, as we shall see, unusual twists occur sometimes in how they relate to each other.

The first article debating the subject appeared in 1865, and its influence proved a lasting one. Flórián Mátyás found six pre-1300 examples where *aszó* was explained (with the usual *vulgo dicitur* formula) as "valley." Seven years after this, a Transylvanian bishop wrote a short note on the local usage of the word, namely that it referred to the type of valley which had a watercourse running through it in wet seasons but was otherwise dry. Still before the turn of the century, in 1883 and ten years later, Bernát Munkácsi published two articles, in which he ably demonstrated that *aszó* had something to do with water, as it appeared in connection with rivers. Given the amount of material written, it may be surprising that until 1969 these were the last significant contributions as far as the meaning of the word is concerned. With rather unusual honesty, the *Historical-Etymological Dictionary of the Hungarian Language* (*MNyTESz*) admitted general ignorance in 1967. The subject was taken up two years after this by the ethnographer Lajos Takács and ten years later by the linguist Antal Nyíri. With their efforts, although still with some doubts, we now have a clearer picture of the word *aszó*.

Several origins were suggested, the latest opinion – and seemingly the most authoritative since it was accepted in the *Etymological Dictionary of Geographical Names* (*FNESz*) – was expressed by Nyíri, who argued that *aszó* (similarly to *asszú*) derives from the verb *aszik*, which means "to dry," and is of Ugric origin.[14] With this, it is beyond doubt that all possible meanings must be connected, however loosely, to dryness. Nyíri was the first to come up with a theory that would bring together all known elements: valley, water and dryness. He carried out a careful linguistic analysis and – for the first time since 1893 – found new data to support his ideas. *Aszúpatak* was defined in two late-medieval examples as "dry stream" or a "stream that flows in wet times."[15] Based on these and on some other analogies, he established the following sequence of changes in meaning: 1. temporary stream 2. its dry bed 3. the valley where this is located 4. the whole area. This would indeed account for all meanings previously asserted, and would make a coherent story. It cannot, nevertheless, explain why the "valley" meaning should appear first in the sources and the "river" meaning later. That, of course, may be by accident, but it seems too consistent. We should remember that five out of six known pre-1300 examples refer to "valley."

Why the word is attractive to the woodland historian was laid out in a 1969 article by Lajos Takács, and further

[10] Pyne is positive that the "Hungarian Plain … is a creation of anthropogenetic fire and grazing." Pyne 1997, 298.

[11] For a few examples, see Csőre 1980, 134–135.

[12] For example Csőre 1980, 38.

[13] In the following, I shall not refer separately to individual articles. Every piece of writing I mention is listed here. Page numbers refer to those parts of the articles where an analysis of *aszó* appears. Mátyás 1865, 189–192; Kriza 1872, 383–384; Szinnyei 1879, 102; Munkácsi 1883, 497–498; Munkácsi 1893, 182; Szamota 1895, 132; Gombocz 1900, 53–54; Hefty 1911, 160; Pais 1912, 391–401; Takács 1969, 120–123; Nyíri 1979, 147–162.

[14] Earlier versions included Turkish origin, and *aszó* not being a derivant of *aszik* but comparable to Finnish *laakso*, meaning *valley*. Mátyás even suggested that it was similar to *oszló*, meaning "something that parts and continues in different directions," obviously an effort to account for the valley meaning, although this looks rather "medieval" even to my linguistically little trained eyes. Nyíri's version is not original, it was Szamota who first published it, albeit with no apparent explanation.

[15] "siccus fluvius" "rivulus tempore flutoso fluens" *Patak* is Hungarian for stream.

elaborated on in his 1980 book on ancient clearing techniques. Takács claimed that *aszó* refers to ring-barking, the "drying" of trees. He demonstrated this with much eighteenth-century material mostly from confessions in boundary trials. The choice of time is obvious: the eighteenth century, the *reconquista* of the country after driving out the Ottomans, was the time when ordinary folk talked much about their land at court. The material, however, was not extended back to the Middle Ages. Therefore we cannot tell whether this meaning may be adequate in a medieval context or not.[16]

Does this above meaning exclude the one proposed by Nyíri? The answer, provided by Nyíri himself, is no. He argued that Takács' conclusions were correct (they are in fact very convincing) but the situation was a mere coincidence. Both versions were relevant because they both derived from the same *aszik*, but that was all they had in common. He continued by asserting that because the "ring-barking" meaning was only demonstrated for the eighteenth century, any occurrence of *aszó* earlier than that must necessarily mean "dry riverbed." This logic, however, is faulty because in most place-names neither meaning can be verified. The case is probably easier with the word *aszaló*, which comes from the same stem and apparently refers to ring-barking. There is at least one medieval example where it is explicitly connected to wood clearance,[17] but it is not yet known whether *aszó* and *aszaló* are free variations of each other in this context, and if so, which one came first.

In sum, our present knowledge of the word *aszó* is that it can mean a valley or area where a stream temporarily runs, or a territory cleared of trees by ring-barking. The former can be illustrated to have been present in the Middle Ages, whereas the latter cannot. What we have in most cases is a desolate place-name, the actual meaning of which is impossible to decide on.[18]

The basic prerequisite for drawing a distribution map of such place-names was to identify the forms (some of them rather inconspicuous) in which *aszó* was preserved. This, being a purely linguistic matter, has been accomplished with fewer question marks in the end. After Hefty had collected many examples, Dezső Pais analysed the development of different forms of *aszó* and established which ones may truly be taken into account. Every publication since 1912 has been based on his data collection. The distribution of these toponyms forms a characteristic pattern (*Fig. 8*)[19]. They are concentrated in hilly rather than mountainous areas and are entirely missing from the Duna-Tisza Interfluve, the flattest and driest middle part of the country. This region does not lack temporary streams, therefore it seems that the word is more closely connected to valleys than Nyíri's theory suggests. Takács, on the other hand, interpreted this map as a clear justification of his own view, because "it depicts the land where in the first centuries after the Conquest – from which age most *aszó* constructions are available – most of the clearance was done in order to foster further settlements."[20]

There can be no doubt that if Takács is right, then we have a powerful tool to help us with establishing which parts of the country were more wooded in the Árpádian period, an otherwise poorly documented age. However tempting this is, I cannot accept it. First, *aszó* does refer to a temporarily dry stream and its valley as well, and it is not possible to distinguish this meaning from ring-barking. Secondly, *aszó* only refers to one type of clearance, which is not very suitable for low-lying areas with flood-plain woods. We do not know how woods were grubbed out in these regions: research so far has focused heavily on upland areas. Thirdly, the fact that clearance is carried out in one region but not in another does not necessarily mean that the region without documented clearance was less wooded.

4. 4. Medieval Woodland Clearance

Thus far, I have discussed how woods may have been grubbed out. An equally significant aspect is how much woodland disappeared in different periods. As I have argued above, it is too early to make definitive statements about the amount of woodland in prehistory. The available samples clearly demonstrate that when the Hungarians arrived in the Carpathian Basin, woodland was limited, but *how* limited, we cannot yet say. It is difficult to see into the very poorly documented centuries preceding the Conquest, but it may well be true that the low population densities triggered secondary woodland formation on some scale.[21] Even this way, the credit Hungarian historians traditionally give to their medieval ancestors in the destruction of wildwood, or woodland in general, is far too much. There are two main reasons for this. Firstly, because the Middle Ages are the first time when written sources of clearance appear, and this type of evidence matters more than anything else in the eyes of historians. Secondly, because the nineteenth-century ideas about the creation of a "nation" and a "kingdom," which tended to underestimate or altogether disregard those who went before, are still with us today. Here, I shall outline what can be confidently argued about medieval times.

Undoubtedly, much clearance work went on in the Middle Ages. However, systematic evidence is restricted to the higher mountain regions, mostly what is today Slovakia and Ukraine. There were references to grubbing out an acre or two in many other places,[22] but only in the north can we speak about anything on a larger scale (not in clearance but in sources). In these regions, clearance was initiated by new settlers, lead by professional *locatores*. It was often included in the privilege issued to a nobleman that he should "grub out the woods, transform them into fertile arable land, bring

[16] *FNESz* states it may not.

[17] 1455: "versus asalosyrtwan" Dl 36 935. Hungarian *irtvány* means "clearance". *Oklevélszótár*, 35.

[18] A recent article, in an acknowledgement of this problem, claimed that making a distinction is "very easy," but, alas, did not go into details as to how this should be done. Rab 1993.

[19] Reproduced from Takács 1980, 144.

[20] Takács 1980, 142–143.

[21] As argued for example by Bartha 2000, 12. For the archaeological part, see Lőrinczy 1993; Bálint 1991; Vaday 1999; Bóna 1993.

[22] Csőre 1980, 136–137.

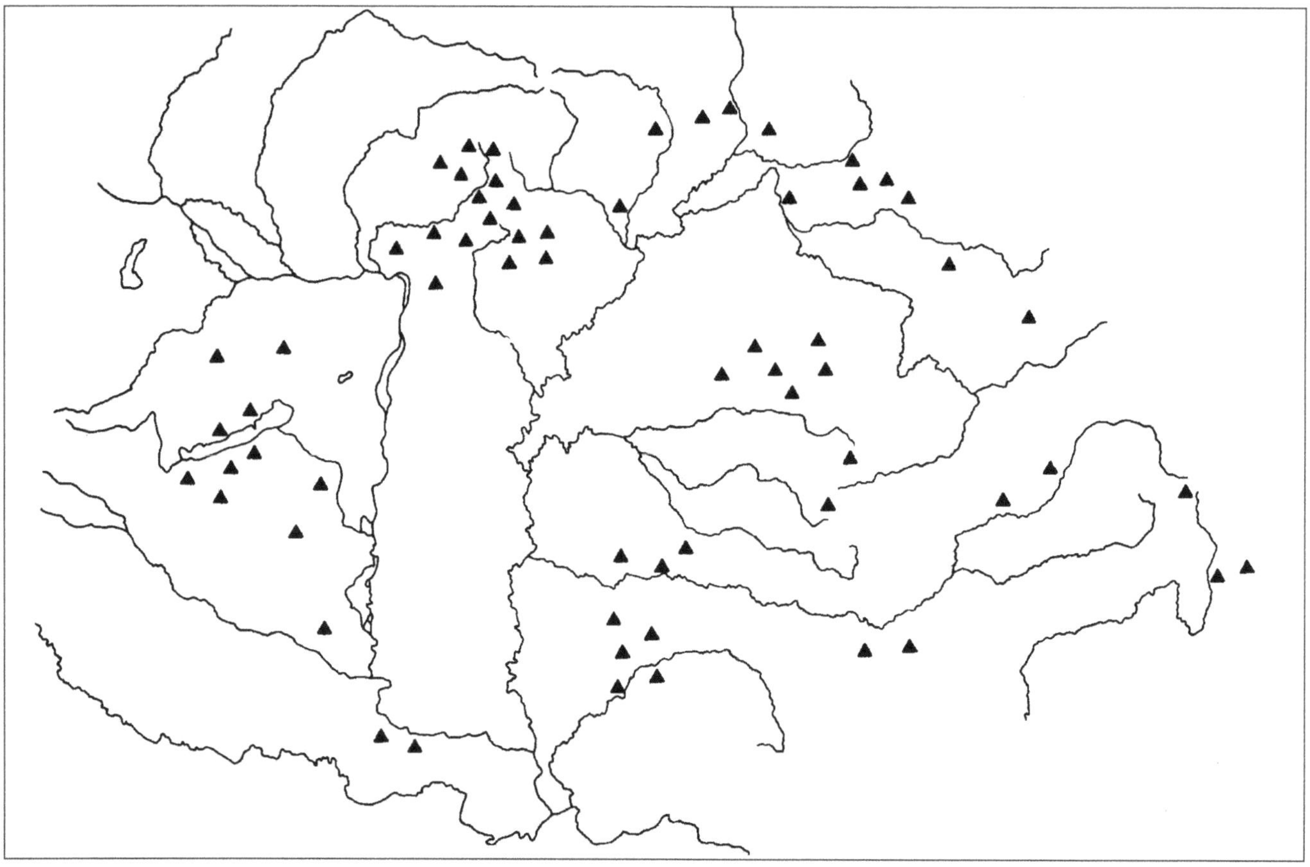

Fig. 8. Distribution map of geographical names containing aszó.

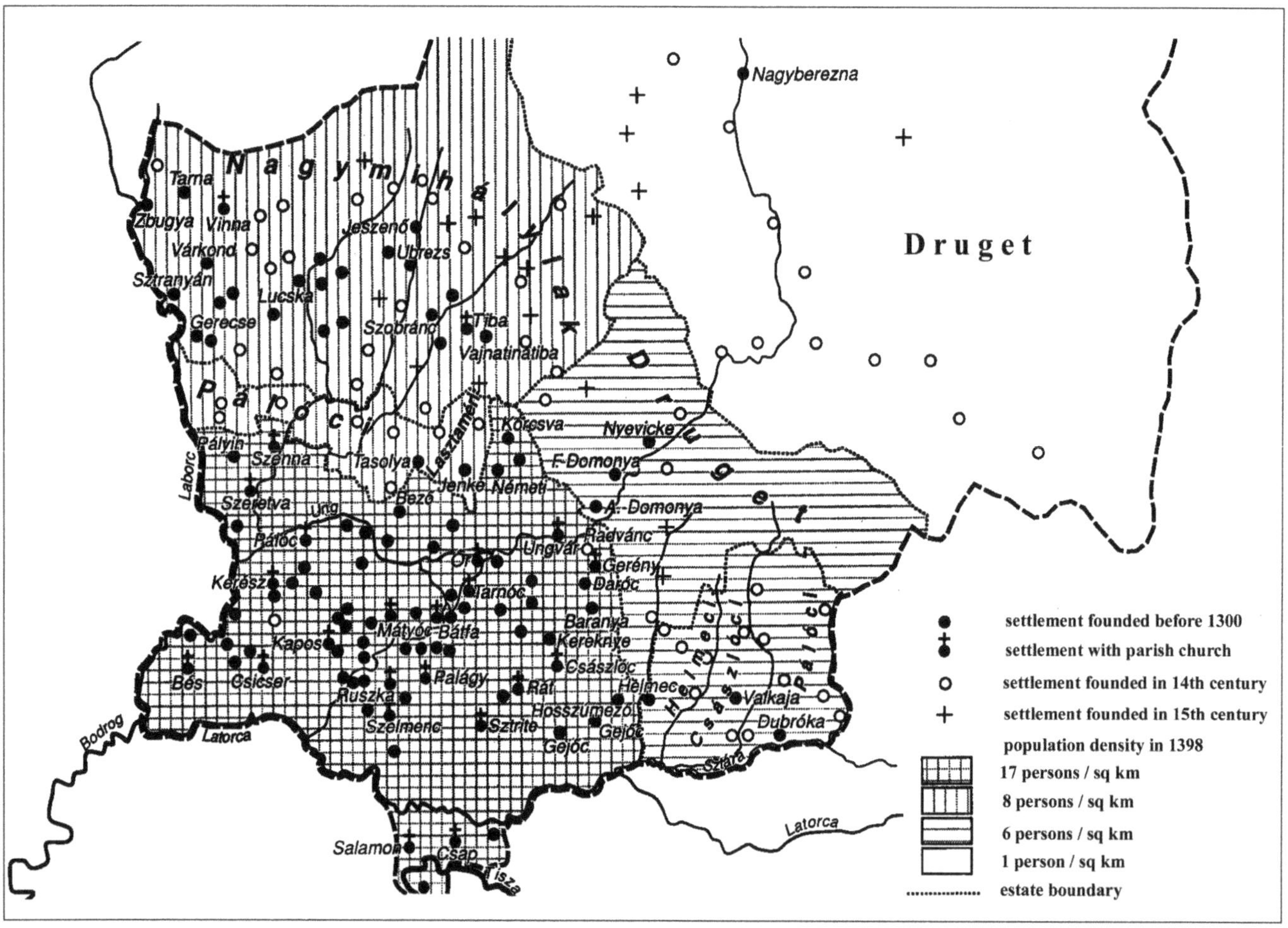

Fig. 9. County Ung in the fifteenth century. Was this a typical border county? There were many settlements in low-lying areas and few in the mountains.

there people, and make … villages."[23] The process started in the thirteenth century, and lasted until approximately the fifteenth. It is rather well-documented, and the expansion of the settled areas can be followed in substantial detail.[24] The new foundations had a characteristic social structure under the leadership of their *scultetus*.[25] Planned villages were also typical of these regions.[26] It is not my task here to present details of this grand enterprise, the reader is referred to the book of Adrienn Körmendy, which also compared the northern Hungarian case with the Polish one.[27] As we shall see in Chapter 5, no statistical data is available to follow the process of clearance in medieval Hungary. The northern region is one of the few examples where – although mostly indirectly, through the expansion of settlements – one could get rough-and-ready estimates of how much woodland was grubbed out in the three hundred years preceding 1526.

4. 5. County Ung: A Case-Study

The best documented example is the county of Ung (in the north-eastern part of the Carpathian Basin) where two conscriptions enumerate settlements, nobles, and peasants (1398 and 1427). In addition, a major estimation of the lands of the Druget family survives (1437) that covers about half the county.[28] Ung has three basic geographical parts: the lowland region, the hilly parts at the foot of the Carpathians, and the high Carpathian Mountains down to the border.[29] By the thirteenth century, only the southern part of the lowland region was populated. Around 1300, 75–80 % of the county must have been covered by woodland. Within two hundred years, the northern lowland region and the hilly parts were moderately populated and, at most, half-wooded. The higher mountains were almost completely without inhabitants until modern times. The 1437 estimation reinforces the impressions gained from the settlement pattern. This document describes the northern half of the county, where woodland was extensive even in the fifteenth century. The villages recorded were mainly in the hilly region, with a few tiny settlements up in the mountains. They contained approximately eighty ploughs of arable, meadow, and managed woodland plus "woods for 340 pigs." As opposed to this, the vast woodland "until the border of the Ruthenians" was estimated to be one thousand ploughs. There was no real surveying, of course, and this number appears to be a synonym for "very much" rather than anything else. The map of county Ung tells us that this great wood occupied about half the county, which leaves the other half as a mostly cultural landscape with not much woodland.[30] Thus, if the northern part of Ung (with the Druget lands) was very much wooded, while the southern one was not, we arrive at a figure of somewhere around 60–70 % woodland in the first half of the fifteenth century. We must not fail to note, however, that the dividing line between inhabited and uninhabited areas was rather sharp, and was probably already established in the eleventh century (*Fig. 9*)[31].

[23] *CD*, vol. 5/2, 598. 1279: "ut ipsam exstirpent, et in agros fertiles transmutent, ibidem populos convocent, et … villas … aedificent."

[24] See for example the maps attached to Engel 1998; Szabó 1937; Mályusz 1922.

[25] Körmendy 1974; Szabó 1966, 109–118.

[26] Földes 1978.

[27] Körmendy 1995.

[28] Engel 1985; Engel 1998. The estimation is Df 234 235. Ung is a fortunate county, not only because many sources survive there, but also because the material was handled by such a first-rate historian as Engel.

[29] This repeats the conditions of the whole kingdom on a small scale.

[30] This estimation unfortunately records woodland and arable together, with the exception of acorn-bearing and axable woods. See also Chapter 5.

[31] Reproduced from Engel 1998, 175, therefore place-names appear in Hungarian.

Chapter 5

The Proportion of Woodland in the Countryside

This chapter will deal with a simple question: How much woodland was there in Hungary in the Middle Ages? However, before we approach this problem, it is worth taking a look at the present situation.

5. 1. Woodland in the Twentieth Century[1]

"In 1991, forests covered about eighteen percent of Hungary," reads the official statistical account of Hungarian forestry.[2] What does this sentence mean? First, not all "forests" are woods. These eighteen percent include plantations as well as woods. "Hungary" refers to the present-day country. In the years preceding 1920, the kingdom was twenty-five percent wooded. After the Trianon peace treaty, only twelve percent of the new country was covered with woodland. This is explained by the fact that almost the whole of the Carpathian Mountains, together with their large woods, belong now to other countries, and what was left is the lower lying regions, where agriculture traditionally dominated. The eighteen percent of land covered with "forests" in 1991 is the consequence of the creation of large-scale plantations, which began after the Second World War.[3]

Native woods have been disappearing in the twentieth century in two main ways. The first, "traditional" way has been to convert them to arable. Secondly, foresters themselves have destroyed woods to replace them with plantations of foreign or native species. The proportions of these two are unknown at present. Forestry records do not differentiate between native woods and plantations of native trees, thus – except for the very present – one can only say how much of the "forested area" was certainly not woodland. The twelve percent of 1920 undoubtedly included plantations as well, and the amount of native woodland has been shrinking ever since. In 1948, with no significant increase in the overall "forested" area, about one fourth of Hungary's "forests" were either black locust or pine, that is planted trees, reducing woodland to a maximum of nine percent.[4] For the present, we have a clearer picture thanks to János Bölöni, who worked out the approximate amount of native woodland from the forestry records.[5] He arrived at 6.2 % which means that people of the twentieth century destroyed about half of the woodland still there in 1900.

5. 2. The Proportion of Woodland in the Middle Ages

To be able to identify the proportion of woodland in the medieval Hungarian countryside, we must have written evidence of quantitative nature. General remarks and travel accounts (such as Pliny's notorious *Pannonia glandifera*, Priskos Rhetor's remark about the total lack of trees where the Huns live, or Arnoldus Lubicensis' description – accompanied by a fancy illumination – of Frederick Barbarossa's 1189 crusade)[6] are not much help, because they reflect mere impressions, and "very wooded" or "devoid of trees" are impossible to translate into percentages. I would be able to design a foreigner's journey today so that he or she would think that Hungary is full of trees or vice versa; similarly much depended on the itinerary in the past: different parts of the Carpathian Basin have been unequally wooded for a long time. Furthermore, "wooded" does not mean the same for an Italian as it does for a German. Contributing to the problem, some historians tend to interpret medieval charters as if they were travel accounts. Csőre claimed that the Great Hungarian Plain must have been very wooded in the Middle Ages, because a great number of charters mention woods there.[7] His list contains some fifty woods, most of them with their own names as reference points in perambulations. Fifty (seemingly not very large) woods for a territory as big as the Alföld are close to nothing. They tell us about the existence of particular woods, but have little to offer in terms of the proportion of woodland in the landscape at large.

There are two types of sources that may provide us with relevant information: conscriptions and estimations. Conscriptions are unfortunately only a theoretical possibility: they are almost entirely missing from the source material.[8] Why I still mention them is because of the existence of the rather mysterious 1229 conscription of the holdings of the chapter of Fehérvár (Co. Fejér) in Co. Somogy, which will be discussed later.

5. 3. Estimations

I have already presented a short introduction to estimations in Chapter 2. Little are we acquainted with the history of this institution. There are numerous cases from the fourteenth century, which, however, usually did not give details, but rather named the sum that one's property was worth.[9] In other

[1] The following concerns present-day Hungary. An overall picture of the Carpathian Basin was beyond my abilities.

[2] Halász 1994, 27.

[3] Halász 1994, 27–28.

[4] Halász 1994, 44.

[5] Bölöni 2001. Strange as it may be, this has been the first ever attempt to answer such an apparently basic question.

[6] For the text, see Gombos 1937, 305. Arnoldus Lubicensis was originally published in *MGH* SS, 21, 115–250; the illumination is reproduced in Zolnay 1983, 160.

[7] Csőre 1980, 97–101.

[8] Which is why they were not included in the introductory chapter on source types.

[9] For example in 1355, the chapter of Veszprém estimated four villages at sixty-eight marks without any further information. *AnjOkm*, vol. 6, 368–370. Similarly, in 1345, four villages together were estimated at five hundred and fifty marks. *Sztáray*, vol. 1, 181–183.

cases, types of land-uses were not distinguished.[10] In contrast to these, fifteenth century estimations are more detailed.[11] Obviously we witness a development in the procedure, sometimes we even catch a glimpse of rivalling local and central customs, as in the case of the village of Záhorská Ves (Sk; Co. Pozsony), which was originally estimated "the usual provincial way" but the nobles insisted that the procedure be repeated "not the provincial way but according to the customs of the kingdom."[12] Parts of this development are difficult to understand today. If, for example, arable and woodland represent the same value (three marks a plough), why differentiate between them and record them separately? Why should certain woods be estimated based on their annual income, and why should this not happen to mills which had easily computable incomes? Why is pasture left out even from the official "know-hows" of estimations,[13] and then why is it still recorded sometimes? Without further examples, it is clear that the history of estimations would deserve a thorough analysis in its own right. Here I shall only discuss those details that concern us directly.

The best-known and finest example of medieval estimations is the filial quarter case between the Gara and the Szécs families (1478).[14] Why this document should be significant is explained by the importance of the Gara family, whose possessions were to be estimated. The Gara were one of the richest families of the age, thus the very detailed charter amounts to forty-six pages. It did not get the attention it certainly deserves until 1986, when András Kubinyi analysed it very carefully, although the source itself remains unpublished.[15] The Gara lands, Kubinyi calculated, had comprised 203 extant and deserted settlements and 93,887 hectares.[16] Had a couple of dozens of such estimations been carried out and come down to us, I would have no serious problems in producing a reliable picture of the proportions of different land-uses in fifteenth-century Hungary. This, alas, is not the case. Presently, we are aware of only three estimations comparable to the Gara-Szécs.[17]

The rest of the material available will not cover more than a handful of settlements each, however, this is not the main problem. The greatest obstacle faced by the researcher is that it is impossible to search for these charters systematically. They lie hidden in documents of various nature and nothing in catalogues or registers refers to their existence. The only attempt to collect them was made by István Szabó, an eminent student of medieval villages, but even with his expertise and doubtlessly unparalleled experience, he could present data for not more than thirteen villages.[18] The only way to locate the evidence would be to look through every charter (approximately two hundred thousand) one by one.[19] This, unfortunately, was unrealistic. What I was still able to do was browse through most of the published sources. This gave me an opportunity to examine large numbers of charters – either by looking at them separately or using the indices of such volumes as *Zsigmondkori Oklevéltár* – yet the acute problem of Hungarian source publication – namely that very little is available for the second half of the fifteenth century – ensured that I did stand much chance even this way of running across most of this material. There was one other volume that proved to be of great practical help: István Bogdán's monograph on medieval linear and land measures.[20] This is a treasure mine of information. Much of its carefully annotated data comes from estimations, which I could relocate and put to my own use.

	arable[21] (ha/percent)	meadow (ha/percent)	woodland (ha/percent)	pasture (ha/percent)	total (ha/percent)
Gara-Szécs	45,382 / 48 %	5,492 / 6 %	42,151 / 45 %	862 / 1 %	93,887 / 100 %
Other	76,824 / 37 %	24,371 / 12 %	60,368 / 28 %	47,397 / 23 %	208,960 / 100 %
Total	122,206 / 40 %	29,863 / 10 %	102,519 / 34 %	48,259 / 16 %	302,847 / 100 %

Fig. 10. Proportions of different land-uses in fifteenth-century Hungary as found in estimations.

[10] *Sztáray*, vol. 1, 198–204.

[11] It does happen, nonetheless, that fifteenth or even sixteenth-century estimations simply communicate the final sum. For example 1450: Dl 67 435; 1523: Dl 47 514.

[12] 1343: Dl 104 692. "more provinciali consueto," "non modo provinciali sed regni consuetudine" The result, by the way, did not change.

[13] The Domesday Book, for example, also almost completely ignores pasture.

[14] On the filial quarter, see Rady 1999; Banyó 2000.

[15] Kubinyi 1986. The charter is Dl 18 145. Kubinyi did not separate the different types of woodland in his calculations, so I had to do this myself.

[16] I do not include the 1794 hectares that were given in combinations of land-uses.

[17] The estates of Makovica (Sk; Co. Sáros) including five settlements in Co. Pest. 1492: Dl 3 022, the above mentioned lands of the Druget family in Co. Ung and Co. Zemplén, and the estates of Császárvár (Cr; Co. Varasd) in 1489: Steiermärkisches Landesarchiv, Graz, Urkunden No. 8502. The last one was discovered and analysed by Kubinyi 2001. As observed by Kubinyi himself, it is only partly useful for us. In only 21 settlements were woods and arable separated and given in royal measures. Out of these I could use 19, because in one, the amount of woodland was unreadable, and in another, there was a wood that had no specified territory.

[18] István Szabó 1975, 11–14.

[19] A computerised database would be of much help, but would not produce *pret-a-porter* results. Estimations can be sought by type of charter in the regesta available for certain parts of the holdings of MOL. This, nonetheless, often yields dissatisfactory results, since many "charters of estimation" only contain someone's order that an estimation be carried out at a certain place. For example: Dl 67 435, Dl 14 329, Dl 14 523.

[20] Bogdán 1978.

[21] The Gara lands contained huge numbers of vineyards encompassing 8074 hectares. I did not want to disturb the four-type classification of land-uses, thus I added vineyards to arable land, because being a very intensely managed monoculture, they are closest to this type. The original data are: arable – 37,308 ha; vineyards – 8074 ha. I added the vineyards in Appendix 1 also to the arable lands.

The detailed results of my findings are presented in Appendix 1. The data shed light on 157 settlements and 208,960 hectares. These numbers present a picture different from that of the Gara-Szécs document at least in one significant respect: the average settlement in the former occupies a territory more than twice as large as in the latter. The reason for this may be that the Gara lands were almost exclusively in the hilly Transdanubian region and south of the Drava river. As opposed to this, my findings include several settlements from flat areas, where by this time desertion and nucleation resulted in villages with huge fields.

The size and proportion of different land-uses on these approximately 300,000 hectares are summarised in *Figure 10*.

How far can these numbers be considered representative for the whole of the kingdom? The Carpathian Basin covers an area of approximately 300,000 square kilometres, that is, thirty million hectares. All my data will not provide information for more than one percent of this huge geographical unit. This is in no sense adequate, but at present this is much as we have. If industrious research reveals five times as many estimations as are presently known, which is rather unlikely, the sample will still amount to a mere five percent of the whole. Sources will not permit anything much better than this until the eighteenth century. We should also note that estimations almost exclusively recorded the land of the nobility, therefore we have no idea how wooded royal and ecclesiastical lands were in the fifteenth century. For the sake of convenience I suppose that they were all similar, however, there is no guarantee that this was so.

The other side of the coin is that the data are largely scattered geographically, which still gives us an opportunity to cast a glance at many different parts of the country. Out of the seventy-five counties of fifteenth-century Hungary,[22] thirty-three (44 %) are illuminated by the estimation of at least one village. This has permitted me to draw conclusions about the kingdom as such with far more confidence than if most data came from just a handful of counties. *Figure 14* clearly demonstrates, nonetheless, that there is a pattern in the distribution of those counties where some data at least are preserved.[23] While the west (Transdanubia) and the north (especially the north-east) are fairly well-represented, the central parts (the Alföld) and the east (Transylvania) are almost empty. In the case of the Alföld the reason is obvious: most documents perished during the Ottoman occupation. These counties are notorious for having little source material. Transylvania, however, is rich in surviving medieval charters, and thus is a promising target for future research.

Figure 15 depicts the proportion of woodland in each county. Somewhat unexpectedly, there is a clear pattern. Woodland below twenty percent was recorded in the inner parts of the kingdom. It is worth noting that the counties where the Great Hungarian Plain adjoins the hills (Counties Komárom, Pest, Heves, Szabolcs), which have probably the most fertile soils in the Carpathian Basin, form a line of low woodland percentages. It is precisely in this region that twentieth-century botanists had the most trouble finding at least *one* ancient wood to study its vegetation.[24] Up to forty percent woodland is characteristic for the regions in the central parts of northern Hungary and the south. Even more wooded were the border counties, and one cannot fail to notice the stronghold of woodland in western Transdanubia (Counties Zala, Veszprém, and Vas). How does all this relate to population density? A comparison to the map compiled by András Kubinyi reveals a relationship that is anything but straightforward.[25] The densest populations (except for the *medium regni*) of the kingdom lived in south-western Transdanubia and in the northern parts of central Hungary. Low population percentages were characteristic for the northern border counties and certain parts of the Great Plain. For the latter we lack written sources, but in the north much woodland and few people came hand in hand. However, in Transdanubia, high woodland percentages coincide with a high population density. In contrast, the relatively well-populated central northern regions had little woodland. Did colonisation work in different ways in these places? Did they have different economies? Further studies may bring us closer to answering these questions.

5. 4. Alföld – The Great Hungarian Plain in the Middle Ages and Before

The fact that there are no written sources for the Great Plain implies more than the simple fact that certain parts of the kingdom will have to be covered by data extrapolated from other regions. The Alföld is so different from the rest of the Carpathian Basin, and in fact from the rest of Western and Central Europe, that percentages from west beyond the Danube must not be used here. Given the lack of written data, one has to turn to archaeological and botanical observations, albeit with caution. Many scientists have discussed the development of the vegetation of the Great Hungarian Plain, and some of these studies have important consequences. I will certainly not arrive at any new conclusions concerning this question, but a generally accepted theory is yet to be constructed in any case. An overview of what has been argued so far will help in understanding how the ecology of the Alföld operates and thus, help in avoiding some of the traps that often catch careless historians.

The Alföld is, for a start, a bit of Central Asia in Europe. It is flat as a pancake, but is ecologically very varied. There are sand-ridges, flood-plains, alkali lakes, fertile loess regions, and many other features. As opposed to most of north-western Europe, the natural vegetation of the Plain is not woodland but woodland-steppe.[26] Most scholars would agree on this point, but *what exactly* is "woodland-steppe" has

[22] I counted those counties that appear in Engel 2001c.

[23] For a detailed list of the proportions of land-uses in each county, see Appendix 2.

[24] They succeeded in the 1950s: Zólyomi 1957. (Wood of Kerecsend, Co. Heves)

[25] Kubinyi 1997.

[26] The usual English term is "forest-steppe," which, in order not to contradict myself in the distinction between Forest and woodland, I shall not use. One also finds "wooded steppe" and "parkland" in some publications. The latter is of no use for my purposes because "parks" mean something entirely different in a medieval context. "Wooded steppe" would be a better option, however, it suggests that the area is essentially steppe that has some trees on it. This may be true in some regions, but in the Alföld, woodland and steppe are of equal importance. Therefore, although I am aware of the uncomfortable and confusing consequences of creating new vocabulary, especially in a language that is not my own, I shall still use "woodland-steppe."

been subject to debate for at least one and a half centuries.[27] One of the most influential figures in this debate was Rezső Soó, the creator of an immense description of the potential vegetation of Hungary. He claimed that the Alföld belonged to the climatic woodland-steppe zone. Thus, the modern woodless landscape was an entirely cultural phenomenon.[28] This was later accepted by those who succeeded Soó in the leading positions of palaeobotanical research: Bálint Zólyomi and Mária Járai-Komlódi.[29] Recently, ecologists have started what the German would call *hinterfragen*, which resulted in a general survey of woodland-steppe sites and a small book on the subject.[30] It appeared that although no ink had been spared on the topic of woodland-steppe, a workable definition of what it was did not, and perhaps might not exist. What all agree on is that a woodland-steppe is a "separate vegetational region in the transitional climatic zone between closed woodland and steppe. In woodland-steppe, more or less closed woods alternate with grassland of usually dry habitat,"[31] and that "the two parts of vegetation are dynamically linked."[32] This is rather vague, and the same difficulties are reflected in the database attached to Molnár and Kun's book. Surviving patches of woodland-steppe were collected and described, however, the tables contain a three-scale categorisation on the probability of the territory in question in fact being woodland-steppe.[33] This, needless to say, is not the ecologists' fault. The remnants of woodland-steppe – the only direct evidence that can aid us in imagining what the Alföld was like before the coming of cereal cultivation – are greatly varied in form and origin. For some, juniper-poplar regions are woodland-steppe; for others they are closer to deserts (*Fig. 11*).

There exists another problem, connected to the origins of natural or artificial woodland-steppe. It might seem unreasonable to speak of artificial woodland-steppe when one is concerned with natural vegetation, yet reality once again overtakes theories. First, we must clarify the meaning of "natural" vegetation. This obviously refers to vegetation from those times when people had no more influence on the environment than any other group of animals. Nevertheless, we must be aware that this vegetation was never static. In the Holocene, temperature and climate fluctuated considerably, and so has vegetation. The present climate – tightly interpreted – is thought to be not more than three thousand years old, and even this was accompanied by such phenomena as the "medieval climatic optimum" or the "little ice age," increasingly recognised by historians.[34] This becomes significant if we consider that the Alföld has been populated since at least the Neolithic, in other words for some eight thousand years. Thus, "natural" vegetation is, in several cases, unrealistic: when climatic conditions would have formed the latest phase in the development of plant communities, people had already transformed the landscape to their own liking. In sum, although much of the woodland-steppe in the Alföld would have grown into this form by itself, a fair proportion of it is artificial: people grubbed out the woods to make room for animals or themselves. The great difference between clearings in the other parts of the country and in the Alföld is that here they are far more capable of maintaining themselves; in other words, once trees are gone, they do not necessarily return.

What is the reason behind all this: why are there parts of the Great Hungarian Plain where trees would not grow? The answer to this question is far less obvious than one might think and has important consequences for the historical ecologist. Most importantly, the two most palpable reasons: a too cold climate and inadequate precipitation can be excluded with ease. The fact that most major tree types of woodland-steppe would grow on territories with even a cooler climate on the one hand, and the dominance of the pedunculate oak (*Quercus robur*) – the oak with the greatest water demands – in woodland-steppe on the other, unquestionably show that neither rainfall nor temperature can be blamed for the absence of trees.[35] Recent opinion has it that the answer lies partly with the dynamics of the large, unregulated rivers, once so characteristic of the Alföld.[36] Before the nineteenth century, a large part of the Plain was flooded more or less regularly. It is difficult to tell how large a territory we are talking about, but some claim that about one third of the Alföld belonged here. In these areas it was too much water that prevented tree-growth. Periodic floods washed away much of the topsoil, resulting in soil whose surface dried out all too easily. This made it difficult for oak seedlings to grow. Some trees (like alder or willow) like water, but a combination of water-logging and high underground-waterlevels resulted in treeless patches in the landscape. This is a simplistic explanation, but it should nevertheless be considered by historians who have a tendency to think that flood-plain woods are absolutely closed woodlands, more like rainforests than woodland steppe.

Those territories without regular floods were not entirely wooded, either. Furthermore, new results from palaeoenvironmental studies have demonstrated that Soó may not have been right in 1931 when he disregarded the possibility that there was natural treeless vegetation in the Alföld. Pál Sümegi and his colleagues argued – based on interdisciplinary studies with methods not to be discussed here – that the most famous of all Hungarian treeless plains (*puszta* – this word will be familiar to all those who have ever been tourists in this country): the Hortobágy, ca. 2200 km^2, which had previously been thought to be the result of the regulation of the river Tisza, may, in fact, have acquired its present, treeless form thousands of years before the Neolithic.[37] They also mentioned a probably important, though almost utterly

[27] Early studies include for example Kerner 1863; Borbás 1881.

[28] Soó 1926; Soó 1931.

[29] For their views, see for example, Járai Komlódi 1987; Zólyomi 1952. More recently see his potential vegetation map in Pécsi 1989, 89. Soó, Járai-Komlódi, and Zólyomi all wrote many articles, however, listing too many bibliographical references would be pointless here.

[30] Molnár and Kun 2000.

[31] Molnár and Kun 2000, 7.

[32] Molnár 1996, 243.

[33] Molnár and Kun 2000, 42–48.

[34] For an overview, see for example Lamb 1982; Rácz 1993. A recent book is Wefer, Berger, Behre and Jansen 2002.

[35] More on the same topic: Molnár and Kun 2000, 16–17.

[36] Molnár and Kun 2000, 18.

[37] Sümegi, Molnár and Szilágyi 2000.

Fig. 11. Tilos wood in Újszentmargita: this is what woodland-steppe looks like after millennia of interactions with people. All management is banned here now, hence the dead trunk lying in the foreground (January 2002).

unstudied aspect of this phenomenon, namely the effect of large mammals. American examples demonstrate that the grazing of bison herds is one of the factors behind the continuity of the prairie; the Hungarian Plain was similarly crowded first by aurochs, European bisons, and then by the herds of the domestic animals of the different peoples who lived there. These millions of animals must not be left out of consideration, yet studies of their actual impact still need to be conducted.[38]

In conclusion, we have to accept the fact that the Alföld, even in its last natural form, had large treeless tracts. Hasty extrapolations, however, must be avoided because the Plain is ecologically varied and contains many different habitats. Environmental archaeological reconstructions of as many places as possible would be needed to arrive at a reliable picture.[39] As for woodland, I would not dare transform the available information into percentages,[40] which, due to the problems connected to "natural" vegetation, would be theoretical in any case. Nevertheless, it cannot be overemphasised that the formation of the cultural landscape started out with open vegetation and not closed woodland.

People introduced agriculture into this landscape. In large parts of the Alföld the soil is very fertile (although by no means everywhere).[41] Thus, it was not only the lack of closed woodland but also this fertility that encouraged early settlers. As I have mentioned above, population has been dense since the Neolithic. I shall not go into details about the different peoples that populated the Great Plain before the coming of the Magyars, however, one aspect must not be left out of the discussion here: the well-known defence ditch (in parts called the Csörsz ditch, a name probably of Slavic origin meaning the "Devil's" ditch) dug in the AD fourth century by the Sarmatians.[42] We are, in fact, not concerned with one single ditch and its accompanying bank, but several parallel ones, at times as many as four. These substantial earthworks (the ditch is 6–8 metres wide and 2–3 metres deep, the bank is 7–10 metres wide and 2.5–3.5 metres high) encircle the Alföld from Vác (Co. Nógrád) to Nyíregyháza (Co. Szabolcs) and down south through Debrecen (Co. Bihar), Arad (Ro; Co. Arad) and Timişora (Ro; Co. Temes), turning back west along the Danube. The ditch was on the outside of this circle (north, east, or south) and was intended to protect whoever was

[38] Sümegi, Molnár and Szilágyi 2000, 216. Further information in Molnár and Kun 2000, 21–22. For an ethnographic summary for the Carpathian Basin, see Szabadfalvi 1970. A famous attempt at assessing the impact of animals on woodland is Vera 2000. This book challenges the view that wildwood was closed woodland and envisages it rather like woodland-steppe. Given the special position of the Great Plain among European ecosystems, the testing of Vera's theory on this area would undoubtedly be a most rewarding project to undertake.

[39] A good example of one place in one period is *Szentkirály* 1996; Pálóczi Horváth 1995.

[40] Csaba Mátyás did the same and declared the question to be "a current task of research." Mátyás 1996, 146.

[41] For a short description, see Stefanovits 1981, 313.

[42] A recent investigation of these ditches is missing. Earlier works: Patay 1969; Balás 1961; Balás 1963; Soproni 1969. A short report on a particular site is in *MRT-6*, 27.

within. Regardless of all other considerations, one thing is clear: these ditches would have been a nuisance in a wooded landscape. First, it would have been hard to dig them out: no one would make a 7-metre-wide ditch in a wood. If someone would still have ventured to grub out the trees and dig a ditch, it would have been pointless: if one's enemies can hide behind nearby trees, the earthwork has no protective value. This is clearly demonstrated by the well-preserved parts of the ditch in Nagyerdő (lit. Great Wood) near Debrecen.[43] Another question in connection with the Csörsz ditch is whether it was topped by a wooden structure or not. Nowadays archaeologists are convinced that earthwork forts (similar structures with similar functions on a much smaller scale) were always reinforced by such a structure. If the existence of this is at least theoretically justified, then further changes in the environment must be assumed.[44] All in all, given the length of these ditches (ca. 1500 km), and the fact that several parallel lines of them existed, it is reasonable to argue that the Alföld in the fourth century was open enough for such a defence system to function properly.[45]

By the AD tenth century, when Hungarian tribes arrived in the Carpathian Basin, the Alföld was already mostly a cultural landscape. Many estimates have been put forward as to how much woodland this actually meant and these generally range around fifteen-twenty percent.[46] This should be stressed, because the publication that influences most historians is largely misleading. In 1994, György Györffy and Bálint Zólyomi published a joint article entitled "The landscape of the Carpathian Basin and Etelköz a millenium before."[47] This contains a widely cited and recycled map, which describes the "Potential vegetation of the Carpathian Basin and its surroundings around the turn of the first millenium." The contradiction is apparent. A map must be *either* potential *or* pin-pointed in time.[48] This particular map depicts potential vegetation, suggesting the undoubtedly false notion that around the year 1000 the Carpathian Basin was untouched by humans. This is all the more dangerous given the general trend among many historians to think of the history of the Carpathian Basin as starting with the Hungarian Conquest.[49]

If already in the tenth century the Plain was little wooded, then medieval people certainly further reduced the amount of woodland, or at least allowed little space for the formation of secondary woods. This is somewhat speculative, however, palaeoecological studies testify without exception that from the tenth century the vegetation of all regions studied experienced the effects of substantial human impact.[50] One historical aspect may also support this idea. Several scholars have observed that when in the early eleventh century the king "monopolised" large intact woods – later Forests – none of these were to be found on the Great Hungarian Plain. Although I myself am largely sceptical of this idea, even a partial acceptance of it strongly confirms that the Alföld had few woods as early as the eleventh century.

There is also an archaeological angle to the problem. Until recently, we knew very little about medieval water regulations and about the topic of land-use in flood-plains in general. Some studies investigated eighteenth- and nineteenth-century forms of flood-plain management,[51] but no direct evidence of what the situation was like in the Middle Ages had been discovered, let alone analysed. Then an archaeologist, Károly Takács, started examining some seemingly random and obscure banks and ditches in the west of the country, which turned out to be parts of a whole system of water management. Ten years of archaeological excavations and comparison of this data with written evidence resulted in a series of studies of the highest importance.[52] Takács demonstrated that certain flood-plain areas were closely regulated in the Árpádian period. One particular area (Tóköz, between the rivers Rába and Rábca), which he studied in great detail, had an amazing network of canals of several types, whose water-regulating functions Takács could clearly illustrate (*Fig. 12*). This was the ultimate cultural landscape: woodland was an integral and necessarily small part of it. The territory of this targeted survey was small, but the author added that "remains of double and triple canals have been discovered in all research areas, which stretch from the eastern side of the Alföld to the western Transdanubia."[53] He concluded that *all* flood-plains were regulated this way during the Árpádian period, which would have amounted to up to half a million kilometres of canals. If this holds true – which will have to be demonstrated or rejected through fieldwork – we are facing a situation in the medieval Alföld that so

[43] For an illustration, see Patay 1969, 411. This, by the way, also means that Nagyerdő is ancient secondary woodland.

[44] In another wood near Debrecen (Fancsika), a short section of the ditch was reconstructed. This has a wattle fence on top and two lines of smallish beams in the middle and at the bottom. Vaday 2003, 266.

[45] It does not strictly belong here, yet I should note the highly interesting series of experiments and fieldwork carried out by Bálint Zólyomi in connection with Patay and Soproni's investigations. The surviving parts of the ditch, as places mostly unsuitable for agriculture, were established as the last reserves of the natural (though secondary) vegetation of a region, all the more important in loess areas, where ancient woodland is virtually absent. Zólyomi 1969.

[46] See for example Molnár and Kun 2000, 23. Based on cartographic and soil evidence, that is, an entirely different approach: Németh 1998. Soil patterns are certainly not as clear indicators as Németh wishes to have us believe. More recently the forester Dénes Bartha produced an article where he came up with a figure of sixty percent woodland in the territory of present-day Hungary for the year AD 895. He arrived at this conclusion based on potential vegetation maps and some haphazard estimates, and without consideration of regional differences. Bartha 2000.

[47] Györffy and Zólyomi 1994.

[48] Although potential vegetation can be interpreted dynamically, that is, different time periods have potentially different vegetations. In that sense, the Carpathian Basin had a different potential vegetation in the year 1000 compared to the year 2000. I owe thanks to Zsolt Molnár for these ideas.

[49] This does not at all apply to early medievalists, but sadly so to many dealing with later periods.

[50] See for example Sümegi 1998a, 377.

[51] Andrásfalvy 1973; Andrásfalvy 1975. These studies, following the unfortunate custom of ethnographic research, extrapolated their material to the Middle Ages. Although there is some linguistic evidence (the presence of *-fok*) to suggest that medieval Hungary did not lack something comparable to that described by Andrásfalvy, researchers should not say more than this. Nonetheless, Andrásfalvy's generous conclusions were adopted by countless others, which resulted in the birth of a whole pseudo-history on medieval flood-plain management. Some examples: Győri 2000; Bellon 1996. All of this has recently been criticised in Takács 2000c.

[52] Takács 2000a. With more or less the same material: Takács 2000b. In English: Takács 2003.

[53] Takács 2000a, 60.

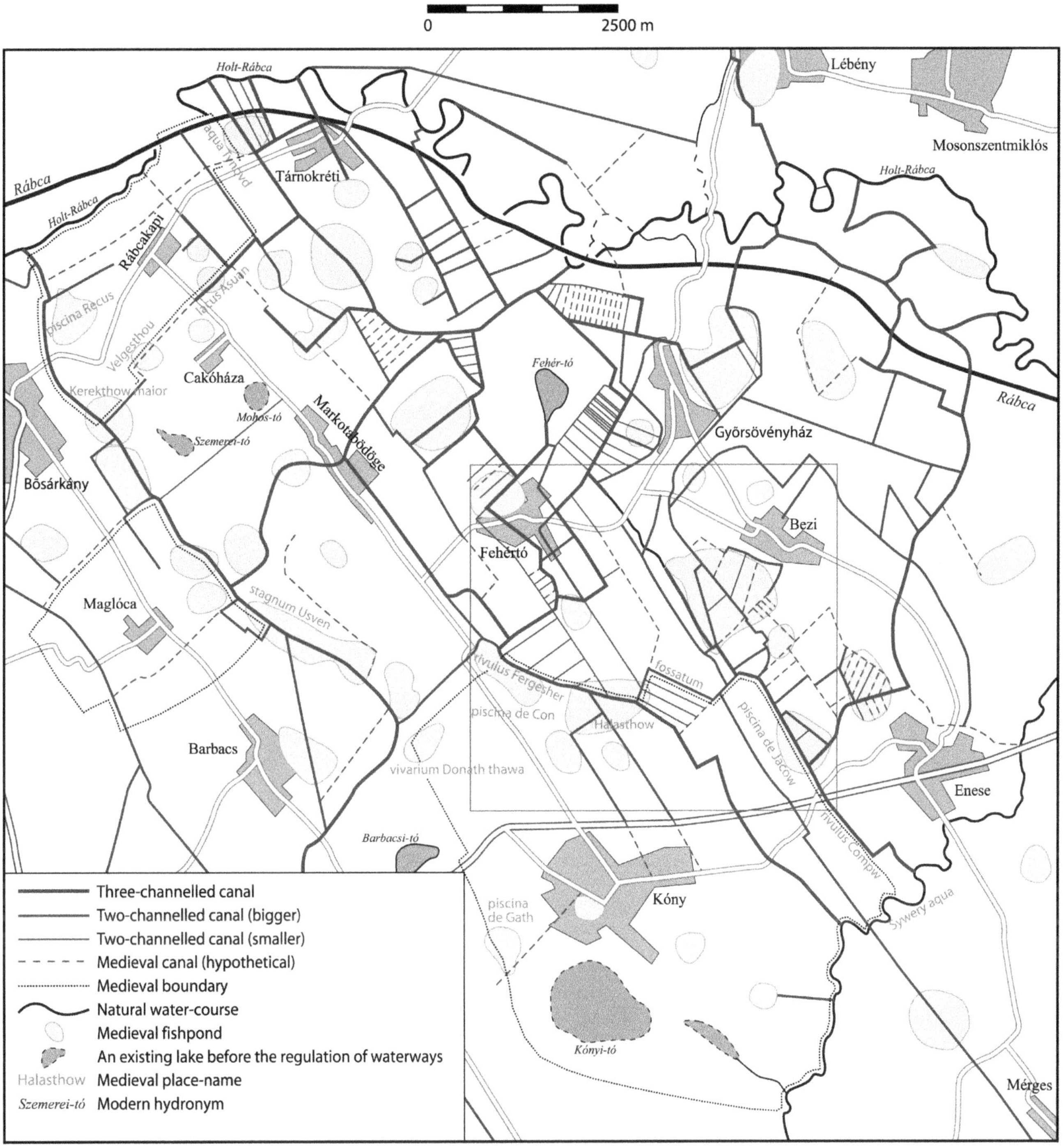

Fig. 12. The Tóköz region with the Árpádian period canals.

wildly contradicts previous beliefs that, I am certain, it will be opposed by many. The natural vegetation in flooded areas would matter far less in this case than thought previously. Large tracts of the Alföld should be thought of as woodless, but not because of vegetational preferences but rather the activities of medieval peasants.

Estimating the amount of woodland in the Alföld in the Middle Ages is impossible. At best, one can make an educated guess. We have to consider that the Plain had few woods before the Conquest and was then densely populated and ploughed in the Middle Ages. By the fifteenth century, taking into account even the depopulation of certain areas associated with the social turmoil of the thirteenth to fifteenth centuries, *my* guess is that about 10–15 percent may have been covered with trees.

5. 5. *The Somogy Tithes*

Before coming to any conclusions, it is worth examining a conscription from the early thirteenth century. I have stated above that conscriptions are almost entirely missing from the Hungarian source material, but this is not exactly true. If descriptions of possessions in connection with ecclesiastical foundations are placed in this category, then several surveys exist from the eleventh century onwards. Conscriptions in the classical sense of the word started to be compiled on church property, as in other parts of Europe, in the thirteenth century. The best known example is that of the abbey of Pannonhalma from 1234–40, better known as the "Albeus conscription."[54] This document, nonetheless, is a good

[54] *ÁÚO*, vol. 2, 1–26.

example of why these sources are useless for the present purpose. Let me quote a typical entry:

"In Beled *predium* these are the names of the *uduarnici*: Creus, Haga, Menke, Moko, who with their relatives are XX two households save the children and the young. And these are similar in services to the *uduarnici* of Hymud. This village has, according to the common estimation of trustworthy men, five ploughs of land surrounded by many boundary sings, and is bordered by these villages, that is by Grabuch, Macha, Ladamer village and a second village called Beled, which belongs to the *uduarnici* of the King."[55]

(*Predium* is a type of settlement, *uduarnici* are bondsmen on royal property supplying agricultural produce.)

Land here is given in sum-totals with no details provided. Sometimes a perambulation follows, but that is all we can learn: the proportions of different land-uses were not considered important. This, by the way, is characteristic of surveys until the time when they are replaced by *urbaria*, which, in turn, concentrated only on individual peasant plots, and ignored the territory of settlements as such.

In sharp contrast to all these is the conscription of the holdings of the chapter of Fehérvár (Co. Fejér) in 1229.[56] It was the result of a long debate between the said chapter and the monastery of Pannonhalma over the tithes in Co. Somogy. Such court cases took place in other periods and places as well, however, a description of the land is included only in this one.[57] The reasons for this are unknown, but are certainly hidden in the practical aspects of the procedure: this was the most suitable solution in this particular case, but not in the others. This important document has not been analysed in sufficient detail, although many are aware of it and refer to it.[58]

The conscription itself is rather similar to estimations in the sense that it concentrates on the properties of one landowner ignoring villages as units. In this case, just as with estimations, we have to hope that the possessions of the chapter sufficiently represent the proportions of land-uses in the whole county, *pars pro toto*. This accepted, the document, which describes parts of sixty-five villages,[59] still presents multiple problems. First of all, calculating absolute proportions is impossible, because meadows and vineyards are given by the piece without any reference to their size.[60] Woodland and arable are computable, thus, their proportions relative to each other can be established: arable is 51,462 *iugera* and woodland 9233. This means a relative proportion 1 : 5.6 in favour of arable, which is certainly surprising. Furthermore, considering the existence of vineyards and meadows, the absolute proportion of woodland was even lower than this, somewhere around ten percent (*Fig. 13*)[61].

Why is this so? Allowing for the fact that these settlements represent the county as a whole, one more distortion is possible. It is often argued that until the fourteenth century most woodland was in common possession.[63] Since the conscription only lists whatever belonged to the chapter of Fehérvár, extensive common woods may be supposed to have existed. This, however, appears unlikely. In two places the document mentions that beside the listed woods of the chapter, common woodland also existed, which means that in the other cases it did not.

Comparison with fifteenth-century estimations is unfortunately not possible, because Co. Somogy is missing from the database. In sum, I can only say that the 1229 conscription depicts a landscape with little woodland, but whether this was the actual situation or the result of some distortion in recording is undetectable today.

5. 6. *A General Estimate*

At the beginning of this chapter, I presented the available written data on the amount of woodland in the Hungarian Kingdom in the fifteenth century. Results demonstrated that those counties that had sources were approximately one third wooded. Botanical, archaeobotanical and archaeological considerations suggested that the region where no written data are available – roughly the territory of the Great Hungarian Plain – had far less woodland. From these I estimate that Hungary at the end of the fifteenth century was not more than 20 percent woodland, excluding the higher regions in the Carpathians. This number will certainly sound too low for most historians. It is my hope that future studies will come up with more data to correct my calculations, although I also suspect that the numbers I ended up with will be, if anything, lowered.

number of villages	arable (*iugerum*)	meadow (piece or *iugerum*)	woodland (*iugerum*)	vineyards (piece or *iugerum*)
65	51,462	151 ps + 617 iug	9233	686 ps, 1266 ps cul, 610 ps inc,[62] 30 iug

Fig. 13. Different land-uses in the villages of the chapter of Fehérvár in 1229.

[55] *ÁÚO*, vol. 2, 12. „In predio Beled hec sunt nomina uduornicorum: Creus, Haga, Menke, Moko, qui cum sua cognatione preter pueros et iuuenes sunt XX due mansiones; et isti pares sunt in seruicio cum uduornicis de Hymud. Hec autem villa secundum communem considerationem bonorum virorum habet terram ad quinque aratra uallatam multis metis, et est conterminalis cum hiis villis, scilicet Grabuch, Macha, villa Ladamer et secunda villa Beled uduornicorum Regis."

[56] *ÁÚO*, vol. 6, 466–477.

[57] See for example the tithes in Co. Zala: Df 200 841, and 17 other charters in the Archives of the Chapter of Veszprém, under the title U 402 SZALAD COTT DEC.

[58] For example Csőre 1980, 95; Kristó 1988, 254, with bibliography.

[59] More, in fact, but I only counted those where computable particulars were given for arable and woodland.

[60] Comparison with documents where the sizes of individual vineyards or meadows are provided might be of help, but the author of the monograph on medieval Hungarian measures found this inexpedient. Bogdán 1978, 240–244.

[61] For a detailed list, see Appendix 3.

[62] "Cul" stands for "cultivated," "inc" for "incultivated." As the table demonstrates, this distinction was sometimes made when vineyards were recorded.

[63] See for example *KMTL*, s. v. "erdő".

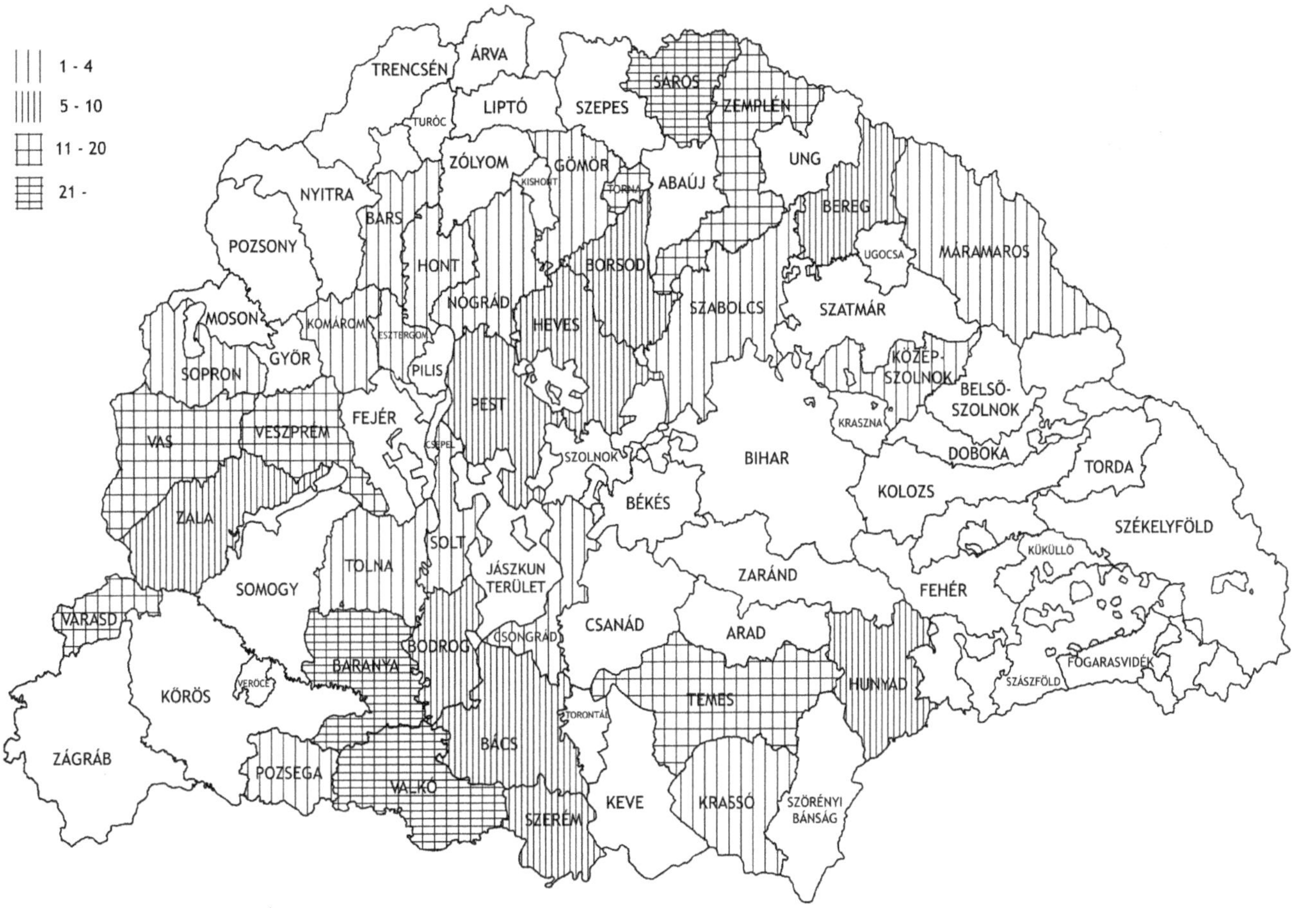

Fig. 14. Number of settlements covered in estimations in each county. No data from counties left blank.

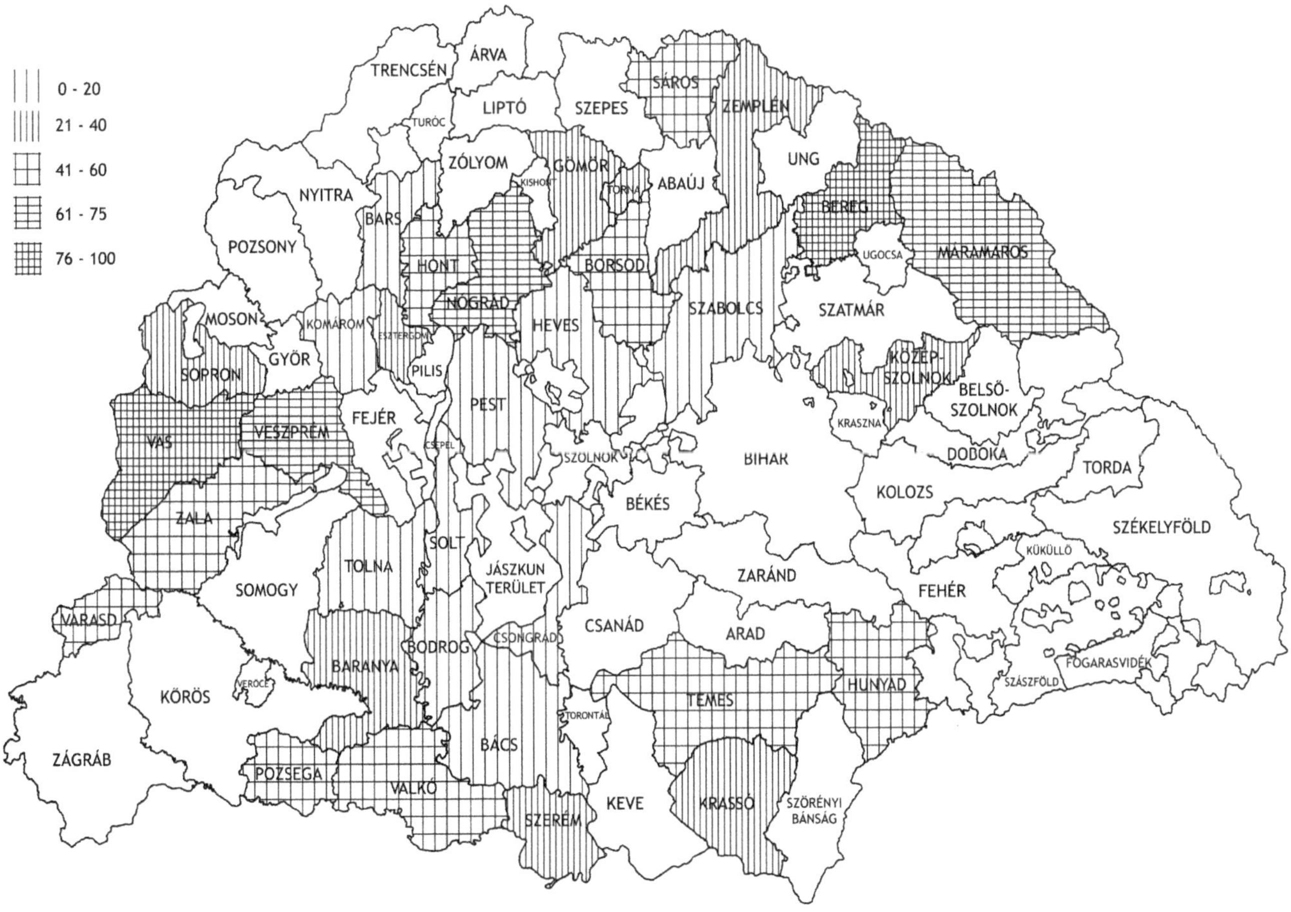

Fig. 15. Percentage of woodland recorded in each county. No data from counties left blank.

Chapter 6

Different Types of Woodland and Their Management

Medieval Hungarian sources refer to several types of woodland. The most general term is certainly *silva*, but one often comes across *silva glandinosa*, *silva dolabrosa*, *silva usualis*, *silva communis*, *silva prohibita*, *permissorium*, *rubetum*, *virgultum*, or *nemus*.[1] All these terms (probably with the exception of *rubetum / virgultum*) referred to different types of land and were not synonyms. It is common to find two types adjoining and having a boundary running between. For example in 1413, a part of a wood was described as having on its four sides two roads, an arable field, and a certain *nemus* or *rubetum*.[2]

Other than occasional bits and pieces of information in general charters and perambulations, the only possible way to decipher the meaning of different types of woodland is through the guidelines provided for estimations. These were compiled to help the work of estimators and mainly consist of lists of how much money different types of land (and other property) were worth. In Chapter 2, I have discussed the estimation list in Werbőczy's *Tripartitum* and others. By and large, they all tell the same story. They differentiate between four types of woodland (including wood-pasture): *silva communis*; *silva permissionalis*; *silva glandinosa, sub dolabro et venatione*; and *rubetum / virgultum*.

6. 1. Silva Communis

"Item *sylva communis*, from which pig tithe or dues in general are not collected, and which has no other fixed income, is estimated like common land, one royal plough to M 3."[3]

The Latin word *communis*, just like the English "common," has a double meaning. It is either "shared by a group" or "ordinary." Werbőczy's wording suggests, for example in this very sentence about common land, that *communis* means rather "ordinary" than "shared by a group."[4] In medieval times, however, it is often very difficult to differentiate between the two. Arable was "ordinary" land, but in the case of many open-fields it was also "shared by a group." Separating the legal and the functional aspects is not an easy task with woods, either. Werbőczy says that a common wood lacks three characteristics that would make it non-common: pig tithe, dues, and fixed income. The first of these defines it as an "ordinary" wood, that is, one without big oaks. The second and the third, however, more likely refer to a wood "shared by a group," namely a group of villagers. Hence the absence of dues (the lord cannot make anyone pay for using it), and the lack of income (he cannot make money out of selling woodland products). Nonetheless, this type of woodland was by no means the wood "where people casually cut what and how much they needed."[5] Common woods had the same value as arable land, in other words they must have been managed. What this management involved, however, we cannot tell.

6. 2. Silva Permissionalis

(alternatively called *permissorium* or *permissoria*[6])

"Item a large wood, which is also called *permissoria*, apt for common work and labour ... to M 10."[7]

All we learn here, except that *permissoria* were more than three times as valuable as common woods, is that a *permissorium* is large and it is apt for common work. As far as the first part is concerned, it was certainly not true. Already in the *Tripartitum* we see a long passage about what to do if a *permissorium* happens to be smaller than half a plough (seventy-five *iugera*).[8] In charters, we find many *permissoria* much smaller than this. In Kopács (Co. Vas), for example, the *permissorium* was 20 *iugera*.[9] The tiniest *permissoria* known to me are the one-*iugerum* wood recorded in Tilaj (Co. Borsod) and the half-a-*iugerum* wood in Kukeč (Slo; Co. Vas).[10] Browsing through the list of estimations in Appendix 1, one gets the impression that *permissoria* were generally medium-size. Leaving apart the five giant woods in Co. Bodrog (450, 3000, 3300, 1500, and 1200 *iugera*), the rest yield an average of 63.2 *iugera* (approximately 127 acres) per *permissoria*. In the Gara-Szécs estimation, again leaving apart a giant wood (1500 iugera), the remaining sixteen *permissoria* arrive at the similar average of 61.6 *iugera*. Let us now return to Werbőczy's description. "Apt for common work and labour," the second part, is a rather general statement, which does not claim any particular meaning. The question still remains: what type of woodland was *silva permissionalis*?

---

[1] Woodland terminology is a complicated question. In England, the Latin "silva" was not much used. Woods were rather called "boscus." In Central Europe, nonetheless, "silva" was the term generally used. For Austria, see Sonnlechner 2000, 211–213.

[2] *ZsOkl*, vol. 4, 171. Gyömörő (Co. Zala): "quadam particula silve que metaliter taliter distingitur: a parte orientalis via magna, meridiei terre arabiles, occidentis similiter via, aquilonis vero quodam nemus seu rubetum." See also the 1284 charter of Barnag, quoted in Chapter 2.

[3] "Item sylva communis, de qua decima porcorum, vel tributum generaliter non exigitur, nec habet aliquem certum proventum: aestimatur sicut terra communis, ad unum aratrum videlicet regalis mensurae adjacens; ad M 3." *Tripartitum*, I, 133, 25.

[4] In the previous sentence he talks about "common or arable land" (*terra communis vel arabilis*).

[5] Bartha 2000, 16.

[6] The last one is, in theory, the plural of the second, but this was not taken seriously in the Middle Ages. *Permissoria* was often interpreted as a singular noun. In this book, the grammatical interpretation I attribute to it is similar to the English word "data": it may be singular or plural.

[7] "Item sylva magna, quae alias permissoria dicitur, pro communi opere, et labore apta ... ad M 10."

[8] *Tripartitum*, I, 133, 31.

[9] 1402: *ZsOkl*, vol. 2/1, 217–218.

[10] *Károlyi*, vol. 2, 104–106; Dl 92 876.

Unlike other types of woodland, *permissoria* had a special medieval Hungarian name: *eresztvény*. This word is not part of the modern language and nowadays only appears in place-names, albeit there abundantly. It is a typical Hungarian (agglutinative, that is) construction in the sense that it is made up of a simple root and multiple layers of suffixes. The reader may remember from Chapter 1 that the Hungarian word for woodland (*erdő*) originates in the verb *ered* which means "to grow by itself" or "to spring." The stem of *eresztvény* is *ereszt*, which, so linguists argue, is very similar to *ered.*[11] They share a common root: *er-*, but have different suffixes. It is usual for verbs ending in *-ed* to have an *-eszt* ending version, which often has a factitive meaning.[12] Therefore, *ered* means "to grow," which implies that something appears on the ground apparently out of nothing just like flowers or grass. *Ereszt*, on the contrary, implies that someone is making something grow out of something that was already there.

The same tendency can be observed in the second layer of suffixes. *Erdő* has *-ő*, which is one of the most common suffixes in Hungarian for forming a noun from an active verb.[13] *Eresztvény* has *-vény*, which presents a number of unresolved problems. Linguists do not agree on its origins and meaning. Some argued that it was a phonetic variant of *-vén*, an adverbial participle.[14] Others claimed it to be a compound in itself, made up of *-v* and *-ény*.[15] What is interesting for us is that in this version *-v* would be an ancient reflexive suffix. A third option is that *-vény* is in fact a compound, but *-v* is not a reflexive but an adverbial suffix.[16] Consequently, what meaning *-vény* implies is also subject to debate. I shall certainly not make order in what does not seem clear even in the heads of specialists, however, there is one article that offers a handhold. Kálmán Szily discovered that *-vény* had had more and less active periods in its history, and its meaning had also changed.[17] To study these changes, he collected the *-vény* ending words from the first active period of the suffix: the thirteenth and fourteenth centuries. All, including *eresztvény*, are strikingly similar. They refer to something that is the result of human activity: *ásvány* = "a channel that was dug,"[18] *sövény* = "a wattle-fence that was woven," *töltevény* = "a bank that was piled up," and so on. In this logic, *eresztvény* should mean "something that was made to grow."

Thus, on linguistic grounds, two separate elements in the name of this type of woodland suggest that people were involved in its creation. It is also extremely important to understand that although Latin sources spoke about *silva permissionalis*, in the vernacular *erdő* and *eresztvény* could not appear together.[19] The latter was not a subtype of the former, but an equal variant with a different meaning. A place where trees were growing was **either** an *erdő* **or** an *eresztvény*. I shall come back to these observations later.

General charters do not reveal much about *permissoria*, but certain peculiarities must not be left unnoticed. As a direct consequence of their value, they were often differentiated from other types of woodland in estate divisions. While the latter were often left in common use, *permissoria* were, in most cases, divided between the parties. In extreme examples, such as one in Opatovská Nová Ves (Sk; Co. Hont) in 1409, only the *eresztvény* was divided by boundary signs, all "the other woods, arable lands, meadows and utilities were rendered to common use."[20]

Some charters contain surprising verbs in connection with *permissoria*. In 1411, the villagers of Spišská Nová Ves (Sk; Co. Szepes) complained that although "a certain *permissionalis* wood in a certain colonial place of theirs, within the just boundaries of their settlement, which has newly been *created*," was exclusively theirs, it was also used by others.[21] The same concept was repeated in Kállósemjén (Co. Szabolcs) in a lawsuit from 1399, where the defendant allegedly occupied a piece of land, and "created" a *permissoria*.[22] A similar word was used in a case of peaceful agreement in Co. Trencsén, when common woods, *rubeta*, and *nemora* were given in compensation for other lands, but it was noted that two *permissoria* did not form part of the agreement and remained with their previous owners (again typical for the outstanding position *permissoria* had among different types of woodland). We learn the names of these woods and also the fact that one of them "was *constructed* beyond the parish church in the village ... called Wgrogh."[23] All these refer to something artificial, something man-made. This is fundamentally different from other types of woodland, and might also explain why *permissoria* were so highly valuable. Without attempting to answer what this "creation" involved, let us consider some further aspects.

There are a number of cases when *permissoria* (or the Hungarian *eresztvény*) were equated with vineyards. These do not, as one might expect, seem to be mistakes or misunderstandings. The one from Gyulakeszi (Co. Zala) is almost verbose: "Vineyards, which were planted by the said nobles last year and were left to grow, and which are called herezthuen ..."[24] At first this does not make any sense, but, as we shall see later on, if interpreted as one aspect of a broader context, it can be accounted for.

---

[11] *MNYTESz*, vol. 1, 782–785.

[12] Simonyi 1881, 262–263. For example *reped* ("to crack") and *repeszt* ("to make something crack.")

[13] My own name, Szabó, originally a common noun, is equipped with a variant of the same suffix. "Szab" means "cut out," consequently "szabó" means "tailor."

[14] Mészöly 1908. *-Vány* and *-vény* are phonetic variants; to avoid repetition, in the following I shall only use *-vény* but shall mean both.

[15] Bárczi, Benkő and Berrár 1978, 332.

[16] Beke 1913; Pais 1933. See also Bárczi 1951, 174–175.

[17] Szily 1919.

[18] The present meaning of the word ("mineral") is a recent development.

[19] Thus *eresztvényerdő*, a usual term in modern literature should not be employed.

[20] "ceteras autem silvas, terras arabiles, prata ac utilitates ad communem usum deputassent." *ZsOkl*, vol 2/2, 242–243.

[21] "quedam silva permissionalis in quodam loco ipsorum colonicali intra veros cursus ipsorum metarum metales de novo *procreata*," *ZsOkl*, vol. 3, 140. Italics mine.

[22] *ZsOkl*, vol. 1, 663.

[23] *Podmaniczky*, vol. 1, 235–238. 1492: "supra ecclesiam parochialem in possessione ... Wgrogh appellata *constructa*." Italics mine.

[24] *Oklevélszótár*, 197. 1255/1415: "vinee vero de anno preterito plantate ab ipsis nobilibus crescendum dimisse que hereztuen dicuntur ..." (I am not quite sure what is meant by "crescendum dimisse.") Another case of *permissoria* meaning vineyard is found in *Sztáray*, vol. 2, 438–447. A third case from 1450: Dl 14 314.

As far as management is concerned, information is again sparse. I know of a charter from 1407 that grants two cartloads of firewood every week for two years from a *permissionalis* wood in Ordzovany (Sk; Co. Szepes).[25] This suggests large-scale coppicing. Some other instances demonstrate a close relationship between *permissoria* and mining. The most interesting of these is an estate division in Milaj (Sk; Co. Szepes). Here the parties divided, among other things, the *permissoria*. They agreed that if they found a mine on either's land, they would use that mine in common. If the *permissoria* needed for mining were not sufficient for one party, they should buy the wood from the other party. If, however, both had sufficient wood, they should supply it on equal terms.[26] Since there were many other woodlands mentioned in the charters, it is beyond doubt that only *permissoria* were apt for the purposes of mining.

Let me now sum up the pieces of information gathered so far. Linguistic evidence suggests that an *eresztvény* was a place where trees grew out of a permanent base involving human activity. Charters reveal data about the special position of *permissoria* among different types of woodland. The suspicion of human activity is further strengthened by such words as "create" or "construct" in connection with *permissoria*. *Eresztvény* was sometimes equated with vineyards. Finally, there is also written evidence for the connection between mining and *permissoria*. The only way to account for all these is if the meaning of *permissoria* is "coppice wood." The word *eresztvény* describes a phenomenon rather similar to coppices. It is man-made in the sense that people cut trees and create permanent bases (stools). Shoots are "made to grow." "Create" and "construct" also then make sense in connection with *permissoria*. They most probably refer to the construction of ditches, banks, permanent and temporary fences, or any other means to protect the new growth. The special position and value of *permissoria* also becomes understandable. If they were to be protected, that involved capital investment. Why vineyards should be called *eresztvény* is also less of a puzzle. Vineyards operate much the same way as coppice woods. The stock is permanent, just like the stool, and the young shoots appear every year (although coppice woods are cut less frequently). This is reflected in terminology as well: the medieval Hungarian word for a stool or bolling was the same as that used for a vine-stock (*tőke*).[27]

My conclusions are justified by early dictionaries. Even in the nineteenth century it was plain that an *eresztvény* is "woodland that is cut and is growing new shoots from the stools."[28] In fact, even some source editions translate it with the proper modern Hungarian term "sarjerdő," which literally renders as "spring-wood." However, some scholars argued that an *eresztvény* was a "young, springing wood."[29] I consider this a meaningless projection of modern forestry terms, where such denominations exist, back to the past. In the Middle Ages, no one would have defined a wood as "young" only to rename it after a decade when it was not "young" anymore. Let us also remember that *eresztvény* was not a type of *erdő*, these terms are on the same semantic level. *Eresztvény* woods were a stable part of the countryside, often mentioned at greater time intervals. Their management made them more than three times as valuable a property as ordinary arable land. The meaning of *eresztvény* has become obscure partly because writers of forestry history, who tried to explain it, were unfamiliar with the basic concept of coppicing or were reluctant to accept it as a separate management practice.

One remaining problem is the connection between the Latin and the Hungarian terms. *Permissorium* appears to be the translation of *eresztvény*, but the stem of the former is *permissio*, the equivalent of the English "permission." This has nothing to do with trees and has puzzled scholars for over a century. Some of them conjectured that a *permissorium* was a wood where everyone had the permission to go to, but already Tagányi observed that access to *permissionalis* woods had usually been prohibited.[30] I propose that *permissorium* is merely a bad translation of *eresztvény*, based on a misunderstanding. *Ereszt* today has another meaning, "to let someone do something." This must have been the same in the Middle Ages, whereby those who had to translate the term *eresztvény*, and were probably unaware of woodland management practices, understood that it had something to do with permission, and translated it likewise.

Were (coppice) woods in fact protected by boundary ditches in Hungary? The answer is yes, but the earliest written evidence for this dates from the Early Modern Period. The best-documented example is that from Debrecen, a town that acquired seven smaller woods from the fourteenth century onwards.[31] Mentions of the digging and maintenance of ditches began to appear in the sources from the middle of the eighteenth century, but it is explicitly said that these ditches followed the paths of older ones made in some distant past.[32] Another such case is the sophisticated system that can be read about in the management guidelines of Somos wood near Carei (Ro; Co. Szatmár) from 1712:

> "One of them [the woodmen] has to walk around the enclosure of our wood every day, and has to look hard for any holes, intrusions, and trespasses. If they catch someone he shall be fined 24 *forints*, half of which sum belongs … to the woodmen together with the gardeners, the other half to the lord… . They have to keep the gates locked at all times, and similarly the smaller gates. They have to go through the gate which is facing the town, but even that they have to keep locked at all times."[33]

Most of the written evidence of woodbanks is confined to the Great Plain, probably because the shortage of woodland there induced earlier or better protective measures than

[25] *ZsOkl*, vol.2/2, 12. Also compare Rackham 1996, 45.

[26] 1405: *ZsOkl*, vol. 2/1, 514–515. Also in 1415: *ZsOkl*, vol. 5, 327–328.

[27] See also Chapter 7.

[28] Czuczor and Fogarasi 1864, vol. 2, 386.

[29] *FNESz*, s. v. "eresztevény."

[30] *MEO*, vol. 1, xiii.

[31] Penyigey 1980.

[32] Penyigey 1980, 339–342.

[33] Merényi 1900, 416.

elsewhere in the country.[34] There are numerous examples like the ones quoted above, but none of them answer the crucial question of whether boundary ditches were already present in the Middle Ages or not. There is no written evidence for their existence. The only ditches mentioned as having some function are those around vineyards and meadows,[35] and there is one case when it seems that two ditches bordered a deer-park, much in the same way as in England or Sweden.[36] Only archaeological research can bring us definitive results in this problem. Ditches, in general, are a much-neglected field in Hungarian archaeology. We know that in the Árpádian period most buildings and settlements were surrounded by ditches and banks, but it is rather telling that in the Alföld, where they were the most abundant, close to nothing is visible on the surface of these thousands of ditches.[37] Most of them are only accessible through excavations, which have not been quite so numerous as to form a chronology of ditches. The only medieval ditch-type of which there is firm knowledge is the irrigation canals of the Árpádian period on flood-plains.[38] The structure of these was also clarified through excavations. The same holds true for the already mentioned Sarmatian defensive ditches. Thus, when I considered the possibility of conducting research on woodbanks, I came to realise that this will have to be a separate project, which cannot be accomplished in this work. I am aware, nonetheless, that without the study of boundary banks, the history of woodland will never be complete. Here, I shall propose a line of investigation that can be followed by future landscape archaeologists.

First, more written evidence of the Debrecen and Carei type has to be gathered, mostly from the eighteenth century. Then all places of interest have to be investigated, working first from old maps and aerial photography. Where written sources demonstrate the continuity of woodland from the Middle Ages to the present, fieldwork is needed to identify possible ditch remnants. Selected sites also need to be excavated to clarify the original structures. Then, if several types of woodbanks are detected, they can be connected to different historical periods in two ways. Firstly, British analogies can be utilised.[39] Secondly, where a particular wood can be demonstrated to be of medieval origin based on written sources, the bank itself might be granted that age. Special attention should be paid to *eresztvény* woods, which can also be gathered and studied individually. This jigsaw puzzle evidence may finally form a coherent picture.

6. 3. An Excursus: Tilos Wood.

Another way to examine the archaeology of ancient woods is to start out from field evidence and seek clarification in written sources afterwards. This I shall demonstrate on the finest example of a coppice wood known to me in Hungary: Tilos wood in Újszentmargita (Co. Szabolcs).[40] Here we are not directly concerned with the post-medieval history of individual woods in the Great Plain, but a short description of Tilos wood may bring considerable insights into woodland history. One has to keep in mind that *no* wood in the Carpathian Basin has ever been studied using environmental archaeological methods.[41] I shall not present a full account of the history of this wood,[42] but highlight the features that make it similar to any other ancient wood in Europe.

Újszentmargita lies in eastern (present-day) Hungary, west of the major city of Debrecen. It is also quite close to the Tisza river. It is a rather new settlement ("új" means "new") being one of those hundreds of villages reinhabited after the Ottomans had left at the end of the seventeenth century. The location of the original village is unknown, but it was certainly of medieval origin.[43] West of the modern settlement we find Tilos wood, which is also one of the best-preserved remnants of woodland-steppe and as such, is protected. Its name – although it does not date in itself, and even if it did, it would only provide a *terminus ante quem* – appears to be the characteristic product of the transition era from the Middle Ages to the Early Modern Period, that is, around 1500. "Tilos" means "forbidden," the *silva prohibita* of the sources. Prohibition is not, as is often assumed, an allusion to management, but simply a demonstration of the landlord's claim to a formerly common wood. There are many woods with this name scattered all over the country. Whatever name this wood may have had earlier is lost.

The wood (ca. 60 ha, just a bit bigger than the average size of medieval *permissoria*) consists of patches of woodland, swampy areas, and steppe-meadows.[44] Most of the trees are oak (dominantly *Quercus robur*, but *Quercus pubescens*, *Q. petraea*, and *Q. cerris* all occur). The steppe-parts comprise *Peucedano-Galatelletum punctati* and *Artemisio-Festucetum pseudovinae*, which, in everyday language, mean alkali soils with plants like fescue, artemisia, and milkweed. This mosaic is wholly natural.[45] However, despite all the treeless areas (approximately one quarter of the whole) the place

[34] However, Transdanubia does not lack sources. For the banks of Dudlesz wood (Co. Sopron), see Frank 2001.

[35] Without pushing the similarities too far, let us remember that *permissoria* are sometimes equated with vineyards.

[36] 1274: "… supra hortum nostrum ferarum, inter duo fossata …" *CD*, vol. 5/2, 187. For English deer-parks, see Stamper 1988. For Sweden, Andrén 1997.

[37] Méri 1962; Pálóczi Horváth 1996, 12; László Szabó 1975; Laszlovszky 1982.

[38] For a more detailed account of these ditches, see Chapter 5.

[39] Rackham 1996, 114–118.

[40] I owe the most sincere thanks to Zsolt Molnár, who called my attention to this wood and provided access to maps and aerial photos.

[41] There is a good historical ecological analysis of one place: Nagykőrös wood. Molnár 1996.

[42] Given the lack of background research, this could comprise a separate book.

[43] Medieval Szentmargita appears first in a reconstructed document of 1009, as the property of the bishopric of Eger. *DHA*, 177. At the beginning of the sixteenth century it still belonged to the same institution, and was recorded in the account book of Bishop Hyppolit as yielding minor income from tithes. Nyári 1870. In 1458, king Matthias donated his "predium Zenthmargyth in Co. Szabolcs" ("predium Zenthmargyth in comitatu Zaboch") to Stephen of Bajom and Sebastian of Gecse. *Károlyi*, vol. 2, 318–319. This is usually thought of as referring to the same village (*Csánki*, vol. 1, 526; Németh 1997, 177.) However, why it should have been the king's property (was not it the bishop's?) why it should have been deserted (*predium*), and if all this is explained, how it got back to the bishop awaits explanation.

[44] Tilos wood had long been a subject of research by what was then called the Botanical Research Institute of the Hungarian Academy of Sciences. Starting in 1966, and lead by Imre Máthé and Bálint Zólyomi, the geobotany, climate, vegetation, and ecology of the wood were studied for almost a decade. The results were published in some twenty articles. Zólyomi, Máthé, Précsényi and Szőcs 1972. A detailed bibliography is on pages 42–43.

[45] Zólyomi, Máthé, Précsényi and Szőcs 1972, 38.

still appears to the modern non-specialist and likewise to the medieval peasant, as a *wood*.[46] The landscape features are most remarkable. Tilos wood abounds in giant oak stools. There are dozens whose diameter is around 180–200 cm. They have not been cut for long decades, and their size suggest an age of at least 300–400 years. It is not rare to find seven or eight shoots stemming from the same stool (*Fig. 16*). The whole wood is surrounded by ditches and banks, the ditch always on the outside. The system is not very complex (*Fig. 17*). On the north, next to the ditch of the modern road we can see a rather massive bank; on the east there is no bank, for the wood seems to have been truncated by a forestry plantation. On the west and south there are two lines of woodbanks. The western and southern lines of the outer bank meet at right angles, thus the whole wood (complemented by the lost eastern bank) takes on a rectangular form. The inner bank, however, cuts off a small triangle in the south-western corner of the above rectangle. As for the appearance of the banks, I have already noted that the northern bank is fairly well-built, especially when compared to the southern and western banks. The latter are rather faint. In better preserved sections their height should come around half a metre. The inner and outer banks are identical in size and form. All these banks are very straight (*Figs. 18* and *19*).

Fig. 16. Giant coppice stool in Tilos wood (January 2002).

There are a number of questions that should be raised at this point. Firstly, the straightness of the banks would suggest, at least in England, a relatively late time of construction. One, however, must be cautious. There is no chronology for ditches in Hungary, and what seems a late ditch in England can still be an early one in Hungary, or vice versa. The oldest tree standing on the inner bank is a turkey oak (*Quercus cerris*),

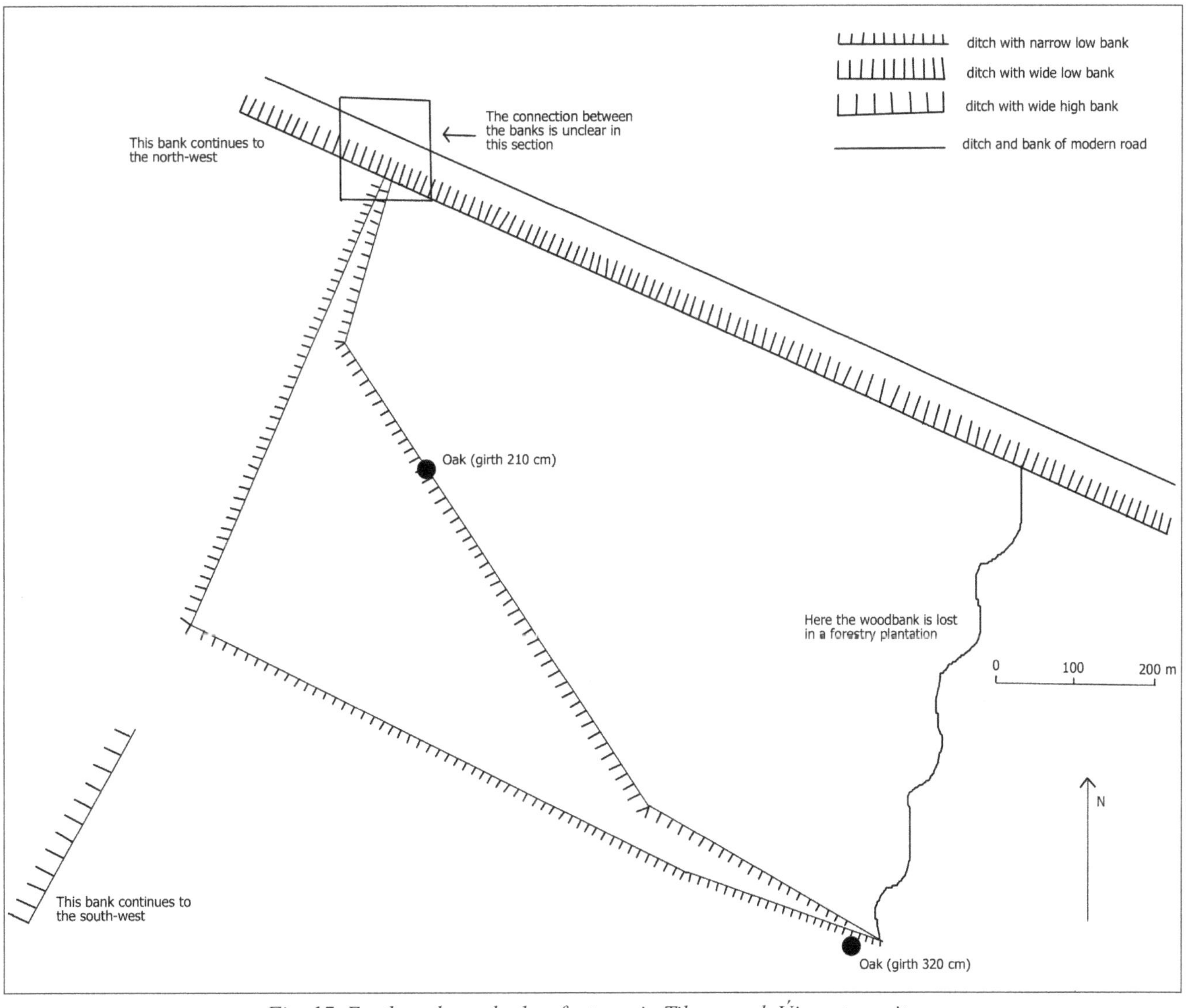

Fig. 17. Earthworks and other features in Tilos wood, Újszentmargita.

[46] Compare Molnár and Kun 2000, 23.

Fig. 18. The outer woodbank in the south-east (January 2002).

Fig. 19. The northern woodbank with the bank and ditch of the modern road (January 2002).

with a girth of 210 cm. This would put the digging of the ditch back in the nineteenth century. At the south-eastern end of the outer ditch one finds a bigger oak (girth 320 cm), but even this may not be older than 200 years. Both trees are unpollarded. Secondly, it would be interesting to know whether the wood had once been smaller and was subsequently enlarged, or it had been bigger originally and the territory between the two banks was given up later for pasture.[47] The vegetation offers little help. It is true that the oak dominated wooded parts conform to the line of the inner bank in the middle of the wood, but this would still make both versions possible, and we should remember that the pattern is thought to be natural. The solution could probably be deciphered from the places where the banks adjoin. However, the south-eastern corner is all blurred because of the plantation, while in the north-west access is impossible for the moment[48] because of the thorns. The banks and ditches of the wood form part of a larger system. The northern bank, for example, does not stop at the north-western corner of the wood, which might explain why it is significantly different from the other banks. There is another bank with a ditch on its western side that begins at the south-western corner of the wood, although there is a discontinuity of some fifty metres. This bank is also conspicuously larger than the southern and western banks of Tilos wood.

This wood then is a typical "European" coppice wood with woodbanks around. Both the deserted medieval village with which it is associated and its name suggest that it is at least 500 years old. This is confirmed by the giant stools and the vegetation itself, which appears mostly natural. The origin of the woodbanks is unclear, and so is their line of development. Archaeological and archival studies will undoubtedly reveal more about the history of Tilos wood, but the basic outlines of its ancient management are already clear.

There are many more woods that are surrounded by ditches in Hungary, for example Nagyerdő in Ócsa (*Figs. 6* and *20*). As far as dating possibilities are concerned, the most promising study area is the Tóköz, south-east of Vienna, around the city of Győr (Co. Győr). The system of ditches in this region was thoroughly studied by Károly Takács and it is hoped that woodbanks, when placed into this system, will be provided with a relative chronology. There is also accumulating evidence in the form of stubs and coppice stools on woodbanks. In Lébény wood north of the river Rábca, for example, a mighty stool stands on a well-built woodbank; in the Bakony mountains, near the village of Olaszfalu (Co. Veszprém), a present wood-pasture site has a large bank with a giant lime stub on it (*Fig. 21*).

6. 4. *Silva Dolabrosa, Glandinosa, et Sub Venatione*

"Item, a larger wood, for example an axable or an acorn-bearing one, which is 'under the axe' and is used for hunting, and which is apt for every work and profession ... is estimated one royal plough to M 50."[49]

Reading the *Tripartitum*, one sometimes has the impression that Werbőczy had little idea of what he was writing about. This rather confused sentence is one of these cases. "Axable," although it sounds somewhat awkward in English, is the best available word to translate *dolabrosa*. The expression apparently refers to a wood where maiden trees overwhelm coppice stools. Just as a *permissoria* was not "a young wood," an axable wood was not "an old wood" in the sense that once the trees were cut it would have ceased to be an axable wood. This type of woodland was also the result of deliberate management, which consisted of not cutting maidens in order to produce high-value timber trees. Some may think that a *silva sub dolabro* was some sort of a wildwood, but this is only true in one aspect, namely that management here was rather extensive. Its value contradicts all other wildwood features. If there had been large tracts of this type of woodland all over the country – the remaining "wasteland" – then it could not have been worth more than ordinary farmland. An axable wood was, on the contrary, seventeen times as valuable as arable or common woods, and five times as valuable as coppice woods. The reason why axable woods were so highly valued is most probably because they were rare and the timber in them was expensive. It is hard to say whether such a wood was closer to a modern "high-forest," that is, a wood without coppice stools, or to a "coppice-with-standards." The presence of timber trees did not exclude stools.

[47] At first sight, the area between the two banks is significantly less wooded than the area within the inner bank.

[48] January 2002.

[49] "Item sylva major, puta dolabrosa, et glandifera, seu sub dolabro et venatione existens, ac pro quolibet opere et artificio valens ... quodlibet aratrum aestimatur ad M 50." *Tripartitum*, I, 133, 29.

Fig. 20. Giant alder stool in Nagyerdő, Ócsa (Co. Pest). It is three metres across and must be ancient, because cuts on neighbouring stools show annual rings of one millimetre or so.

"Used for hunting" is much more difficult to interpret than "axable." The primary meaning is self-explanatory, yet if it means anything more than that is hard to say. *Sub dolabro* and *sub venatione* do not necessarily occur together. In a simple interpretation *sub venatione* was a bonus, just like the acorns of a *silva glandinosa*. Woodland beasts, so this argumentation would continue, were abundant in this type of habitat. However, we know that most "woodland" beasts live perfectly happily in other types of woods, and nowadays they inhabit plantations as well. Some of them, such as the wild boar, are originally not even woodland animals. The other option is to look for some legal meaning in *sub venatione*.[50] Very little, however, is known about medieval Hungarian hunting rights. Csőre argued that hunting was not a royal right, but rather a territorial one. He also claimed that basically all members of society had the right to hunt until the 1504 royal decree, which prohibited peasants from hunting. There is a contradiction in this argument. All land belonged to someone, thus peasants must have hunted one someone else's estate. What difference would it have made to a nobleman if his own peasants had hunted around his village or those of another nobleman? I shall not try to clarify this problem here, but shall return to those aspects that concern Royal Forests in Chapter 9.

Glandinosa (acorn-bearing) is a type of woodland that had been differentiated from other types since at least Roman times.[51] As is the nature of the subject, this should be an occasional, well-profitable bonus. If the wood contained oaks (for acorns) or maybe beech (for beechmast), it had a computable income based on the number of pigs that it could render for pannage. Sometimes these estimates were included in the descriptions of woods, but certainly as management-plans rather than actual pigs in actual years.[52] We do not know if this involved management of trees as well, but we can presume that they were seldom cut in a *silva glandinosa*.[53] Coppice woods were not much good for pannage, because they were too intensely managed and lacked large enough oaks.[54] Acorn-bearing woods were recorded in very high numbers in estimations. In other words, they do *not* appear in the sources as rarely as one might expect. Why this may have been so is explained further on, when I examine the amount of woodland recorded in estimations.

Fig. 21. Lime stub on a bank near Olaszfalu.

6. 5. Rubetum or Virgultum

"The same [as about common woods] is to be known of *rubeta* and *virgulta*."[55]

This laconic sentence of Werbőczy only reveals that *rubeta* and *virgulta* had the same value as arable land or common woods. As far as the two words are concerned, they appear to be interchangeable, very much like *fenetum* and *pratum*. Thus far, I have not been able to identify anything that would suggest one was different from the other.

The meanings of these Latin words make it clear that we are concerned with a piece of land dominated by shrubs. *Virgultum* itself may mean "shrub" and takes its origin from *virga* meaning "twig." *Rubetum* is linguistically less general; it comes from *rubus* meaning "bramble" in classical Latin. Rather surprisingly, the Hungarian equivalent of these terms cannot be traced in the sources. In modern source editions, it is customary to translate them as *cserjés*, which broadly covers

[50] For a short overview of medieval hunting rights, see Csőre 1980, 85–88. See also Zolnay 1971.

[51] Montanari 1979, 35.

[52] For example *Zichy*, vol. 8, 150–153. Széphely and Kétremete (Co. Temes) had a common wood where 500 pigs could be pannaged "in times of acorns" (*tempore glandium*).

[53] The only, rather indirect, evidence for this is an obscure sentence in the so-called *Glosses of Kolozsvár* (ca. 1550), a German-Latin dictionary annotated in Hungarian. The first part of this says: "Sylva est autem glandifera, quae non ceditur …" ("An acorn-bearing wood is one that is not cut …") Berrár and Károly 1984, 195.

[54] Compare Rackham 1980, 119–120.

[55] "De rubetis etiam, et virgultis idem est sentiendum." *Tripartitum*, I, 133, 26.

the concept of *virgultum*.[56] However, I know of only one single example where this equation was made in a medieval source, and even that is more than questionable.[57] *Rubeta* and *virgulta* (including combinations) make up approximately fourteen percent of all woodland recorded in my Database. Why we cannot find the Hungarian name of such an important type of land, I cannot suggest.

Earlier writers proposed that *rubeta* were scrubs formed as a result of "irregular use of woodland" – meaning short coppice cycles.[58] This, however, does not happen in reality. Short coppice cycles do not destroy trees, but form giant stools. The dense shade of the shoots prevents thorny bushes from invading the territory. On the other hand, coppice woods had their own name, *permissoria*, as we have seen above. The real question is whether scrubs can be part of the natural landscape of the Carpathian Basin or whether they are the – intentional or otherwise – result of land management. If the latter is true, we also need to know how they were maintained. I have discussed in Chapter 5 that, save the highest parts of the mountains, the natural vegetation of the Carpathian Basin is woodland or woodland-steppe. Even the latter of these does not qualify as *virgultum*, because, although is it open enough, there is no reason to suppose that its vegetation would be dominated by thorny bushes. Therefore it is most probable that *virgulta* are connected to management, namely grazing.

Scrubs are "thickets of hawthorn and young trees with ancient trees embedded in them," wrote Rackham about the scrubs of Hatfield Forest, Essex. He added that they "bear witness to the uneasy tension between grazing and trees." This he explains later: "Land, unless cultivated or grazed, turns into scrub in a few years and into woodland in a few years more. The scrubs in the Forest result from a *temporary* lack of grazing."[59] Even if medieval Hungarians associated scrubs with bramble rather than hawthorn, there is no doubt that *virgulta* were similar. The *Tripartitum*, when referring to vineyards, adds the following: "Vineyards in general are estimated, according to the ancient customs of our kingdom, like *rubeta* and *virgulta*, because once their cultivators disappear, or they neglect their job, vineyards quickly and easily turn into *rubeta* and thorns (*vepres*)."[60] However, scrubs are not very stable by nature, and turn into woodland "in a few years more." In this case, a perambulation like the Barnag example,[61] where the boundary runs between a *virgultum* and a wood, would not make any sense. *Virgulta*, however, were a stable part of the landscape, and the only thing available to stabilise them was grazing. Thus, *virgulta* and *rubeta* were medieval Latin technical terms for *wood-pasture*. This is sometimes explicitly stated, as in connection with the village of Béc (Co. Zala) in 1373: "rubeta seu silva pascualis."[62] With this in mind, we can account for the almost entire absence of expressions referring directly to wood-pasture, although this type of land use undoubtedly existed, as attested by the medieval presence of pollards (see Chapter 7). It may also explain why pasture is not mentioned at all in the guidelines to estimations. The estimations themselves also disregarded *pascuum*, with the exception of Co. Bodrog. The Gara-Szécs case records pasture, but as a mere 0.9 percent of the entire area. If this reflected a real lack of pasture, then one type of substitute was wood-pasture in the form of scrubs: *virgulta* or *rubeta*.

If one wants to see a very much medieval-looking *virgultum* which is still in use, one does not have to venture too far from the capital. The closest example known to me is the wood-pasture (as it is called in Hungarian[63]) next to the village of Páty (Co. Pilis), south of the Budai Mountains (*Fig. 22*). Here, on a few acres, one can find a plain spotted with hawthorn and bramble (mostly the common *Rosa canina*). Sheep still graze here, and thus, trees cannot return. Near a spring there stands a mighty black poplar, and, most surprisingly, among the bushes one finds a biggish pollard sweet-chestnut. This was obviously planted here. Another fine example of such a landscape is south of the village of Pusztavacs (Co. Pest), where the same bushes cover parts of the otherwise grassy ground accompanied by five large pedunculate oaks[64] and a wild pear, all apparently unpollarded. This *virgultum* is bound to disappear not because of the lack of grazing, but because on one side the village is encroaching upon it while on the other side a forestry plantation is gaining ground.

Why and how were such wood-pastures maintained? First, they come into being by grazing either in woodland or on land that had been cleared of trees and used for some other purpose. The first case takes much longer time and more continuity, because once grazing slackens off, nearby trees can quickly fill the gaps. In the second case, the bushes establish themselves during "a *temporary* lack of grazing." It takes little time for thorns to grow big enough to make them unpleasant for animals to consume. These bushes, then, protect young trees during the first few years of their existence, so that pollards can be established. This was, and still is, an essential element in any wood-pasture system. It should also be mentioned that the shrubs themselves were important sources of firewood. Modern writers without exception despise scrubs, but the medievals valued them as highly as arable land.

6. 6. Woodland in Charters

How were the above distinctions reflected in the actual sources, both estimations and others? Let us first examine the different types of woodland recorded in estimations.

Figure 23 depicts those types of woodland of more than one percent overall. In half of the cases, the source simply refers to *silva*, which leaves the researcher no chance to establish what kind of woodland was in question. The full table (Appendix 4) includes twice as many categories, but most of these are not

[56] *Cserje* means "shrub."

[57] *Oklevélszótár*, 125. 1495: "Quibusdam virgultis wlgo tharyanoscherrye" The author himself was undecisive whether to interpret this as "cserje" or "cser" + "je," a possessive structure involving "cser," one type of oak.

[58] Csőre 1980, 105.

[59] Rackham 1989, 3, 12, 234–236.

[60] "Nam vineae generaliter, tanquam rubeta et virgulta, de regni nostri antiqua consuetudine aestimari solent, ex eo; quod ubi, et postquam cultores vinearum aut defuerint aut earum labores praetermiserint, vineae cito de facilique in rubeta et vepres convertentur." *Tripartitum*, I, 133, 45.

[61] Quoted in Chapter 2.

[62] *Zala*, vol. 2, 68–69.

[63] *Fás legelő*, literally "pasture with trees."

[64] The largest is six metres around.

Fig. 22. Virgultum in Páty. Note the lynchet on the opposite hillside and the ditch in the middle. A pollard sweet chestnut hides in the shrubs on the left (February 2002).

new types, but rather cases where the officials listed the types of woodland present and recorded their territories together. Noteworthy exceptions are *virgultum*, *silva dolabrosa* and *silva sub venatione*, comprising 0.6 %, 0.4 %, and 0.4 %, respectively. The first one, especially bearing in mind the marked presence of *rubetum*, suggests that *rubetum* was the "official" term, for which *virgultum* was only an occasional substitute. The third one reinforces my conjecture concerning the nature of *sub venatione* woods, namely that they had little to do with management. It is much more difficult to explain the almost complete lack of axable woods.[65] Were they in fact so rare? Or were they recorded under some other name? If the second answer is true, the obvious choice is *silva glandinosa*. This type of woodland represents no less than 16.5 % of the whole, which is 31 % of all *specified* woods.[66] It is possible that acorn-bearing woods would have become axable woods before felling, that is, "axable" and "acorn-bearing" were simply two different ways to look at the same trees. For one reason or another, "acorn-bearing" was preferred in recording. In theory, *glandinosa* woods should have been recorded by computing their incomes from pannage and then multiplying it by ten. In practice, however, this was rarely done.

Apart from odd combinations, there is one type of woodland that reoccurs in estimations but is missing from the guidelines: *nemus*. This woodland type comprises 3.5 % of all woodland, that is 6.6 % of all specified woods.

Nemus was usually translated into Hungarian as "berek."[67] The linguistic background of this word is unclear. Some argue it

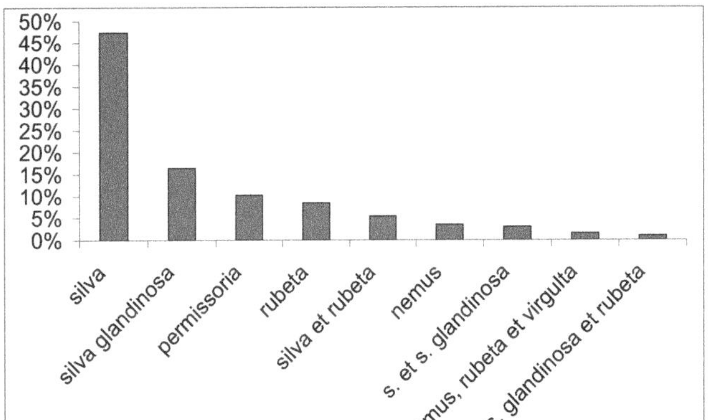

Fig. 23. The percentage of different types of woodland in estimations.

is Finno-Ugric, and is based on "bere" meaning "mud, marsh," which was given the diminutive suffix "-k." Others claim that it is of Slavic origin and means "bank of river." Neither of these explanations, so the *Etymological Dictionary* goes, is very convincing.[68] In any case, it seems clear that "berek" implies some sort of a wetland environment. What it was used for I cannot decide with certainty. There exist, however, fragments of information suggesting that it may have been a form of wood-pasture. Firstly, several present-day landscapes with historical connections to the word still retain a traditional pastoral way of living. One of these, Co. Bereg,[69] carries the memory of such a place in its name.[70] Secondly, a short sentence in the

[65] Even if we add up all those types where axable woods are included, the result is a mere 1 %.

[66] Acorn-bearing woods including all combinations comprise up to 21.6 %.

[67] It is easy to find counterexamples: "silva seu rubetum Wagothberek" 1446. *Kolozsmonostor*, vol. 1., 331. "quoddam rubetum Zyluasberek dictum" 1361. *Oklevélszótár*, 66.

[68] *MNyTESz*, s. v. "berek." It should be mentioned that there are cases when the "bank of river" translation is present in medieval documents: *Oklevélszótár*, 66.

[69] Most of the medieval county is in Ukraine today, together with the capital, Berehove. "Bereg" and "berek" are variants of the same word.

[70] For animal husbandry in the region, see Bodó 1992.

above-mentioned *Glosses of Kolozsvár* (ca. 1550) informs us that a "berek" is "where cattle are grazed and kept warm or pass the winter."[71] Finally, we know that in Austria, *nemus* signified wood-pasture in the Middle Ages.[72] The analogy is not necessarily relevant, but it is certainly worth noting.

There is, however, one aspect to the problem that does not fit the overall picture. *Nemus* is sometimes equated with what appear to be small orchards and not wetland wood-pasture.[73] The list of estimations also contains a certain "nemus pirorum," that is a *nemus* of pear trees.[74] These types of *nemora*, nonetheless, were never translated as "berek." Thus, at present, it appears that the Latin word *nemus* was used both as a translation for "berek" – one form of wetland wood-pasture – and also as a word for smaller fruit tree orchards. In many cases the two are impossible to distinguish.

Students of medieval charters will know about two other recurrent types of woodland: *silva usualis* and *silva prohibita*. These occasionally feature in estimations as well, especially the former term. I have not discussed them above because these types of woodland do not involve management. *Usualis* is, in fact, a synonym for *communis* in the sense that these woods had extensive common rights. It was in contrast to common woods that some time in the fifteenth century prohibited woods appeared. Local noblemen all over the country realised that woods were marketable property and tried to make their peasants pay for what used to be free common rights since time immemorial.

Combining the evidence from estimation guides, the estimations themselves, and other sources, I propose the following list of woodland and wood-pasture types for medieval Hungary:

1. common wood – *silva communis*
2. coppice wood – *silva permissionalis – eresztvény*
3. acorn-bearing wood – *silva glandinosa*
4. wood-pasture – *rubetum / virgultum*
5. wetland wood-pasture – *nemus – berek*

6. 7. Times of Woodcutting

An additional question about medieval woodland management is in which part of the year trees were felled. This is of some importance, because the plants themselves would require cutting in the period when they are at rest: from late autumn to early spring. Cutting outside this time can damage trees. This fact is well-known but little respected nowadays; foresters cut trees all year round.

The obvious source to find an answer to this question would be month-to-month account books of woodland-owners, but these do not exist for medieval Hungary. I know of some documents that imply that trees were cut all year round,[75] but these are privilegal charters that contain what the grantee had the right to do, not what he or she actually did. The only type of document in which woodcutting appears in a statistically significant number is the one that describes a complaint about someone else illegally cutting down trees in the plaintiff's woodland. Here is a typical example from 1406:

> "Sigismund, by the grace of the Lord, King of Hungary ... It was reported to us in person by master Ladislaus, son of Ladislaus of Bathmonostra that Salamon of Chepch, Michael of Zemere and Johannes ..., officials ... of Paul, abbot of the monastery of Batha, by the consent, order and mandate of the same, together with Blasius of Damanch, [here follows a long list of peasants] around the past feast of saint Michael archangel entirely cut down and annihilated a certain wood of the said Ladislaus that belongs to his above-mentioned settlement called Bathmonostra. The cut wood they took away and had it taken away as it pleased them ... *potentia mediante*, causing great damage to him [Ladislaus]."[76]

(Note that this was all happening within one settlement. *Potentia mediante* is a judicial term to be explained in the next paragraph.)

These documents appear some time around 1400, and there is an ever-increasing flow of them from all over the country into the sixteenth century. Why they are not present earlier is hard to say. A possible explanation is that it was at this time that woodland owners started to take the matter to court instead of managing it some unrecorded way. These cases are one type of the criminal act called *acta potentiaria*, which – judged by the little that we can say with certainty – were a way of conflict solving among noblemen.[77] For those years where all documents have been published (1400 – 1420),[78] I collected all such cases and recorded their number in a monthly distribution (*Fig. 24*).

There are far more examples, especially at the beginning of the sixteenth century, however, gathering them would hardly be possible. It is only the said two decades that allow us to eliminate problems of biased data-collection. Here I can argue that I work with virtually all recorded cases,[79] in other words, this is as far as sources can take us.

The chart signifies a largely "nature-friendly" timing for woodcutting. The majority of cases were recorded for the period of November – April, with the exception of June, which has three times as many cases as May or August. Nonetheless, even without having to account for June, several doubts arise about the validity of the chart. First, in most cases we have no information as to whether the actual woodcutting was done shortly before it was reported or not. Appendix 5, which provides detailed information about every document, shows that it was

[71] "Berek, ubi pasci et aestuare vel hyemare pecudes solent." Berrár and Károly 1984, 195.

[72] Sonnlechner 2000, 212–213.

[73] An example from Pilis is *AnjOkm*, vol. 2, 16. "nemus prunorum, nucum et aliorum fructuum"

[74] *Zichy*, vol. 11, 337–342.

[75] For example the one I have quoted in connection with coppice management: *ZsOkl*, vol. 2/2,12.

[76] *Zichy*, vol. 5, 487–488. "Sigismundus dei gratia rex Hungarie ...Dicitur nobis in persona magistri Ladislai filii Ladislai de Bathmonostra quod Salamon de Chepch et Michael De Zemerey ac Johannes ... officiales ... Pauli abbatis monasterii ecclesie de Batha, ex consensu, iussu et mandato eiusdem, unacum Blasio de Damanch ... circa festum sancti Michaelis archangeli proxime preteritum, quandam silvam ipsius magistri Ladislai ad possessionem suam Bathmonostra vocata predictam pertinentem penitus succidi et annihilari, lignaque succisa quo eorum placuisset voluntati deferri fecissent et facerent ... potentia mediante, in preiudicium eiusdem valde magnum."

[77] *KMTL*, 255; Engel 2001a, 219; *Tripartitum*, II, 67, especially footnote 7 on pages 335–336.

[78] *ZsOkl*, vols. 2–7.

[79] Virtually all, because the indices of *ZsOkl*, however well-done, are not perfect.

not uncommon that half a year had passed between the act and the production of the charter. In other instances, however, it was said that only a few days had gone by. Unfortunately, for most events there is no indication of how much time passed between felling and recording the event. Given this situation, I had to take it for granted that such affairs were reported soon after they had taken place. The other disturbing factor was that the monthly distributions of these charters might reflect the schedule of courts rather than actual events. In other words, documents might have been produced when courts were in session (which was not continuous) but not at other times of the year. To try and test this, I counted all documents for six sample years (1415 – 1420). Their monthly distribution is quite even, with something of a decreasing tendency towards the end of the year (*Fig. 25*).

The peak is in June, which perhaps explains why there should be so many cases of illegal woodcutting recorded for that month. Further data, however, support the idea that the monthly distribution of illegal woodcuttings is not influenced by the rate of overall document production. In the latter, May comes in second, and even July has more charters than any of the months that follow. In fact, December, which has the highest number of illegal woodcuttings, has the lowest number of overall records. In sum, with the exception of June, there does not seem to be anything in the larger mechanism of charter production that could account for the distribution pattern of illegal woodcuttings.

One other thing we should not forget is that these cases were, after all, about thefts. It would be only too logical that people should steal firewood during cold weather, not warm. This presumption leads us in a number of possible ways. One the one hand, it could serve as evidence of the otherwise hard-to-prove point that we are concerned with firewood, that is coppicing. On the other hand, implying firewood for *heating* leaves open the question of the other uses (equally important) of firewood. Did these people have enough wood to feed their ovens, forges, etc all year round and needed to commit a crime only when they were cold? The answer, as yet, lies hidden in the documents.

Keeping in mind that the evidence, all in all, is rather weak, but also that this is what we must make do with, *Figure 24* tells us two interconnected things. First, most of the reported cases were probably not different from the usual coppicing, except that they happened in the wrong places.[80] Second, medieval people were well-aware of the fact that the best time to cut trees in this climate was the cold season, which they mostly respected.

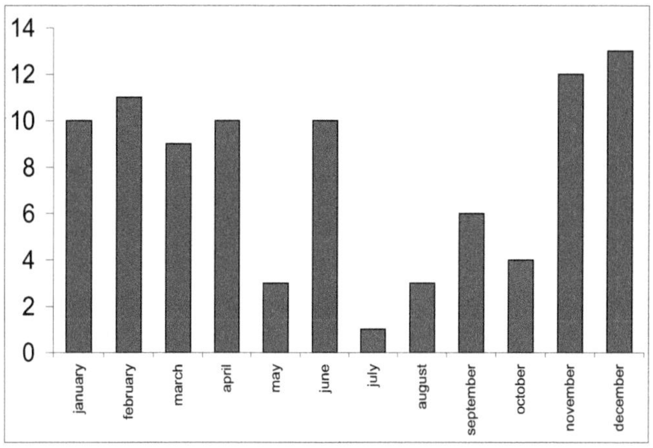

Fig. 24. Monthly number of illegal woodcuttings 1400 – 1420.

6. 8. Timber and Underwood

The distinction between timber and underwood clearly existed. Terminology, however, was not very consistent. The Latin word most often used to denote dead trees of all sizes and types was *lignum*. This could mean giant timber trees, as in 1351, when master Wezzeus donated "one hundred big trees called borona, which are thirty *ulne* in length" for the construction of the church of Saint Adalbert at Esztergom.[81] It is unclear what was meant here by *ulna*, but thirty *ulne* are at least nineteen metres.[82] Firewood was also called *lignum*. Its standard unit of measurement from the earliest records on was the *cartload* (*plaustrum* in Latin, usually *Fuder* in German)[83] although no data survive to tell us how much wood this meant.[84] It seems that there were several different names for different kinds of timber in the vernacular (such as *borona* in the above example) while underwood was simply called "wood" (*fa*).[85] In writing, a clear distinction between timber and underwood was made only when necessary: then, however, rather scrupulously. In privilegal charters that involved woodcutting rights, it was always specified whether firewood or timber (or both) were meant. Apart from these charters, one may happen upon a clear expression of the distinction only by accident. In a court-case from 1322, for example, the people of Spišská Kapitula (Sk; Co. Szepes) got

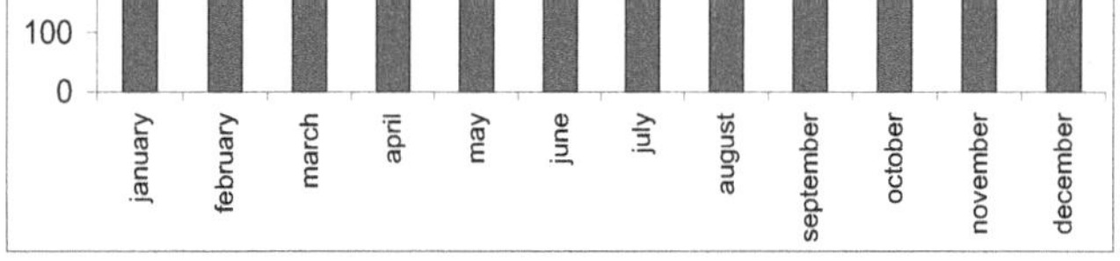

Fig. 25. Monthly distribution of the number of documents preserved in Hungary 1415 – 1420.

[80] Sometimes they are taken as woodland clearance, but that view, knowing that trees grow again when cut, does not deserve more than a footnote. What is more interesting is the implications within the chart concerning *acta potentiaria*, namely that they were a bureaucratic, "officialized" way of managing petty conflicts.

[81] *MES*, vol. 4, 51. "centum magna ligna borona dicta in longitudine triginta ulnas in se continentia"

[82] Bogdán 1991, 87–102.

[83] Csőre 1980, 68–69. For different types of carts, see Domanovszky 1979. A possible reconstruction of a *plaustrum* is in Vörös 2000, 119.

[84] The only data until 1874 (!) is from 1696, when a cartload of firewood was 1.7 cubic metres. It should be noted, however, that the early modern standard unit of measurement of firewood was the *öl* (Latin *ulna*), which was well-defined. Bogdán 1991, 415–416, 421.

[85] There is some discussion about the existence of *tűzifa* ("firewood" literally, the proper modern term) in the Middle Ages, but whatever the palaeographic background, it only occurs once, and therefore has not much relevance. Csőre 1980, 68.

into an argument with the local chapter about a certain piece of land. They had to swear to leave the territory alone so that "the *virgulta*, the pastures, and the wood for firewood and timber be conserved intact."[86]

We should also be able to find separate references to timber and firewood in account books, because timber was more expensive and was sold by the piece rather than by the cartload. In general this is so, however, at present we cannot go further than simply noting this. Account books were rare and came into use from the fifteenth century on. The little material they contain remains unpublished (and unused). Information is available only for some bigger towns such as Sopron (Co. Sopron), Bratislava (Sk; Co. Pozsony) and Bardejov (Sk; Co. Sáros).[87] Apart from references to the purchase of timber, which appears to have been an expensive and somewhat extraordinary event, the account books recorded the regular supplies of underwood. The meaning of this was taken for granted, thus the most common entry is something like: "pro lignis 300 den." In the published material there were just a handful of cases when the number of cartloads was also recorded (which allows us to calculate the price of underwood per cartload). Thus, for the moment, account books cannot be utilised in a meaningful way. Given the importance of this source type – not only in establishing the role of wood in commerce, but also in its potential to provide data on coppice cycles[88] – it is an essential task of future research to study unpublished account books.

Underwood and timber were present, if not much preserved, in the ordinary buildings of the age.[89] This book is not the place to go into details about medieval vernacular building techniques, but even a basic survey of research provides important conclusions about tree management.[90] Here, as ever, precautions must be taken in the interpretation of data. Firstly, it has been only recently that archaeologists tried to reach some consensus about the typology of medieval buildings. In consequence, there has been no uniform terminology, and many older publications are rather hard to understand.[91] Secondly, excavations remain notoriously unpublished, nowadays even more so, because of the large-scale motorway rescue excavations.[92]

Ever since the pioneering studies of István Méri, one of the most influential Hungarian medieval archaeologists, we differentiate between Árpádian period and late medieval houses.[93] The former were rather simple and short-lived semi-subterranean houses; the latter, bigger and better built houses sitting on the ground. The time of change in building techniques was the thirteenth century.[94] This, of course, is a gross oversimplification, and leaves little room for internal development within individual types, which themselves consisted of many different variations. Also, there were houses above ground in the villages of the eleventh and twelfth centuries as well,[95] but semi-subterranean houses were probably dominant. We should not, in any case, suppose that late medieval buildings somehow evolved from the semi-subterranean houses. At present, it seems more probable that both types were present and developed individually from the very beginning.[96]

Let us start with the semi-subterranean houses of the Árpádian period.[97] What kind and how many trees were used in the production of the kind of houses where most of the population lived until the thirteenth century? No standing building survives to illuminate this question, although ethnographers have studied and described many similar constructions from all over the Carpathian Basin. Archaeologists excavated pits (rectangles with sides of around three or four metres each) and the postholes of the timbers that held the purlin, but not much else: apparently these houses had no walls, and the "roof" started at ground level, thus leaving no traces. The two or three timbers and the purlin were small: approximately twenty centimetres in diameter, and did not have to be longer than the height of the whole construction (around three metres.)[98] Lately, there have been several attempts to reconstruct such houses.[99] These supposed that the roof was made from around ten pairs of rafters and that the space between them was filled with wattle-and-daub: interwoven underwood and clay. It also turned out that making such a house was quick. Sabján wrote that in the Árpádian period it could be done in two weeks or less.[100] A recent discovery somewhat changed this picture. László Pintér excavated a house that burnt down

[86] *RDES*, vol. 2, 357–358. "virgulta et pascua et silva pro lignis conbustilibus et structuris indempniter conserventur."

[87] Dányi and Zimányi 1989; Fejérpataky 1885.

[88] Compare Rackham 1996, 63–65.

[89] For a general survey, see Tari 2001. Despite the title, the article covers the whole of the Middle Ages. A useful book for comparison and terminology is Grenville 1997.

[90] Castles, manors, and other noble dwellings would probably supply more data, as they are better studied, but their number was never quite high enough to affect woodland management on the one hand, and on the other hand, timber could be brought from greater distances. Such buildings, in my view, represent what their owners could afford rather than what might have been going on in ordinary local woodlands.

[91] For more on this, see Miklós Takács 2001.

[92] For a list of rescue excavations, see Miklós Takács 2001, 16–19.

[93] Méri 1952; Méri 1954.

[94] For a general history of vernacular buildings, see Balassa 1985. Although outdated in many respects, this books remains a standard reference work and the only monograph dedicated solely to medieval vernacular buildings. The latest summaries are Miklós Takács 2001 and Pálóczi Horváth 2001.

[95] Bencze, Gyulai, Sabján and Takács 1999, 96. There is an archaeologist who does not believe in the existence of semi-subterranean houses at all. He claims that the recovered subterranean parts were cellars. Bakay 1989.

[96] Pálóczi Horváth 2001, 222–224.

[97] Some of the many excavations: Kovalovszki 1960; Kovalovszki 1975. For a copious bibliography, see Bencze, Gyulai, Sabján and Takács 1999. See also Balassa 1985 and Miklós Takács 2001.

[98] There are debates about where the roof started and consequently about the height. It used to be thought that the roof started right at the edge of the pit, which, so Tibor Sabján argued, was not possible because the side of the wall would have collapsed and the house itself would have been uncomfortably small. Ethnographic analogies also show that the roof started further away from the edge leaving some extra space for sleeping and storage and resulting in a taller building. Bencze, Gyulai, Sabján and Takács 1999, 131–176.

[99] Bencze, Gyulai, Sabján and Takács 1999, 131–176; Szentgyörgyi 2000.

[100] Bencze, Gyulai, Sabján and Takács 1999, 145.

and preserved the purlin and some of the rafters.[101] Here, the rafters were set more closely than usually supposed; in fact, they were touching and thus holding one another (*Fig. 26*). In sum, we can say that these semi-subterranean houses did not need much building material (a few dozen small trees and an insignificant amount of wattle,) which must have been available to all. We may infer that maiden trees and coppice shoots were both needed (in the form of timber and wattle) but the amounts in question were so small that, without knowing how many such houses had to be built every year, it would be pointless to draw far-fetched conclusions about woodland management.

Late medieval houses were much bigger enterprises. They were meant to last longer, had proper walls and usually several rooms. Apart from stone buildings, which were rare and are not my concern here,[102] there were two basic types. One of them was the "log-cabin," that is, a house built entirely of big timbers laid on one another and joined at the corners.[103] There is actually little direct evidence for the existence of such houses. What archaeologists have found were stone foundation walls without any traces of holes for timbers in them. They concluded, again supported by ethnographic analogies, that this reflected a log-cabin type of building technique. The original idea was that these buildings were wide-spread in the well-wooded western parts of the country,[104] however, the discovery of such a log-cabin construction in the Great Plain (at Nyársapát, Co. Pest) signifies the situation must have been more complicated.

We know of more examples of the other house type,[105] and because in an archaeological sense it left better identifiable remains than the log-cabin type, we know more about its walls as well (*Fig. 27*)[106]. This type, although rather far from an English or German timber-framed house, used similar materials.[107] The structure of its walls consisted of many small timber trees dug into the ground some one metre from one another. The space between was filled with wattle, using underwood. This is inferred by the examination of remains of daub in which the shape of sticks was printed. Such a house required an impressive number of trees in its construction. At Túrkeve-Móric (Jászkun district), I counted around one hundred and thirty holes for timber,[108] in varying sizes from eight to twenty-five centimetres in diameter. Timber larger than this was rarely used, and only in structurally important positions. Allowing for the roof, the overall number of timbers used was reasonably over two hundred. Underwood was also needed in considerable amounts. All this is comparable to medieval

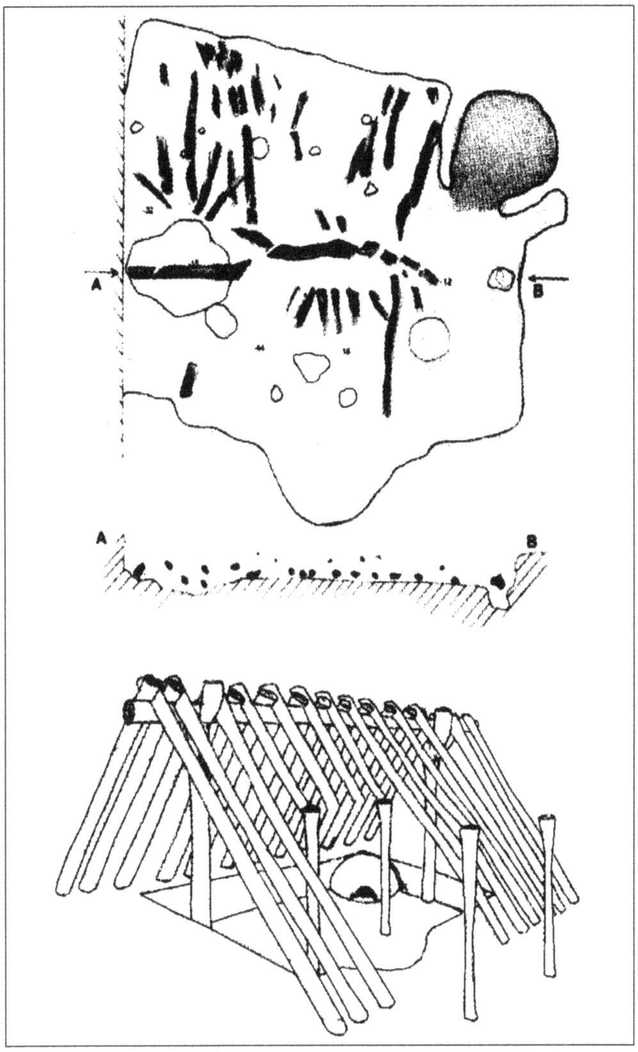

Fig. 26. Charred timbers in a burnt-down Árpádian period house, and a possible reconstruction of the structure of the house.

English houses, both in the number and the sizes of timbers.[109] Just as log-cabins are thought to have been connected to more wooded parts of the Hungarian Kingdom, this type should have been characteristic for the Great Plain. However, the remains of such houses came to light in Csepely (Co. Veszprém), in the supposedly wooded Bakony Mountains.[110] In England, the existence of timber-framed houses is clearly not related to the amount of woodland in the neighbourhood.[111] Csepely

[101] Pintér 2001. The survival of the trees means that the roof was covered with earth, which, when the rafters were half burnt and could not support its weight, fell on the trees and put out the fire.

[102] See for example the stone houses at Nyársapát: Benkő 1980; or Csepely: Kovalovszki 1969.

[103] For example Holl 1979; Holl and Parádi 1982a; Holl and Parádi 1982b; Bálint 1960–1962; Benkő 1980. Some argue that certain Árpádian period houses were built using a similar technique. Fodor 1994.

[104] Holl and Parádi 1982a, 13–16.

[105] Kovalovszki 1969; Benkő 1980; Pálóczi Horváth 2000a; Pálóczi Horváth 1996; Méri 1954. A reconstruction attempt is Sabján 2000. This has highly instructive photographs about standing buildings (not medieval) of a similar type. Such buildings were characteristic of the Carpathian Basin from the Neolithic to the eighteenth century. The huts of the Árpádian period, in fact, represent one of the few exceptions to this general pattern. Similar houses are thought to have been built in medieval Scotland, for example. Crone and Watson 2003, 67.

[106] Reproduced after Pálóczi Horváth 2000a, 149.

[107] There is now accumulating evidence for the existence of timber-framed houses proper, as well. Müller 1972; Holl 1990; Kvassay 1996. For more examples, see Pálóczi Horváth 2001, 240–245. Reconstructions of such houses in Visegrád appear exactly like any other timber-framed houses in medieval Europe. To establish how widespread this type was, however, remains a task for future research.

[108] Méri 1954, house no. 1 on map 1.

[109] Rackham counted some 330 trees in a fifteenth-century Suffolk timber-framed farmhouse, half of which were less than 23 cm in diameter. Rackham 1972.

[110] Kovalovszki 1969.

[111] Rackham 1996, 66–67.

is within a day's walk of Sarvaly (Co. Zala), where the classic examples of log-cabin houses were excavated. This tells us that the choice between the two types was governed partly by principles and fashions invisible in the archaeological record and inexplicable in strictly materialistic terms.

"Building practice adapted to using a certain size of tree implies a woodland practice that regularly supplied it."[112] Late medieval Hungarian "timber-framed" houses, wherever excavated, all used very similar trees: young maidens, around twenty centimetres in diameter, cut at an age that is way below what would be considered "mature" nowadays. Most of these seem, inasmuch as can be judged from the holes, to have been full – that is not split – trees.[113] If remains were found, they were almost exclusively oak.[114] "They imply a practice of woodmanship with a rapid turnover of small oaks and no difficulty in replacing them."[115] Such trees were certainly not to be found in great numbers in a wood anywhere near a wildwood state, and woodmen would not have cut so many young trees if replacement had been problematic. We know from the sophisticated construction of wells that they were capable of fine carpentry,[116] and thus could have made small beams from bigger trees, but chose not to do so. All in all, one imagines that these trees came from managed coppices. Timbers were cut young, and wattle was made from shoots. Some timbers were left to grow bigger to supply purlins and other bigger structural elements. Some excavated tree remains also seem to reflect such management practices. Grynaeus observed that for the well at Szécsény (Co. Nógrád), the small trees had all been cut at the same time.[117]

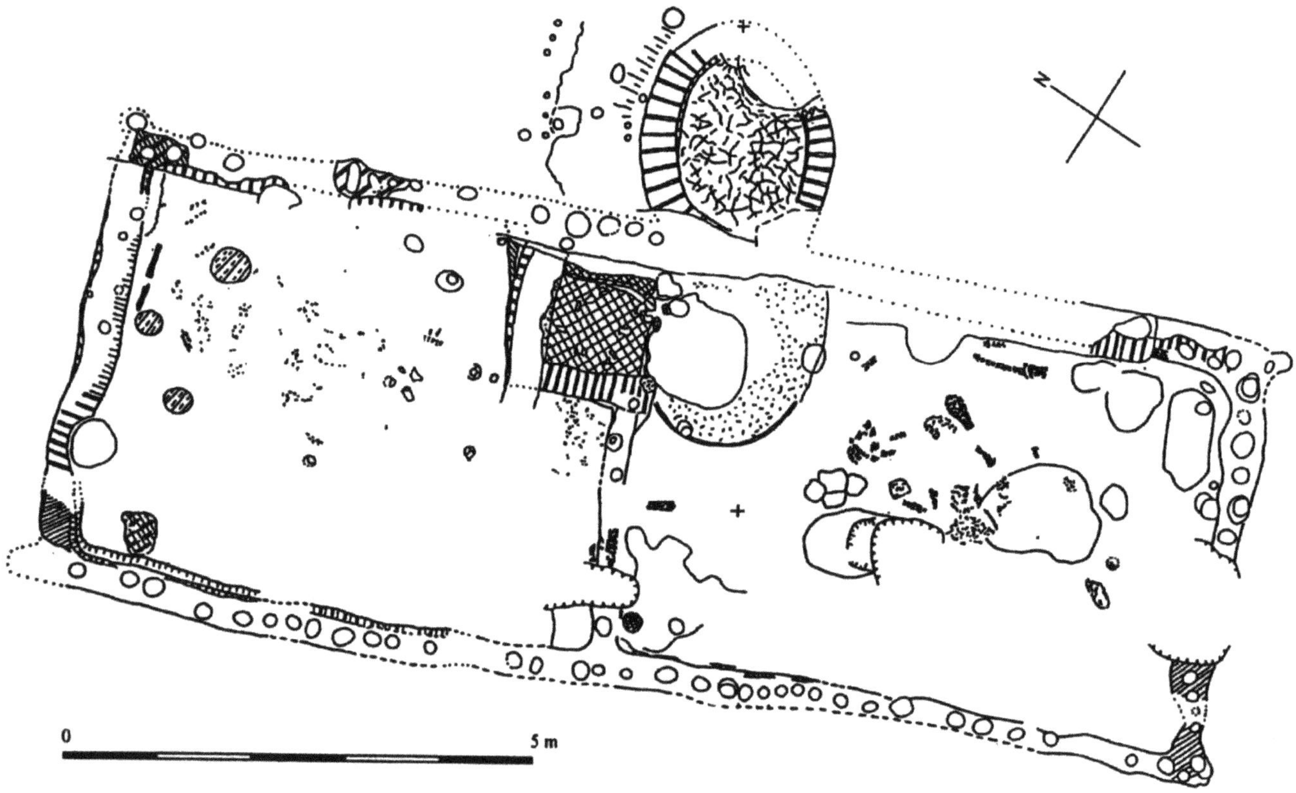

Fig. 27. Groundplan of a late medieval house from Szentkirály (Jászkun district).

[112] Rackham 1996, 66.

[113] Méri 1954, 144 remarked that some of the timbers had been split.

[114] Grynaeus 1997, 114.

[115] Rackham 1986, 87.

[116] For example: Pálóczi Horváth 1987. More on medieval wells in general: Grynaeus 1997, 92–97.

[117] Grynaeus 1997, 113–116. Inordinate attention is paid to a 1426 charter of King Sigismund in this matter. *MEO*, vol. 1, 25. Here the king ordered that a different part of the woods (in Co. Zólyom) should be cut each year and then ploughing had to be forbidden in that part to allow for regeneration. This is commonplace knowledge for any woodman in any historical period.

Chapter 7

Trees in the Landscape

7. 1. Trees in Perambulations

I have already introduced perambulations and their basic functions in Chapter 2. Landscape history – especially in Hungary – relies so much upon perambulations that it is worth having a closer look at them. The ethnographer Lajos Takács, already mentioned in connection with his research on clearing techniques, also studied early modern boundary-walks.[1] His monograph focused on the procedure and on the types of individual landmarks. Perambulations were carried out in Hungary as late as the nineteenth century, and the excellent research of the ethnographic school contains many issues relevant to the Middle Ages as well, even though the precautions I listed in Chapter 4 must be kept in mind.

Medieval perambulations are known to all who read charters. However, basic questions have not yet been addressed, such as: Approximately how many of them survive? What are the landmarks most often used, and what are their proportions in the whole? How many perambulations were written in different periods and what are the reasons behind the changes, if any? All these aspects are important because the individual trees and woods must appear in their proper context. If I read about a certain wood next to some arable land, this tells me very little unless I am able to locate both the wood and the arable in the modern landscape. However, if I know that woods appear twice as often as arable in the whole perambulation corpus of a certain area or period, and, say, that the two put together still do not come near pasture, then I can judge the value of that small piece of information with more confidence. In other words, there are two directions in the interpretation of perambulations. One is the qualitative, micro-level search, where individual landscape features are identified.[2] The other is the quantitative, macro-level approach, which has not yet attracted any scholarly attention in this country.[3] In the following, I shall concentrate on the latter.

A statistical analysis of all medieval perambulations is not unrealistic, but would be so time-consuming that it would have undoubtedly exceeded the limits of this piece of writing. In consequence, I had to choose a smaller, yet well-defined group of charters. I could have used either the *Anjou-kori Oklevéltár*, or the *Zsigmondkori Oklevéltár*,[4] which contain all documents issued in certain years, but that would have ignored time dimensions. I could also have taken published family archives, such as those of the Károlyi or the Zichy families.[5] This would have been relatively quick, and would have allowed for sampling different time periods, however, such collections are inevitably regionalised, connected to the estates of the given family. Furthermore, and this applies to both options, with edited charters one always has to trust the person who transcribed the documents. Palaeographers, undoubtedly skilled in many aspects of medieval life, are rarely experts of landscape terminology, especially of tree types.[6] In the end, the choice I had to make was between databases that were limited either in time or in space. Finally, I decided on the latter option. With an intent to study hitherto unpublished material, the private archives of the chapter of Veszprém seemed a reasonable choice. If I had to choose a regionalised database in any case, it appeared logical to pick one with immediate relevance to the more concrete subject of this book.

The chapter of Veszprém had two relatively rich archives, both of which survive. One of them is the archival material connected to the activity of the chapter as *locus credibilis*.[7] This, however, seemed too specific, concentrating only on one type of legal activity. The places covered are more scattered than in the case of the private archives, and the surviving charters number only 214. The private archives (*Veszprémi Káptalani Magánlevéltár*, henceforth Vkpmlt)[8] then, are simply a collection of the charters of the chapter as an institution. The chapter had lands, villages, different sources of income, just like any other private or public "noble" body. The nearly 1100 charters discussing these lands formed a corpus that was of manageable size yet a sample large enough to serve as a basis for general conclusions. The time range is typical of the Hungarian Middle Ages: very little from the eleventh and twelfth centuries, while from the thirteenth century there is a sharply rising number of sources.

---

[1] Takács 1987. The title is misleading: there was never a "feudal period" in Hungary, yet the tradition is to use the term up until the mid-nineteenth century. The reason for this is that Hungary preserved a social structure seemingly similar to that of the Middle Ages until the reforms of the early 1800s. The ill-chosen title, however, does not affect the scholarly quality of this book.

[2] For the identification of individual features in perambulations in Hungary, see Laszlovszky1986a; Szűcs 1993a, esp. 26–31; Takács 2000a; Major 1959; Zatykó 2003. This also can be done on a large scale, as in Hooke 1999. For a more theoretical overview, see Hooke 1992.

[3] A notable exception is Rómer 1860a. Rómer claimed to have counted trees in the *Codex Diplomaticus*, where oak would have appeared most frequently, however, his data collection is very incomplete. Dénes B. Jankovich published a list of the kind of things to be gathered from charters. Jankovich 1986. No results that I know of followed.

[4] *Anjokl*; *ZsOkl.*

[5] *Károlyi*; *Zichy*.

[6] I do not claim that my readings are perfect, still mistakes herein will be at least *my* mistakes.

[7] *Loca credibilia* are peculiar to the Hungarian Middle Ages. From approximately the beginning of the thirteenth century, they acted as public notaries. Many people and institutions also kept their charters in the chapter-houses. For a quick introduction on how they functioned, see Szentpétery 1930, 121–138. *Loca credibilia* were also the primary source of common knowledge, oral legal tradition. Fügedi 1986. We still use the proverb "Nem káptalan a fejem" ("My head is not a chapter"), if we cannot remember something.

[8] OL Df 200 353, 200 610 – 200 899, 200 902 – 201 203, 201 205 – 201 634, 230 049 – 230 077, 273 681, 283 184 – 283 247. The old reference is U 402. The original charters are kept in Veszprém. For a description of the archives, see Lukcsics 1930; Körmendy 1980. A short English version is in Balázs 1976, 196–197.

The Vkpmlt contains 1096 charters connected to the settlements possessed in one way or another by the chapter, though 199 discuss matters like tithes, *ius gladii*, or the inventory of the library, and one may even find the records of incomes from the holdings of the chapter from around 1500.[9] The majority of the material was catalogued on the basis of the settlements to which they pertained. The number of these settlements is 112. I found altogether 63 documents containing perambulations.[10] One is an obvious early modern forgery, which still has medieval origins. A perambulation, as we have seen, was a strict legal procedure, thus in theory I should have counted every section that starts "Primo incepissent..." ("Firstly, they started..." the formulaic opening of the description) as a separate boundary-walk. Nonetheless there was only one document in which perambulations would have referred to very different places. Some had more than one bounds,[11] but these were carried out in one day in several places in one settlement, therefore I counted them as one. This was also practical for other calculations with the number of charters.

What are the drawbacks of this corpus? In other words, how far does it represent the surviving medieval material as a whole? Every corpus, for a start, must be biased. Representative sampling would require so much effort that it would be easier to work with all 200,000 documents. The Vkpmlt most probably has more perambulations than, for example, family archives, simply because the chapter had more land than most families. Then it is partial towards hilly countryside and woodland as opposed to flat land and pasture. The majority of the settlements mentioned are either in Co. Veszprém or in Co. Zala, with some in Co. Somogy, which appear in the data of estimations as rather more wooded than most of the kingdom (*Fig. 28*)[12].

Finally, we have to take into account that the distribution of perambulations in time is largely influenced by the history of the chapter of Veszprém itself.[13] They happened when the chapter was given some land or had some trouble with an existing settlement. I believe that the low number of perambulations in the fifteenth century is partly due to this phenomenon. All in all, the data and calculations presented in the following are intended to be the first solid ground in the sea of existing material, but it will take much more time and work to have a whole island on which one can set foot without the risk of sinking.

The first question I would like to answer, however feebly, is approximately how many perambulations may exist in medieval sources. In the Vkpmlt, there are 63 perambulations for 1096 charters, which is 5.75 %. Today, we know of the existence of approximately 200,000 documents up to 1526. I shall use this number, and not the 318,000 charters whose texts have been preserved in the 200,000 documents, since the 1096 charters that make up the Vkpmlt also contain transcriptions or mentions of far more charters. The 5.75 % of 200,000 is not less than 11,500, which is an astonishingly high number. One possible means to check how far this might be true is by examining the number of settlements in relation to the surviving perambulations. Here we are concerned with 112 settlements, at least as judged by those who set up the reference system of the Vkpmlt. Although naturally we cannot say that one perambulation always equals one and only one settlement, and even less that every perambulation represented the full boundary for one settlement – very often the perambulation is on the common boundary of two settlements: how can that be inserted into such calculations? – let us still utilise the rough-and-ready percentage of 56.25 % of the settlements of the Vkpmlt that have a perambulation. The total number of settlements known from written sources in the medieval Kingdom of Hungary (deserted or extant) is estimated to be around 20,000.[14] 56.25 % of this number comes to 11,250, which will certainly not prove, but at least strengthen confidence in the number of perambulations calculated above.[15] There exist some attempts at transforming edited perambulations into a computerised database,[16] but, as I shall demonstrate in the following, counting landmarks is not as mechanical as it seems.

The perambulations of the Vkpmlt are unevenly distributed in time. Most of the material comes from the fourteenth century, which is not surprising as opposed to the thirteenth, but it may be so in relation to the fifteenth (*Fig. 29*). It is clear, nonetheless, that the great age of perambulations in Hungary was in the thirteenth and fourteenth centuries. In these two hundred years, the whole system of land ownership and estate structure was transformed. In the thirteenth century, huge royal lands fell into private hands. In the fourteenth century, the previously scattered possessions of landholders were concentrated into compact estates. All these involved perambulations on a large scale. The boundaries of individual units of land did not necessarily change, but they needed to be defined in writing for each new owner or new system. In contrast, the fifteenth century was a time of relative tranquillity.

Counting boundary signs was not as easy as it initially promised to be. The first problem was the recurrent features in the same perambulation, for example, when the officials passed by a road, and then returned to what appears to be the same road.[17] This was sorted out by a fortunate custom of Hungarian medieval latinity: when a feature reappears in

[9] Df 201 634. The source is published: Kredics, Madarász and Solymosi 1997.

[10] For a list, see Appendix 6.

[11] Df 200 902, 200 956, 200 980, 201 061. The first one is the exception, it has two separate bounds, which I count separately.

[12] Reproduced after Kredics, Madarász and Solymosi 1997, appendix.

[13] Lukcsics 1908.

[14] Szabó 1966, 70. This number, although accepted and referred to, is more than problematic. As we shall see in Pilis, and to a lesser degree also in Bakony, in the Árpádian period there were far more settlements (of whatever character) than what written sources suggest. Although nucleation from the thirteenth century resulted in settlements that would be called "villages" today, and written sources mention probably the majority of these, counting *all* remains of medieval settlements in the Carpathian Basin could easily lead to a number twice or three times more than twenty thousand.

[15] There is a statistical analysis of the archives connected to the estate of Szenyér (Co. Somogy.) Here, out of the 340 charters, 32 were perambulations, which is 9.4 %, almost twice as much as in Vkpmlt. Borsa 1984.

[16] Grynaeus and Grynaeus 2001. The Grynaeus' research focuses on plants alone.

[17] Oliver Rackham, when facing the same problem, used a straightforward method: every feature, even those most probably mentioned before, was counted separately. Rackham 1986, 10.

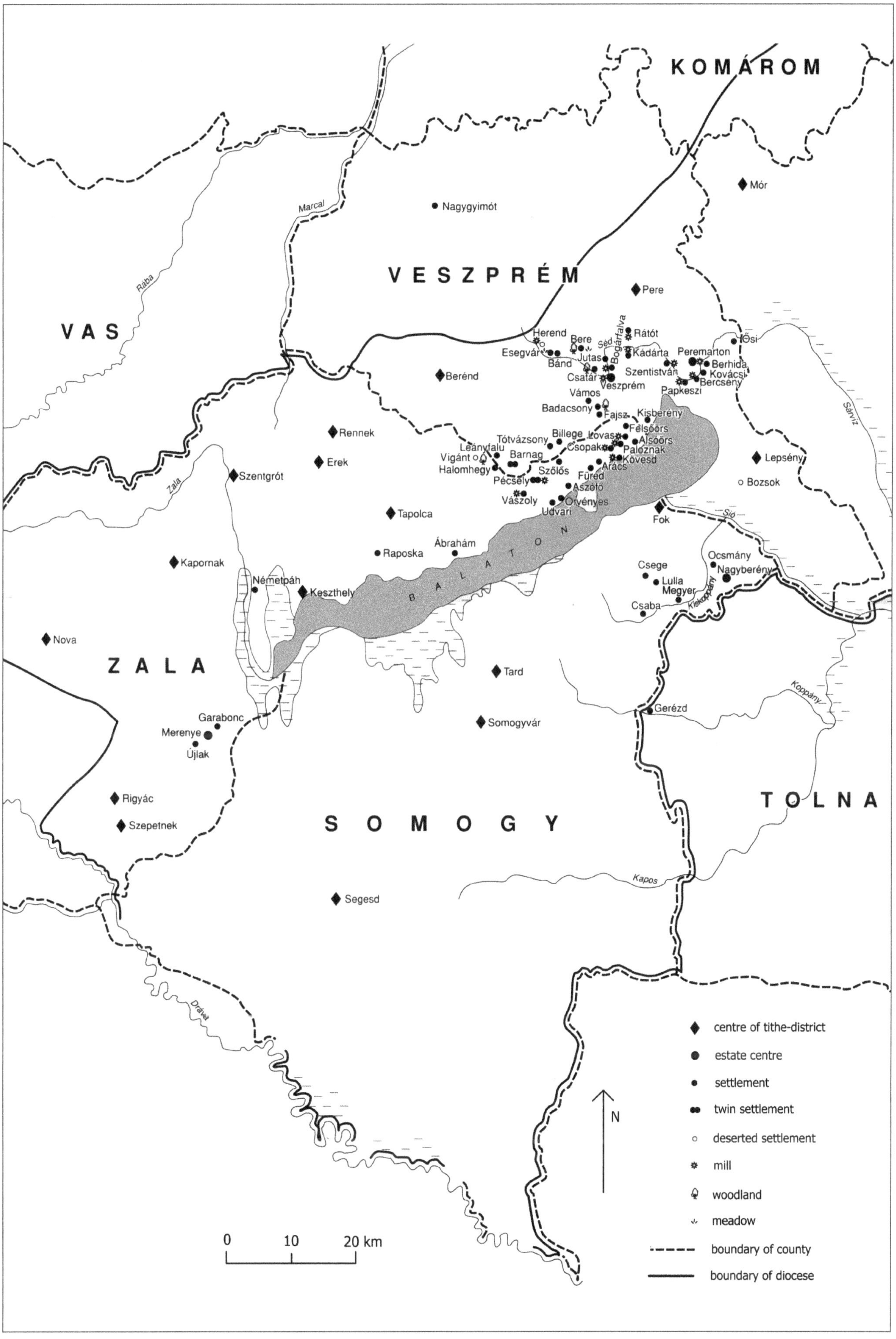

Fig. 28. Holdings of the chapter of Veszprém around 1500.

century	number of perambulations
13th	14
14th	37
15th	9
16th	2

Fig. 29. Number of perambulations in the Vkpmlt corpus.

a charter, this is noted by the help of an adjective, such as *prefatus*, *predictus*, or *prescriptus*.[18] The second problem is connected to linguistics, and to our inadequate knowledge of the medieval countryside. Does, for example, *via magna* denote something characteristically different from *via*, or is *magna* a simple adjective as we would use it today? What is the relation of these two to *via publica*, or *via erbosa*?[19] Is there a difference between *aqua* and *rivulus*?[20] My final choices of what to count as a separate entity are laid out in Appendix 7. Trees were separated in the most detailed possible way, for example a *pirus*, *pirus silvestris*, *pirus et meta terrea*, and *pirus silvestris et meta terrea* are four entries in the table.

The total number of landmarks in the 63 perambulations is 1901, which gives the average charter 31 boundary signs. Altogether 134 types of signs were counted, however, this number is easily modified by merging different types of roads or trees. If we start examining this list at the bottom, we encounter many things that were mentioned just once, among them such curiosities as the nest of a hawk, a foxhole, or a vintage road (*via vindemialis*). Moving up the list, many types of tree are mentioned twice or three times (chestnut, cherry, crab-apple). Items mentioned more than ten times are three types of tree: walnut, pear, and oak, accompanied by meadows, certain types of roads, or mills. The top ten are most intriguing. Woodland, with 35 occurrences, is just outside this limit, whereas both arable and vineyards are within it. Hills and valleys are mentioned altogether 115 times in the charters. The oak-tree surrounded by an earthen boundary (*ilex meta terrea circumfusa*) occupies the fifth place with 60 occurrences. The first three places testify to a striking feature of both the perambulations and medieval Hungary. *Via* is in the third position (129 mentions while all types of roads added up come to no less than 208), preceded by earthen boundary (*meta terrea* – 348), and boundary (*meta* – 466). These two are most probably the same since there are many allusions to the simple *meta* being actually made of earth. The two together are mentioned 814 times – 13 times on average – comprising almost forty-three percent of all boundary signs! This overwhelming dominance signifies at least three things. Firstly, that from the thirteenth century, when earthen boundaries start appearing in large numbers, the descriptions in perambulations concentrate less on natural features and more on one man-made aspect: little hillocks accumulated during the process itself. More and more of the actual countryside, thus, is hidden from the eyes of the modern researchers. Some late charter bounds are nothing more than lists of earthen boundary signs. Secondly, that Hungary in the Middle Ages was a landscape where almost every settlement was surrounded by these earthen boundaries, which, as the logic of perambulations would prescribe, were within sight of one another. Sparse data provide some information about the size and shape of these hillocks. We know that sometimes they were quite big (an early modern charter speaks of one that was 14–16 steps around), but more often about as high as a man.[21] One Vkpmlt perambulation related that a certain hillock might not be a boundary, because it was "not round but elongated,"[22] which gives us an approximate idea of the shape. Thirdly, that a perambulation consisted of some earth moving – although the regular repairing (*renovare*, as it is usually referred to) of these hillocks may not have been such a great task – and thus, must have involved a number of people.[23]

Trees appear in these perambulations in large numbers, mostly as individual, non-woodland plants. They were very often combined with the well-known earthen boundary (*meta terrea circumdata* or *circumfusa*). We find altogether 305 trees, recorded in 75 forms. Out of these, 14 were only defined as "tree" or "shrub." Twenty-three different types were recorded, from the obvious oak and walnut, through hornbeam and lime, to juniper and service. In this sense, the list is definitely representative. There are not many types of trees mentioned in medieval documents that are not included here.

How much botanical precision should be expected from perambulations? Identification to the genus level is usually unproblematic, and if recognition of the species is also possible, it is because the genus sometimes automatically denotes species as well, as with beech (*Fagus sylvatica*), for example. In some other cases, different names for the same species had to be sorted out (*Fig. 30*). I would also like to refer to the scholarly opinion which supposes that names of trees in charter bounds should not be interpreted straightforwardly. Thus, it is more advisable to speak about "a beech, so-called in the sources" rather than an actual beech.[24] In my view, this is an overly critical approach. We cannot expect the medieval perambulators to think in Linnean terms – that would be anachronistic – but I do not see any reasons why we should question their ability to tell a hornbeam from a lime. In particular instances one becomes suspicious (as with the single *pinus* in the Vkpmlt corpus), but those are simply inevitable mistakes that will occur in any human affair. Too much scepticism would hinder our work.

[18] In the few other cases features were counted separately.

[19] One perambulation suggests that *via erbosa* might be a footpath. 1258: *PRT*, vol. 2, 307–308. "capitenea meta autem incipit in quadam via herbosa seu pedestri semita"

[20] Fieldwork revealed that in an eleventh-century perambulation around the river Tisza, *fluvium* and *aqua* had had different meanings. Laszlovszky 1986a.

[21] Takács 1987, 80.

[22] "quendam tumositatem terre non rotundam sed longuinam pro meta minime aptam" 1392: Dl 201 106.

[23] Compare for example Winiwarter 1999, where on page 41 the author wrote that "to calculate the amount of labor needed for agriculture in the Middle Ages one has to include the maintenance work needed for infrastructure and the work needed for erection and subsequent dismantling of temporary features such as fences."

[24] Grynaeus and Papp 1977.

Latin in sources[25]	modern Hungarian[26]	botanical Latin	modern English
acer, platanus[27]	juhar	Acer sp.	maple
alnus	éger	Alnus glutinosa[28]	alder
betula	nyír	Betula pendula[29]	birch
(carpinus? ornus?)[30]	gyertyán	Carpinus betulus	hornbeam
castanea	szelídgesztenye	Castanea sativa	sweet-chestnut[31]
cerasus, merasus, prunus	meggy, cseresznye, szilva	Cerasus sp., Prunus sp.	cherry, sour cherry, plum[32]
cornus	som	Cornus sp.	cornel
avella	mogyoró	Corylus avellana	hazel
cinus	galagonya	Crataegus sp.	hawthorn
fagus	bükk	Fagus sylvatica	beech
fraxinus	kőris	Fraxinus sp.	ash
nux	dió	Juglans regia	walnut
iuniper[33]	boróka	Juniperus communis	juniper
pomus (silvestris)	vadalma	Malus sylvestris	crab-apple
pinus	fenyő	Pinus sp.	pine
populus, tremulus	nyár	Populus sp.	poplar[34]
pirus (silvestris)	vadkörte, vackor	Pyrus pyraster	wild pear
ilex, quercus	tölgy	Quercus sp.	oak
salix	fűz	Salix sp.	willow
sambucus	bodza	Sambucus nigra[35]	elder
sorbellus[36]	berkenye	Sorbus sp.	sorb
-	barkócaberkenye	Sorbus torminalis	service[37]
taxus	tiszafa	Taxus baccata	yew
tilia	hárs	Tilia sp.	lime (GB) / linden (US)
ulmus	szil	Ulmus sp.	elm

Fig. 30. Principal trees and shrubs that occur in medieval sources.

7. 2. Problem Cases

7. 2. 1. Oak

Medieval sources contain several words in connection with oak. Some are obvious (*quercus*, or the Hungarian *tölgy*), but some are more problematic. First and foremost, one needs to include *ilex*. This word, the most abundant type in the Vkpmlt corpus, definitely referred to some sort of an oak. There are countless charters where it is explained this way.[38] *Ilex* in this meaning is usually taken to be a Hungarian speciality, but no one has ever examined whether this is in fact so. The other two words that doubtlessly meant "oak" are the vernacular *magyal* and *cser*.[39] Both are also fairly well-represented in perambulations. After the articles of Camillo Reuter, the

[25] Spelling differences, such *pirus* or *pyrus*, *quercus* or *kercus*, are omitted.

[26] These are not the scientific but the common Hungarian names. Competing versions can be found in Bartha 1997, 320–331. Also a useful, but in an historical sense entirely unreliable collection is Prisztcr 1986.

[27] The Latin versions hardly ever appear.

[28] *Alnus incana* is also found in Hungary, but it is rare.

[29] *Betula pubescens* is a less likely, but possible option.

[30] The modern Latin version (*carpinus*) seems to be missing from the sources. *Ornus* (today present in the name of *Fraxinus ornus*) might be the contemporary name for hornbeam. Berrár and Károly 1984, 213. In most cases, however, only the vernacular *gyertyán* appears in the sources.

[31] Horse-chestnut (*Aesculus hippocastanum*) has little to do with its name-sake, and is a post-medieval addition to the landscape.

[32] There is much confusion here. The reason is that these originally wild species have been subject to cultivation since time immemorial. *Prunus* in the sources seems to refer to *szilva*, that is one of the many types of plums, most probably originating from *Prunus spinosa*. *Cerasus* stands for *meggy*, sour cherry (*Cerasus avium*) or any domesticated version of it. *Cseresznye*, cherry, is usually only given in the vernacular. Identification problems are mostly due to modern botany and Latin. The three simple categories (plum, cherry, sour cherry) are easy to tell apart, but all the variations are difficult to regulate in botanical terminology, furthermore, different countries have different traditions (*cerasus* vs. *prunus*). One should never attempt to extract more from a charter than the difference between the three basic categories.

[33] In some cases *iuniper* equals *fenew*, that is *Pinus* sp. *Oklevélszótár*, 239. They are both gymnospermae, therefore juniper might be mistaken for some sort of pine. If the scribe knew the Latin name, it seems unlikely that he would not know the proper Hungarian version. It is more probable, if the two cases we know of are not mere accidents, that juniper, based on its appearance, was considered to belong to the family of pines.

[34] *Tremulus* probably refers to *Populus tremula*, aspen, but that cannot be proven.

[35] Above 500–600 metres *Sambucus racemosa* also occurs.

[36] The fact that *sorbus* occurs far less frequently is noteworthy.

[37] It is likely that *Sorbus torminalis* is different from other types of *sorbus*, especially because the vernacular name is never accompanied by a Latin equivalent. It should also be noted that different *sorbi* intermingle very freely and produce numerous types within the genus.

[38] For example: *Oklevélszótár*, 1005. 1368: "arborem ilicis vulgariter tulfa vocatam"

[39] See *Oklevélszótár*, 122, 606. In modern botany *magyal* refers to *Ilex aquifolium*, holly. *Cser* today is the turkey oak (*Quercus cerris*). There is, however, no guarantee that this was the same in the Middle Ages.

mysterious Hungarian *haraszt* is also understood as "oak,"[40] however, despite the author's efforts, there is not one source that would demonstrate this explanation.[41] Reuter also claimed that medieval Hungarians had used different words for different species of oak. *Tölgy*, *ilex*, and *haraszt* would have referred to the pedunculate oak (*Quercus robur*),[42] while *magyal* and *quercus* to the sessile oak (*Quercus petraea*). Reuter's theory was based on the fact that in his database the occurrences of *ilex* were in most cases connected to wetland environments, home mainly to the pedunculate oak. This, then, appears to be a classical case of too broad conclusions based on a too narrow dataset. Reuter worked on the perambulations of Co. Baranya, in the south-west of the country, a landscape dominated by the Drava river. As opposed to this, we can see that *ilex* is just as dominant in the Vkpmlt perambulations, that is, Co. Veszprém and Co. Zala, upland regions with a natural dominance of the sessile oak. As for the positions of the individual trees – that *ilex* grew in wet valleys and *quercus* on hills – it is all too easy to find opposite examples.[43]

There is, in fact, little that one can say for certain. Identification of different oak species is more of a modern botanist's dream than the reality for medieval sources. The following words, however, all refer to the oak genus: *tölgy* (*tul*, *thul*, *tulg*), *magyal*, *cser* (*cher*), *ilex*, *quercus*. Haraszt is a tree that we cannot identify.

7. 2. 2. Jegenye

This word today refers to a cultivated type of black poplar, imported from Italy in the eighteenth century.[44] Medieval sources often refer to it, quite clearly meaning some poplar.[45] With the apparent contradiction, what tree the medievals had in mind remains unknown.[46]

7. 2. 3. Gyümölcsény

This is a misleading word, because it sounds like modern Hungarian *gyümölcs* ("fruit.") Although *gyümölcsény* as such does not exist in present-day Hungarian, for a while it was understood as some medieval form of *gyümölcs*, thus the tree was interpreted as a "fruit tree."[47] This is certainly not true. In all cases when some extra information is added, *gyümölcsény* is always a (thorny) shrub.[48] It is sometimes called "white thorn" (*alba spina*), which lead to the tempting hypothesis that *gyümölcsény* might mean "hawthorn."[49] Although hawthorn undoubtedly has white flowers and thorns, this is true for a number of other shrubs as well.[50] Furthermore, we know that hawthorn was called *galagonya* in the Middle Ages, just as today. Naturally, there may have been two words for the same plant, but in that case one should explain why no connection was ever made between them.

7. 3. Frequency of Different Trees

A glance at *Figure 31* reinforces the impression gained from the perambulations that most trees mentioned were free-standing. It would otherwise be impossible to account for the position of wild-pear and walnut. The latter is not native to this country and does not grow in woods. The question, then, is how to interpret oak.

Oak is the most nationalist tree in Europe. Every country is proud of its connections to oak, and perceives its best features as being somehow embedded in this tree. Pride is, however, not associated with all oaks, but only with the pedunculate oak (*Quercus robur*). Hungarians are no different: medievalists are proud that King Béla III once held court under an oak,[51] and – although there is nothing whatsoever to prove this – are convinced that ancient Hungarians venerated certain oaks.[52] The above list also clearly signifies the attention people paid to oak. The tree is recorded as many times as the rest of the top ten put together. Is this because, as is often argued, "our woods mainly comprised this tree?"[53] The answer must be no, for we have seen that perambulations tended to record non-woodland trees. In contrast to the modern landscape, there must have been a large number of *free-standing* oaks scattered around the medieval countryside, which were particularly apt

type of tree	occurrences	type of tree	occurrences
oak	131	"shrub"	4
wild-pear	50	ash	3
willow	19	plum	2
walnut	16	cherry	2
beech	9	hazel	2
sorb	9	*gyümölcsény*	2
elm	8	hawthorn	2
hornbeam	7	*troncus*	2
crab-apple	7	"fruit tree"	1
"tree"	6	maple	1
lime	6	alder	1
poplar	4	service	1
juniper	4	pine	1
chestnut	4	*rubusculum*	1

Fig. 31. List of trees mentioned in the Vkpmlt corpus.

[40] Reuter 1965; Reuter 1975.

[41] *Haraszt* occurs many times, yet it *never* appears together with its Latin equivalent.

[42] *Haraszt* should be a south-western dialect word.

[43] For example, *ZsOkl*, vol. 7, 370. 1420: "ad partem occidentalem declinando ad quendam monticulum … ad quandam arborem ilices"

[44] Gencsi and Vancsura 1997, 356.

[45] *Oklevélszótár*, 428.

[46] Perhaps it meant the most distinctive native poplar, the black poplar (*Populus nigra*)? I owe thanks to Oliver Rackham for this suggestion.

[47] *MNyTESz*, s. v. "gyümölcsény."

[48] For a list, see Csőre 1980, 120–121.

[49] Csőre 1980, 120–121; Reuter 1963. The real basis of this is a syllogism. Once *gyümölcsény* was called "alba spina," in another case *galagonya* was "alba spina," thus *gyümölcsény* should be the same as *galagonya*. "Whitethorn" in English is another name for "hawthorn." Not a trace of this, however, is found in Hungarian.

[50] Blackthorn (*Prunus spinosa*) for example.

[51] Forster 1900, 344.

[52] Csőre 1980, 24.

[53] Rómer 1860a, 305.

for boundary signs, because they lived long, had characteristic shapes, and were in general venerable trees.

To check how characteristic the Vkpmlt list may be for the country as a whole, I counted the trees in the fifty-five perambulations carried out in the years 1417–1420.[54] The results show a pattern rather similar to what we have seen above (*Fig. 32*).

type of tree	occurrences	type of tree	occurrences
oak	59	sorb	3
wild-pear	25	cherry	2
willow	13	beech	2
haraszt	8	hornbeam	2
"tree"	7	alder	2
walnut	6	maple	2
elm	5	elder	1
lime	5	service	1
crab-apple	4	chestnut	1
troncus	3	pine	1
ash	3	"shrub"	1
poplar	3	"fruit tree"	1

Fig. 32. List of trees mentioned in perambulations from 1417–1420.

Oak, wild-pear, and willow again occupy the first three places. Walnut falls somewhat behind, however, it should be noted that all eight occurrences of *haraszt* are found in one single document. Also remarkable is the fact that with a similar number of perambulations (63 and 55), the Vkpmlt documents contain almost twice as many trees as the charter bounds of 1417–1420 (305 and 160, respectively). Is this because fifteenth-century perambulations are mostly concerned with "earthen boundaries," and fewer trees appear in them than in earlier documents, or because the first list concentrates on a smaller area with many trees, whereas the second one on the (less full of trees) country as a whole?

7. 4. Pollard Trees[55]

> "It often so happens that beauty or need encourages the farmer to pollard the oakwood. Here experience has it that moisture always serves the wounds better than other places: in the place of one branch that is cut, in a year there will spring twelve other ... The oaks must be thick, at least as big as a man's thigh, if we want to pollard them for firewood or some other household duty."
> János Nagyváthy *Hungarian Practical Cultivator*. 1821.[56]

In many countries, non-woodland trees are often pollarded. I have introduced the idea of pollarding in Chapter 1: In wood-pasture, normal coppicing is not possible, because grazing animals devour the young shoots and prevent the new growth. Here, trees are cut at a height of about two or three metres, outside the reach of animals. In the previous chapter, I have argued that wood-pasture existed in medieval Hungary (described with the words *rubetum* and *virgultum*). Such landscapes would have involved pollarded trees as well. Pollarding as such has received very little scholarly attention in Hungary. To date, no one has examined whether pollards were present in the Middle Ages or not.

The history of pollards has been most carefully studied in England, but there is also knowledge of pollarding in other countries, notably in the north.[57] These trees grab our attention mostly because of their appearance. An ancient pollard, of which there are tens of thousands in Europe, usually has a fantastic, grotesque shape. Pollarding became unfashionable about one hundred and fifty years ago;[58] surviving ancient pollards, uncut for a long time, have a huge *bolling* (the permanent trunk) two to three metres tall and frequently six to eight metres around, the top of which is full of humps and bumps due to the numerous times the branches were cut and the wounds properly overgrown. Nowadays the branches themselves are huge, and often touch the ground whereby they root and start new trees.[59] In sum, if one has some experience and pays due attention, an ancient pollard is rather difficult to miss. The word *ancient* is not used here by accident. Pollarding prolongs the life of a tree on the one hand, and the fact that it is in a place where pollarding seems to pay back the effort ensures its survival on the other. Its life is prolonged by pollarding because every time the shoots are cut, growth is slowed down. The tree reaches old age much later, when material made by the leaves is no longer sufficient to provide for the amount needed to grow a new ring over the whole surface of the trunk.[60] Pollards are not usually found in woods, unless the territory of the wood previously used to be a wood-pasture or a field. Woodland trees rarely survive long: when they are considered to be "mature" (in other words when foresters think they do not grow fast enough and straight enough to be marketable, which is around one hundred years nowadays) they are cut. Consequently, a high proportion of ancient trees are pollards. Pollarding, similarly to coppicing, is not entirely artificial. The best-known examples to demonstrate this are the so-called stag-headed oaks. The topmost branches of these trees start to dry, which is usually taken as the first sign of ultimate decay. The oaks then, more often than not, in fact die, because they are considered dangerous and are felled. Oliver Rackham discovered that

[54] These are easily accessible in volumes 6 and 7 of *ZsOkl*. Reference numbers are: *ZsOkl*, vol. 6, no. 18, 371, 440, 451, 465, 563, 629, 665, 687, 751, 903, 940, 945, 973, 1068, 1204, 1288, 1396, 1405, 1583, 1741, 1915, 1947, 2113, 2220, 2563, 2679; *ZsOkl*, vol. 7, no. 6, 158, 169, 329, 372, 440, 444, 655, 669, 833, 837, 843, 860, 930, 1261, 1305, 1377, 1592, 1694, 1762, 1765, 1802, 1871, 1885, 2257, 2276, 2333, 2387.

[55] For more on this, see Szabó 2001.

[56] Nagyváthy 1821, 285–286.

[57] Hæggström 1998; Hæggström 1983; Austad 1988. In 2003, Helen Read completed a European tour to study pollarding techniques in different countries. Her diary is available at the Ancient Tree Forum website (http://www.woodland-trust.org.uk/ancient-tree-forum). Through her experience, it seems that if ones looks hard enough, pollards can be found in every country in Europe.

[58] This differs from country to country.

[59] This new growth is spontaneous. When people do the same to trees that otherwise would not grow this way, it is called *layering*.

[60] A tree has to grow a new ring over its whole surface (branches, trunk, roots) every year. The material for this is made by the leaves. In connection with the ratio between the amount of wood to be produced and the amount of leaves, a tree has three major phases in its life. When it is "young," the crown expands, and the tree has more energy than it needs. When the crown stops growing, the tree is "middle aged." As time passes, it has to grow more and more wood with the same foliage. When this is no longer possible, the tree becomes "old," and starts to lose branches. Rackham 1996, 11–12.

stag-headed trees are actually not dying. There are trees which have been in that condition for more than fifty years and the lower part of their foliage is still in excellent health.[61] Stag-headed trees "pollard themselves": they get rid of a number of branches, so that the amount of wood they have to produce in the form of annual rings is reduced and the leaves may still provide enough energy for the tree to survive (*Fig. 33*).

Trees are pollarded for a number of reasons. Firewood is only one of these, though probably the most important. In northern countries, where winter is severely cold, their main purpose is the production of leaf-fodder, but this does not hold true in the British Isles, where winters are much milder. In Hungary, as in most of Europe, many riverside willows are pollarded in order to grow withes for basket making. These trees are treated in such a way not because animals would eat the shoots – although that is also possible – but because the head of the bolling has to be above water even when floods occur, so that the tree does not "drown." A later tradition is the pollarding of mulberries, especially the white mulberry (*Morus alba*). The leaves of this tree were the most important fodder for silkworms from the seventeenth century onwards in Hungary.[62] Although mulberries do not live particularly long, one can find fairly old ones. Close to Veresegyház (Co. Pest), on the bank of Sződligeti stream there stands a pollard mulberry with a girth of 300 cm, it is hollow inside but the foliage is in fine condition.

Fig. 33. Stag-headed oak at Nyirád in Bakony. This is not a dying tree (April 2002).

Finally, it is worth mentioning "beauty," as understood by Nagyváthy in the paragraph quoted above. It is very often aesthetic reasons that dominate pollarding, especially in the case of alleys. The only Hungarian deer-park that survives intact both in physical reality and in function (Gyarmatpuszta, Co. Esztergom) has a row of pollard horse-chestnuts along its redbrick wall. One of the largest of these trees is 430 cm around and also hollow inside.

7. 4. 1. The Hungarian Case

In England, the tradition of pollarding goes back at least to the Middle Ages. Crete has pollards from Byzantine times, and such trees are not unusual elsewhere in the Mediterranean.[63] Hungary also has many pollards, yet the initial question I had to pose was how ancient the practice of pollarding actually was. This question was ultimately answered by the trees themselves. In the catalogue of ancient trees there are a number of pollards,[64] among which probably the most spectacular is a lime at Szabolcsbáka (Co. Szabolcs), at least 500 years old (*Fig. 34*).[65] Then, in the Sekler – that is entirely Hungarian – area of Transylvania, near Lupeni (Ro; Székelyföld[66]), I have found a wood-pasture that has hundreds of very ancient pollard beeches, a landscape of outstanding beauty, rarely paralleled in continental Europe. Dozens of the beeches have a girth of 400 or 500 cm. The territory is still grazed by cows, the characteristic browse line (the height up to which the animals can still reach the leaves) is around 150–160 cm. The lower part of the crown of many trees appears as if it were drawn with a ruler (*Figs. 35* and *36*). I have discovered similar landscapes in the Bakony Mountains, which will be discussed in Chapter 15. Mighty oak pollards cover many hectares in Co. Somogy and Co. Baranya, in the south of Hungary. Recently, a list has been published about the known wood-pasture sites of present-day Hungary.[67]

Fig. 34. Small-leaved lime pollard at Szabolcsbáka. Note the immense bolling and the branches that touch the ground and have taken root (November 2000).

[61] Rackham 1996, 12.

[62] See for example Kunoss 1853.

[63] Grove and Rackham 2001, 190–216.

[64] A list of several thousand ancient trees is found in Bartha 1994. Pollarding, alas, is not recorded. Pictures of some ancient trees are accessible in Vajda 1969.

[65] Its girth is 870 cm. Bartha 1994, 92 does not guess at its age, but the immense bolling suggests at least five hundred years. This is supported by my ring count on a relatively small broken side branch, which had more than one hundred annual rings.

Fig. 35. Wood-pasture with ancient pollard beeches at Lupeni, still in use (May 2000).

Fig. 36. Pollard beech at Lupeni, some six metres in girth (May 2000).

Thus, it seemed reasonable to argue, based on living trees alone, that the tradition of pollarding can be traced back to the Middle Ages in the Carpathian Basin. Geographical names also preserved a memory of the pollards that once stood in the area. The Hungarian name for a pollard – *csonkafa* – and the adjective derived from it – *csonkás* – are found by the hundreds all over the country. To strengthen the argument, the same words appear in written sources from the fourteenth century onwards.

Hungarian research has thus far neglected the subject almost completely. The only writer to dedicate a few pages to pollarding was the ethnographer Lajos Takács. What historians and archaeologists know about pollards is taken from his books.[68] Sadly enough, he misinterpreted the phenomenon and took it to be a form of woodland clearance. He was right in the sense that some eighteenth-century sources talk about pollarding (*csonkolás*) when the branches of trees to be destroyed later by other methods were cut down. He did not point out, nonetheless, that if the branches of a living tree are cut, that, *in itself*, will very rarely result in the destruction of the tree. Nagyváthy, quoted at the beginning of this subchapter, described the events with much more accuracy. In fact, all sources cited by Takács speak about pollarding and not the grubbing out of woods,[69] and they are all from the eighteenth century. As we have already seen, both living trees and written sources prove that the tradition is actually much older, dating at least to medieval times.

7. 4. 2. Written Sources – Vocabulary

> "Then it [the boundary] turns back towards the east through a *rubetum* until a ditch, where there is a pollard oak (*detroncata quercus*). From this, descending through a valley it proceeds to the big road, and at the edge of this road there is a great ditch, whence it returns to the said place."

Extract from the perambulation of Ganna (Co. Veszprém), dated to 1171, but most probably written around 1234.[70]

The obvious sources for pollards are perambulations. I have already discussed the many types of trees that occur in them, but I have not yet referred to the details they provide. Sometimes we read about young or old trees, trees with a double or a broken trunk, or even several trees with intermingled branches.[71] Supposing that pollarding existed in the Middle Ages, one would expect pollard trees to appear in perambulations. I do not claim that all pollards were recorded as such, for sometimes the compiler did not even specify what type of tree he was writing about. However, a pollard, especially if the bolling is old and the shoots are young, looks so characteristically different from its standard companions that this feature may have easily made it to the charters.

Csonkafa

Csonka, as already mentioned, is modern Hungarian for a pollard. This term appears in very low numbers in the sources, always in the form *csonkafa* (*fa* means "tree") or *csonka* + type of tree. The *Oklevélszótár* lists nine cases and I have found only one further example.[72] Out of these ten trees three are willows, three are pears, two poplars, one oak, and one is not specified. Pear may seem surprising: why would a fruit tree be pollarded? I do not know the answer, but the presence of a standing tree once again demonstrates that the practice of pollarding wild-pears once existed. On Nagy-Villám Mountain, not far from the castle of Visegrád, there is a biggish wild pear, pollarded and surrounded by stones. Because of the pollarding, the canopy is so dense that the trunk is almost invisible even in winter. Boundary marks were often surrounded by stones (*lapidibus circumdata*) in the Middle Ages, thus, this tree may well be a decent illustration for a handbook of landscape history (*Fig. 37*).

---

[66] This was not a regular county, but a more or less autonomous area of the Seklers.

[67] Haraszthy, Márkus and Bank 1997.

[68] Takács 1980, 160–166; Takács 1976, 46; Takács 1987, 46–47.

[69] He admitted at one point, although with reluctance, that "trunks without branches, thus, may not have been only the result of clearances, but may also have been created to produce the once so important winter fodder, and are in general the result of the woodland management techniques available before the introduction of modern forestry." Takács 1980, 165.

[70] *PRT*, vol. 8, 256–260.

[71] For example: 1415: *ZsOkl*, vol. 5, 82. "ad quandam arborem *zyl* curvatam"; 1302: *Oklevélszótár*, 245. "venit ad arborem tul fyotol dictam" (*tul* means "oak," *fyotol* is "young"); 1300: *HO*, vol. 8, 411. "tres arbores nucum, ubi quarta per ventum percussa fracta esse dicitur." 1296: *ÁÚO*, vol. 10, 235. "ad alias tres arbores fraxsiny, wlmy et ylicis stantes coniunctim, quarum rami inter se mixtim creuerunt"

[72] 1431: *Károlyi*, vol. 2, 135. "pervenit ad quandam arborem chunkafa dictam"

Fig. 37. A pollard wild-pear in Nagyvillám (February 2000).

Troncus, Tőke, Törzsök

If this were all there was, one could argue that even if pollards were a characteristic feature of the medieval landscape, they rarely attracted the attention of the compilers of charters enough to record them. There is, however, another word that appears far more frequently in perambulations and seemingly denotes something comparable: *troncus* (and its spelling variant *truncus*). It is only logical, by the way, to look for a Latin word, since perambulations were written in Latin, with the vernacular version of landscape features only occasionally included. Nevertheless, the case here is not merely that *troncus* would be Latin for *csonkafa*. These two words were never connected, because *troncus* had a different Hungarian equivalent. We shall see later on that in the Middle Ages two, by now forgotten words used to exist which denoted pollards and stools.

Troncus is a formulaic part of many perambulations. When a tree was mentioned, basically three types were differentiated: *dumus* (shrub), *arbor* (tree), and *troncus*. Undoubtedly, the most general of these is *arbor*, but a shrub was not necessarily what one would imagine today. What sense does "oak-shrub," often mentioned in documents, make, for example?[73] In modern languages a ligneous plant is either a tree (one trunk) or a shrub (multiple trunk). At the moment, nonetheless, we are concerned with the medieval meaning of *troncus*. The first question is whether or not it was simply the remaining part of a dead tree, which appeared too difficult to remove and thus was left standing to decay.[74] Several things contradict this. Firstly, the Middle Ages were a period when, for example, in the purchase of a plot the seller had it included in the contract that the newly planted willows were not for sale.[75] (Planting a willow is the easiest thing in the world: one cuts a fresh shoot, sticks it in the ground, and that is it, providing there is enough water.) In another case, the Forest-guards of Bakony would not come to an agreement with the people of the abbot of Pannonhalma until the latter promised not to touch any living tree in the Forest.[76] In 1464, the villagers of Ráckeve (Co. Fejér) obtained the right to gather dead-wood on Csepel island from no less a personage than King Matthias.[77] The list could go on endlessly, and every item indicates that medieval people did not waste wood, or at least that there were areas where firewood (or the access to it) was not abundant enough to let a whole trunk rot away standing instead of chopping and burning it. *Troncus*, on the other hand, appears as a *terminus technicus* all over the kingdom. Secondly, in almost all cases, we also learn what type of tree the *troncus* was.[78] With a dying, leafless trunk, it is hard to believe that among medieval perambulators there was always someone capable of such identification. If a *troncus* had been a dead tree, it would have been much more efficient to record it as such, instead of bothering with identifying it – after all the aim of recording landscape features in a perambulation was to make it easy to recognise them later. Thirdly, I know several cases where charters mention the dry trunks of dead trees, yet the word *troncus* never appears in connection with these.[79] In 1337, the boundary of a settlement in Co. Ung touched upon "some dry trees, which were still standing in their roots."[80] This description could have easily been substituted by *troncus*, had that meant anything like this.

In sum, *troncus* denoted a living tree.[81] Thus, as a restriction of the meaning of *arbor*, it must have had something to do with shape or management. In the charters, unfortunately, there is no explanation of what a *troncus* might have looked like or how it was created. No agricultural treatises survive from the Middle Ages in Hungary, and different ways of cutting trees certainly did not interest those few who knew how to write. The only way to find out more is through the medieval Hungarian equivalent of the word. Glossaries and dictionaries from the fourteenth-sixteenth centuries translate *troncus* as either *tőke* or *törzsök*.[82] I also managed to find both words as vernacular forms in charters, the former three times, the latter once.[83]

---

[73] In the Carpathian Basin, that is. In southern Europe, oaks can grow into a shrub form. Rackham 1998.

[74] This is how the word is usually interpreted. In modern source editions, it is translated as *tönk* (stump).

[75] *Kolozsmonostor,* vol. 1, 201.

[76] 1258: *ÁÚO*, vol. 2, 314–315.

[77] Magdics 1888, 21.

[78] For example 1368: *Károlyi*, vol. 1, 304. "ad verticem cuiusdam montis, in cuius latere pervenissent ad unum truncum arboris vulgo nyar vocate meta terrea circumfusum" (*nyar* is "poplar."); 1412: *ZsOkl*, vol. 3, 641. "penes quandam silvam glandinosam iuxta truncum arboris ilicis"

[79] For example in Uzsa (Co. Zala): *Zala*, vol. 1, 473–475.

[80] 1337: *Sztáray*, vol. 1, 126. "venit ad quasdam arbores exsiccatas, tamen in radice stantes"

[81] This was sometimes emphasised: 1411: *ZsOkl*, vol. 3, 217–218. "sub quodam tronco vivo magno"; 1413: *ZsOkl*, vol. 4, 281. "troncum arboris vive kewrus" (*kewrus* is "ash")

[82] Berrár and Károly 1984. "*Törzsök*" appears in the form "ultra silvam Tursoc" in the *Gesta Hungarorum*, written ca. 1200 by a Hungarian chronicle writer, commonly known as Anonymus. *SRH*, vol. 1, 76, 80.

[83] 1366: *Károlyi*, vol. 1, 272. "postmodum venissent ad unum troncum in latere unius silve ad usum ulmi (!) vulgariter zyltursuk" (*zyl* means "elm") I have not seen the original, but "ad usum" makes absolutely no sense here, thus we certainly have a scribal mistake or a mistake in the transcription and/or edition. The meaning of the sentence, nonetheless, is clear without "ad usum." 1302: *AnjOkl*, vol. 1, 183. "in quodam loco ubi est quidam truncus Byktuke vocatus." (*byk* is "beech") 1293: *HO*, vol. 7, 233. "iungit quendam truncum qui Egurthuke vocatur" (*egur* means "alder.") 1471: Dl 17 193. "inter duos troncos thewke dictos" *Troncus* appears in the Bible as well. Job 14:7–9 reads as follows: "Lignum habet spem si praecisum fuerit rursum virescit et rami eius pullulant. Si senuerit in terra radix eius et in pulvere emortuus fuerit *truncus* illius, ad odorem aquae germinabit et faciet comam quasi cum primum plantatum est." This was translated into Hungarian by Gáspár Károli in the second half of the sixteenth century as "Mert a fának van reménysége: ha levágják, ismét kihajt, és az ő hajtásai el nem fogynak. Még ha megaggodik is a földben a gyökere, és ha elhal is a porban a *törzsöke*: A víznek illatától kifakad, ágakat hajt, mint a csemete." (Italics mine.)

Törzsök has by now almost disappeared from the Hungarian language. The only living, though outdated usage is *törzsökös nemes*, which means a nobleman from an ancient family. *Tőke*, on the other hand, is very much alive today, and has three basic meanings: 1) what remains of a tree above ground when it is cut, and the same when taken out and used for example for chopping wood 2) vine-stock 3) capital, as described by Karl Marx. We are in a lucky position, because the 1533 *Lexicon* of Murmellius also defines *tőke*: "Die lenge ader stam des boms von der wurtzel bis an die este," that is "The length or trunk of the tree from the roots until the branches."[84] This is a perfect definition: if a charter contains mention of a *tőke*, then the perambulators saw a standing tree with a big trunk but only small or no branches, which is exactly what a pollard tree looks like when in use.[85] This, by the way, also explains why *csonkafa* and *troncus* are never considered equivalent to each other. They are two different approaches towards the same subject: *csonkafa* (English pollard) refers to the specific management form, whereas *tőke* (English bolling) to the plant itself. Perambulations, primarily intended as depictions of whatever there was to note in the landscape, tended to record the appearance not the management.[86]

The only connection between the words *troncus* and *csonka* is the verb-form. According to the 1595 dictionary of Verancsics, "truncare: chonketanni" (*csonkít* is the verb derived from the adjective *csonka*; and *–ni*, the suffix that forms the infinitive).[87] At the beginning of this subchapter I quoted a perambulation, where a certain "pollard oak" was mentioned. In the Latin original this was "detruncata quercus." Somewhat more was revealed in a 1364 perambulation from Co. Zala, where the boundary was in "an oak pollarded from above" ("in arbore ilicis desuper truncata.")[88] In 1296, an "oak tree pollarded of its branches" ("unam arborem yliceam ramis truncatam") was noted down.[89] All in all, the activity was called *csonkítás* or *csonkolás* (English pollarding), whereas the result was a *tőke* or *törzsök*. When translating the latter two, the scribes started out from the Hungarian verb, and used its Latin nominal form. This difficulty was not faced by Hungarian clerics alone. In England, for example, the Latin name of pollards was *robur*, which in modern botanical language refers to the pedunculate oak (*Quercus robur*).[90]

The second meaning of *tőke*, vine-stock, is also worthy of our attention. This is also at least medieval in origin, in fact we do not know which meaning came first. The obvious connection between the two is that a grapevine works very similarly to coppice stools or pollards. (May I remind the reader at this point that coppice woods were sometimes equated with vineyards in the Middle Ages.) Every year the stock is cut back – lower or higher depending on the method – and the grapes themselves appear on the young shoots. We do not plant grapevines every spring, and we do not have to (or should not) plant trees either. Just as stools and bollings produce young shoots for firewood, vine-stocks grow branches so we will have wine the following year. The third meaning of *tőke*, capital, is undoubtedly a semantic continuation of this image. Capital in a bank produces interest like vine-stocks produce grapes. Turning back to trees, with pollards it is not the fruit, but the shoots themselves that are valuable. According to Murmellius: "Ramus, quod de ipso trunco arboris: *tőke* aga" ("A branch from that *truncus* of the tree: a shoot.") This is another proof that *troncus* does not refer to a dead stump but to a living tree.

Tőke and *törzsök*: their Latin translation was the same, but the Hungarian originals may have differed in meaning. One solution could be that *tőke* meant a stool, while *törzsök* was a bolling, but, unfortunately there is no evidence to support this otherwise appealing idea. It should be noted that although the two words sound similar, it is most probably by accident. The stem of *tőke* is *tő* (meaning "stem,") which is of Finno-Ugric origin. In the case of *törzsök*, however, linguists do not even know if it is a derivate of the modern word *törzs* (trunk) or the other way round. In either case, the origin is unknown. In theory, then, we are confronted here by two absolutely separate words.[91] The main problem is that the word almost always used in the sources was *troncus*, therefore there is not enough material to differentiate between the Hungarian originals. This difference had either disappeared by the Middle Ages, or had not made it to the Latin of the period. From another point of view, I cannot tell whether a particular *troncus* in a perambulation referred to a stool or to a pollard. Examples suggest that both were possible. In Halam (Sk; Co. Zemplén) a *troncus* was mentioned in the middle of a coppice wood, and this could only refer to a stool unless the territory had been a wood-pasture before.[92] Most *tronci*, nonetheless, appear as individual trees (such as the one in the Ganna perambulation), which makes it most probable that they were pollards (*Fig. 38*).

7. 4. 3. Pollard Trees in Pictures

Pictorial sources are usually rather disappointing for the woodland historian. Pollard trees present one of the few exceptions, most probably because their appearance is so characteristic.[93] The most famous picture with pollards

[84] Berrár and Károly 1984, 710. Murmellius was a German humanist, whose Latin-German dictionary was published with Hungarian glosses in 1533, in Cracow.

[85] We should not forget that every word of a perambulation was translated from the vernacular into Latin. There was no artificial terminology, especially in the beginning, for the actual landscape to be described. Latin, or the Latin of those who had this task, was not very good at expressing what was needed. Early perambulations, such as the 1055 charter from Tihany (Co. Zala), demonstrate the kind of difficulties scribes faced. As time passed, "vulgo dicitur" appears less and less frequently, and later perambulations are much more formulaic. *Troncus*, thus, is not a Latin word whose vernacular meaning needs to be worked out, but rather a translation which may reveal or hide a meaningful original.

[86] It is also possible that *csonka* has only recently acquired its present meaning. In medieval sources it may simply mean a broken or otherwise damaged tree, *arbor fracta*.

[87] Berrár and Károly 1984, 151. Verancsics was Dalmatian by origin, and published a five-language dictionary very much in the form of modern dictionaries: words in alphabetical order with short translations.

[88] *PRT*, vol. 7, 530–531.

[89] *ÁÚO*, vol. 10, 236.

[90] Rackham 1980, 182.

[91] *MNyTESz*, s. v. "törzs" and "törzsök."

[92] *ZsOkl*, vol. 3, 217–218.

[93] Hæggström 1994.

Fig. 38. A pollard in a wood: Égés-bérc in Börzsöny. This lime grew up in open conditions and was probably intended as a landmark (October 2002).

Fig. 40. February in the 1579 calendar from Trnava. The man on the left is pollarding a tree in the middle of a field.

Fig. 39. Legend of St. Ladislaus at Vítkovce. The trees are basic but recognisable depictions of pollards.

– although certainly not because of them – is the one from the *Very Rich Book of Hours* of Duke Berry (fifteenth century) with six pollard willows on the bank of a river in front of a magnificent white castle.[94] The best known Hungarian examples – again not because of the trees – are the early modern engravings of Wilhelm Dilich depicting towns and castles (for example Vác and Fehérvár, both around 1600). To the best of my knowledge, the earliest depiction of a pollard in Hungary is a wall painting in Vítkovce (Sk; Co. Szepes). This fresco was made around the mid-fourteenth century and represents scenes from the legend of St. Ladislaus (*Fig. 39*).[95] The trees in question are very unusual. Wall paintings in most cases depict a generic tree, a plant with no individual features. Species occasionally might be guessed at, but "it is curious how few artists ever portray a recognisable oak or ash tree."[96] The Vítkovce frescos contain three trees that were cut at around two metres (as judged in comparison to the height of the people) with some young shoots that have already appeared on top of the bollings. The trees themselves are not very elaborate, partly because of the simplicity of the style. However, because of the clearness of expression, it may not be entirely unjustifiable to assert that the trees were indeed intended as pollards.

Far less ambiguous is the 1579 calendar from Trnava (Sk; Co. Pozsony), where February is represented by two activities: sowing and pollarding (*Fig. 40*). The type of tree, once again, is not identifiable, but the scene itself is rich in authentic detail. The man is holding a typical medieval axe and the tree stands where pollards very often appear, in the middle of a field. These territories, when cultivated in a two- or three-field system, were from time to time available to grazing animals, thus firewood could only be obtained through pollarding.

---

[94] Longnon and Cazelles 1989.

[95] László 1993, 121–126. The picture reproduced here is a tinted drawing based on copies by József Hanula. (The original wall-paintings are in a very poor condition.) I would like to thank Dóra Sallay for drawing my attention to this picture.

[96] Rackham 1980, 15.

Chapter 8

Conclusions

In the first part of this book I hope to have outlined the basic features of medieval woodland management in Hungary. The picture drawn so far is sketchy, but apparently coherent.

Palaeoenvironmental studies prove that woodland has been cleared in the Carpathian Basin since at least the early Neolithic, although we know little about how this was done. Trees have had their ups and downs for millennia, and in certain parts of the Basin the dominance of open landscape goes back to Celtic or Roman times. I tried to stress that the conquering Hungarian tribes arrived to a largely transformed cultural landscape, where woodland was not quite as dominant as usually believed. The first solid written data about the proportion of woods in the countryside come from as late as the fifteenth century, and even then their quantity is hardly adequate. Reflecting an earlier inheritance and the internal colonisation process of the thirteenth century and later, data suggest that woods covered about twenty percent of the kingdom at the end of the fifteenth century, excluding the High Carpathians. However, the distribution of woodland was rather uneven: some parts were more wooded than others, which feature most probably goes back to pre-medieval conditions. This has the important consequence that woodland was sparse enough in parts of medieval Hungary to inspire conservation measures and intensive management.

All three traditions of medieval tree management were present in contemporary Hungary, although direct written evidence about these is again sparse. The most important clues in written sources turned out to be the names of different types of woodland. Coppice woods, intensely managed and relatively small, were found all over the kingdom and had their own vernacular name: *eresztvény*. Large stools can still be found in Hungary, while woodbanks are certainly present and can be studied through fieldwork, although nothing yet positively proves that they existed in the Middle Ages. Wood-pasture was usually called *rubetum* in Latin, and although it was a land-use in a wide swathe of the countryside, I could not find its Hungarian name. Free-standing trees abounded in perambulations, and, as elsewhere in Europe, oak was most often recorded. Connecting pasture and free-standing trees, pollards were also present. There are some medieval pollards still standing. Their medieval Latin name in Hungary was *troncus*, which was the translation of *tőke* and *törzsök*. These terms may have had somewhat different meanings unknown at present.

Thus far I have only summarised those features that were similar to medieval England, or, to be more precise, to what emerges increasingly as *medieval European woodland and tree management*. However, as I tried to point out in the Introduction, anything that has to do with nature necessarily reflects local conditions as well. This is especially true for the Carpathian Basin, a huge tract of which is ecologically very different from north-western Europe. Furthermore, traditions and customs affect the management of land as much as they do that of society. What were, then, the specifically Hungarian elements in woodland management? First, one has to emphasise animal husbandry. The Great Hungarian Plain probably supported many large mammals in prehistory, and the pastoral tradition has been continuous ever since. The fifteenth century is of special significance in this aspect, because it may have seen the emergence – most probably through deliberate breeding – of the large Hungarian Grey cattle,[1] of which tens of thousands were driven to Western Europe yearly.[2] This might easily have lead to the creation of savannah-like landscapes.[3] In connection with the flood-plains that once were so extensive in the Alföld, I suspect that the Hungarian *berek* and its Latin equivalent *nemus*[4] refer to a local way of simultaneously handling trees, water, and grazing.

Beside cattle, pigs were also important. Although in Western Europe pannage was often no more than a topos connected to woodland, in Hungary pigs really created considerable incomes for the owners of woods. Pannage appeared in estimation guidelines, and we know that pigs were not only theoretically regular visitors to woods, but, as reported in many court-cases, this happened in reality as well. One could, of course, easily overestimate the significance of pannage because it fell within the landowners' sphere of interest and thus left many records. However, one should not forget that almost one third of all woodland specified in estimations qualified as "acorn-bearing."

Other than *silva glandinosa*, local types included *silva sub venatione*, and *silva dolabrosa*. The meanings of these are unclear, especially that of the former. An axable wood seems to be a timber-oriented wood, although not in the intensive management of forestry, but rather in an extensive management. I believe that there is some connection between acorn-bearing and axable woods, which is supported by the fact that axable woods are basically non-existent in estimations. Another type of wood was *silva communis*. Although "common wood" appears self-explanatory, what its management entailed is unknown. The logical solution in a wood commonly used by villagers would be coppicing, but in that case I cannot see the difference between *permissionalis* and *communis* woods.

There is another, equally important way to look at the same set of problems. Thus far I have written about medieval tree management as if it had been a compact and unchanged unit, however, I am aware that it had its own line of development and separate elements in it probably emerged at different times and in different ways. At this stage of the research, nonetheless, I have only questions and suggestions.

[1] Bartosiewicz 1999, esp. 148.

[2] Pach 1970, 223–246. For what this meant in ecological terms, see Hoffmann 2001.

[3] Rackham 1998.

[4] *Nemus*, as the reader may remember from Chapter 6, can have other meanings as well.

How much of the system was adopted from abroad, and how much was the result of internal development?[5] Internal development is inevitable. This type of tree management was not specific to the Middle Ages; we have seen in Chapter 3 that coppicing was already practised in the Neolithic and pollarding is thought also to be a prehistoric tradition. This stems from the recognition that shoots (coppice or pollard) are much better for firewood than standards: anybody who had to chop a timber tree without heavy machinery will know this.

> "I believe that the idea of coppicing woodlands originated from watching beavers and their tree-felling activities. Similarly ... the idea of pollarding and shredding came from seeing the condition of trees perhaps a year after a great wind like 1987. Many trees which have been self-pollarded and shredded by the gale, are subsequently flushed with new growth."[6]

These lines from tree-enthusiast E. E. Green give us an approximate idea of how woodland management may get a start at any time in any society. Coppice woods in Hungary, for example, have their own special vernacular name (*eresztvény*) which may refer to internal development. Also, the fact that the Latin version (*permissoria*) is possibly a bad translation of the same word seems to mean that the tradition was not introduced from abroad. However, it is also possible that clerics were simply inarticulate in Western terminology. There still exist three villages of early French settlers that contain within them the French name of coppice woods (*taillis*): Tállya (Co. Zemplén), Andornaktálya (Co. Heves), and Nagytálya (Co.Heves).[7] Similarly, although *berek* seems to be a local answer to a local question, *nemus* had a similar wood-pasture meaning in neighbouring Austria.

I have already listed a number of problems requiring further research. The most important tasks are, at the moment, a fuller analysis of the available written sources; the study of as many ancient woodland sites as possible with intensive targeted fieldwork; the recognition of differences in woodland management in different parts of Hungary; and attempting to provide research with a vertical aspect.[8] What we already know, however, will provide a solid basis for the second part of the book, as we move on to study two wooded Forests: Pilis and Bakony.

---

[5] This question have been raised in other issues of agricultural history as well, notably in connection with peasant plots. Here, the German settlers and their *lehen* is thought to have had a crucial impact, also reflected in early terminology. Belényesy 1955a, 72–79.

[6] Green 1993, 91.

[7] FNESz, s. v. "Tállya," "Andornaktálya," "Nagytálya."

[8] Equally important would be advances in palaeoenvironmental studies and dendrochronology, but these are the tasks of specialists.

PART II

FORESTS

Chapter 9

Royal Forests

> "We, Matthias, by the grace of God, King of Hungary ... entrust to memory ... that our faithful men ... dwelling in our village of Mothoffalwa ... came to our majesty's sight ... and revealed to our majesty that they ... had, from ancient times, enjoyed these liberties given to them by the sacred kings of Hungary: ... that they are not obliged to any burden of servitude except for those to be performed in guarding the Forests and the beasts in them, and in hunting, ... that whatever beasts they bring to the royal majesty, the castellans of our said castle of Zolon are obliged to reimburse them for their expenses."

Extract from the 1464 charter of King Matthias to the Forest-guards of Mot'ová (Sk; Co. Zólyom) commissioned to replace the documents that perished during the intrusions of the Bohemians.[1]

In Chapter 1, I have briefly discussed the difference between woodland and Forest. The latter, at least in England, was an administrative unit where Forest Law applied. The land itself did not necessarily belong to the king, and was not always covered with trees. This system was imported to England from the Continent. The idea of Forests supposedly originated from the Merovingian Franks, and in fact the only other modern languages that preserve the memory of medieval Forests in having two words to describe wooded areas are the three tongues of the Frankish Empire (French *bois* and *forêt*; German *Wald* and *Forst*; Italian *bosco* and *foresta*). English Forests were the last and clearest phase in the development of the concept (and thus are particularly useful for comparative purposes). In other places in Europe, however, the process went in different directions and brought different results.[2] What is certain, nonetheless, is that everything was connected, from the very beginning, to hunting. *Forestum*, argued Jarnut and also Chris Wickham,[3] started out with territories reserved for royal hunting, but later "slowly ceased to be a property (royal or ex-royal) with specific rights attached, and began to be a right on its own," and finally "kings began to grant forest [Forest] rights over the properties of third parties too – forest [Forest] rights and property were ... decisively divorced."[4] The creation of such Forests had to be reconciled (or at least made enforceable) both with the local nobility and the peasantry. While the former were often left dissatisfied, the traditional common rights of the latter were usually not cut back very severely.[5]

Hungarian – in contrast to the above languages but similarly to Spanish, Norwegian, Romanian, Swedish, Croatian, Finnish, etc – possesses only one such word: *erdő*. Nonetheless, the quotation above demonstrates that something similar to Royal Forests did exist in medieval Hungary. Sources mention *silva regalis*, but it needs to be explored whether this expression was more than a mere attributive compound. While the first part of this book concentrated on woodland, the second part will explore (wooded) Royal Forests.

Medieval sources used one word: *silva* to denote both woodland (from a few acres up) and what seems to be a Forest.[6] Presumably, whenever a *silva regalis* was mentioned, it meant nothing else but a wood that belonged to the king. At times this woodland could be gigantic, but that did not make much of a difference as far as its status was concerned. In part this is true, but an example will illustrate that the matter was not so simple. A charter from 1470 wrote about "the lands and places, meadows and pastures, woods and income of the *honor* or county that is commonly called Bakonispansagh," where a certain John of Esseghwar had ordered the royal huntsmen to stay out.[7] In a near contemporaneous charter (1453) the king discussed his *silva regalis* called Bakony.[8] Is this latter a wood? If so, what is its relationship to the county? Why would the former include many other things besides woods? Why do they have the same name? What does a county do within another one (Co. Bakony in Co. Veszprém), a very unusual phenomenon in medieval Hungary? If the former is included in the latter, how could a nobleman chase the king's people away? The situation is complicated and at the heart of it we may see something that is more than an oversized wood, yet not the same as a Western-type Royal Forest.

9. 1. The Story So Far

The generally accepted theory is that when the Kingdom of Hungary was formed around the year 1000, uninhabited and lordless lands came into the possession of the king.[9]

---

[1] Dl 24 339. "Nos, Mathias dei gratia rex Hungariae ... memoriae commendamus ... quod nostrae maiestatis venientes in conspectum fideles nostri ... in possessione nostra Mothoffalwa ... commorantes ... nostrae exposuerunt maiestati ... quomodo ipsi ab antiquis temporibus hiis libertatibus ... per divos reges Hungariae ... eis ... concessis ... gavisi fuissent ... quod aliqua onera servitiorum obligati non fuissent, praeter quam ad custodiam sylvarum et ferarum in eisdem existentium ac ad venationem faciendam, ... aliquas feras regiae maiestati adducerent, castellani dicti castri nostri Zoliensis eis expensis dare tenebantur"

[2] Rubner 1964; Jarnut 1985; Hennebicque 1980; Montanari 1984, 174–190.

[3] Wickham 1994.

[4] Wickham 1994, 160.

[5] For England: Birrel 1987. For France: Aubrun 1988.

[6] Latin *foresta* does occur, but only four times. See *Lexicon Latinitatis*, vol. 4, s. v. "foresta." One cannot but agree with Csőre, who argued that the scribes had used the word as a fancy expression without really understanding the concept behind it. Csőre 1980, 79.

[7] Dl 102 568. "...terrarum ac locorum, fenetum et pratorum necnon silvarum et proventum honoris seu comitatus quod communiter Bakonispansagh nuncuparetur." (*ispansagh* means "county")

[8] *HO*, vol. 2, 315. "silva nostra regali Bocon vocata"

[9] *ÁMTF*, vol. 1, 45–47; *KMTL*, s. v. "erdőispánság." Mályusz 1922.

Early Hungarians are believed to have been a pastoral and agricultural people, so they mostly settled in the fertile lowlands of the Carpathian Basin. Larger wooded areas, mostly on the edge of the kingdom in the Carpathian Mountains, became royal lands. They were used very extensively, and remained marginal until around 1200, when they were organised into counties proper[10] alongside with a major wave of settling and donation processes. The new counties lacked many characteristic features of "regular" counties, and were thus termed "forest-counties" in modern scholarship.[11] Later on, during the fourteenth and fifteenth centuries, "forest-counties" (for the sake of convenience, henceforth I shall call these "Forests") gradually lost their special appearance, and either became indistinguishable from any other county in the kingdom, or disappeared altogether, their territories swallowed up by neighbouring counties.

How far can this theory be justified? Parts of it are certainly true, but others are not. To begin with, nothing positively proves that larger woodlands automatically came into royal possession because they were entirely empty. We *do* find them in that legal state when they appear in the sources, however, when and how they became crown lands is hidden in the darkness of early Hungarian history. What is more, the only "forest-county" with a known early history appears to be a counterexample. The Forest of Patak, related the chronicler Anonymus around 1200,[12] was deliberately acquired by King Andrew I (1046–1060) through an exchange with the descendants of a certain Ketel. The king wanted the land because it was good for royal hunting, and because his Russian wife felt closer to her homeland in that northern region. This, of course, could be another one of Anonymus' fancy stories, however, as Jenő Szűcs argued very convincingly, it has a fair chance of being true.[13] In this case, then, the territory did not fall into the hands of the king accidentally. He wanted it for some special purpose. It should be noted that by far not *all* major woods went into royal possession. The Vértes Mountains for example, between Pilis and Bakony, were never a Forest. They were the land of the Csák kindred. György Györffy deduced from this that Szabolcs, the forgotten forefather of the kindred, had been member of the Árpád family.[14] Börzsöny, to quote another example, is rather similar to Pilis. They are neighbours of approximately the same size, both bordered by rivers on two sides, although Börzsöny is more rectangular as opposed to the triangular Pilis. All this being true, no one, to the best of my knowledge, has ever bothered to ask why Börzsöny never became anything like a Forest.

The second point in the above theory that cannot be accepted is that all later Forests were marginal land with a rather loose administrative structure and no definite counties. Two examples that contradict this are Pilis and Bakony Forests. Pilis lies in the heart of the kingdom, surrounded by the most important lay and ecclesiastical centres of the Hungarian Middle Ages. In the north-west we find Esztergom, seat of the archbishop of Hungary, and an early royal centre. In the south-east, there is Óbuda (possibly the focal place of early Hungarian leaders[15]) later to be replaced by the most significant of all towns: Buda. Somewhat further to the south-west is Fehérvár, the town of St. Stephen, coronation and burial place of most medieval kings. These were joined in the mid-thirteenth century by Visegrád in the north, a smaller royal town, which, nonetheless, had a symbolic significance and could claim to be the "capital" of the kingdom throughout the fourteenth century. All this was already called *medium regni* by contemporaries.[16] Bakony, as we shall see later in more detail, was also an early occupied region with a very stable settlement structure. Its focal place, Veszprém, was the "town of the queens" and also the seat of a bishop.[17] Furthermore, both Pilis and Bakony had been parts of early (eleventh-century) counties, Visegrád and Veszprém, respectively. The system of counties was undoubtedly established by St. Stephen, however, if one leaves aside tempting hypotheses and relies on the (very few) contemporary sources alone, only a handful of counties can be firmly argued to have been established in the time of Stephen.[18] There is definite knowledge about four in a charter from 1009: Veszprém, Fejér, Kolon, and Visegrád.[19] Out of these, Veszprém and Fejér are still on the map, Visegrád – although important as a town – is never mentioned again as a county, and Kolon has still not been identified in an entirely satisfactory way.[20] It has been argued that "originally the whole territory of the 'forest-estates' belonged to the king ... During the twelfth century, this homogeneity came to an end due to donations of which little is known... This process ... reached the point when it became necessary to transform the administrative model of the territories ... and develop a county, in the first decades of the thirteenth century."[21] What sort of necessity was there, one may ask, to form a "Bakony

[10] Some of them (Torna, Sáros) had been considered to be parts of counties, but this was interpreted rather loosely.

[11] In fact, the expression is medieval. "Erdyoyspanssag" was mentioned in 1454. *MEO*, vol. 1, 26–27. What the medievals meant by the word is not known. A similar expression existed for the territories around Aachen, Düren, and Monschau-Konzen, which were called *comitatus nemoris* from the end the of twelfth century. Kaspers 1957. It is fairly obvious, however, that the Hungarian "forest-counties" grew out of the local county system.

[12] *P. dictus magister*, better known in Hungary as Anonymus, is the most debated figure in Hungarian medieval history. All he told us about himself is that he was the notary of King Béla. Which Béla was meant out of the four possible, however, has been re-examined for centuries. Most scholars would now settle on Béla III, and the composition of the *Gesta Hungarorum* around 1200. For a summary of the historiography on Anonymus, see Csapodi 1978. Anonymus wrote his *Gesta* about the Hungarian Conquest of the Carpathian Basin in the fashionable style of his age. With very little real knowledge about the events to rely on, his work is a strange mixture of fiction and oral tradition. He should be thought of as someone comparable to Geoffrey of Monmouth as far as "historical credibility" is concerned.

[13] Szűcs 1993a, esp. 1–4.

[14] Györffy 1983, 33–34.

[15] This, although usually taken for granted, is in fact based on a theory of György Györffy, not proven beyond doubt. Györffy 1997, 67–71.

[16] *Medium Regni* 1999.

[17] Gutheil 1979; Zsoldos 2000a.

[18] Zsoldos 2001.

[19] For the charter, see in *DHA*, 52–53. For related problems and Co. Visegrád: Zsoldos 1998.

[20] Kristó 1988, 247–248.

[21] Zsoldos 1998, 19.

county" within Co. Veszprém, which remained basically intact from the eleventh century to the 1950s? The fact that large tracts of lands were given out as donations was not at all unusual: that happened in most counties. It appears that the scholarly discourse has been somewhat misled by the marginal geographical position of some of the territories in question. However, it is hard to underestimate the differences between what was to become Zólyom Forest (a huge area, great mountains, and bears) and Pilis Forest (gentle hills one can cross on foot in a day, with deer and the occasional wild boar). In one case, even woodland is a problem. The island of Csepel[22] is rarely counted among the "forest-counties," although it acquired all the characteristic features (if not more) of the others.[23] Scholars are still reluctant to talk about Csepel as a "forest-county," because it was not hilly but flat, and not entirely wooded. The similar histories of all these places cannot be explained by taking the characteristics of some representative examples and treating others as mavericks. We must find the features they *all* shared, and concentrate on those. What causes difficulties in this process is that historians are far from fully understanding the early history of the Kingdom and the changes of the thirteenth century.[24] One of the least researched areas is forms of land ownership, which would be the essential background to study Forest history. As a consequence, my observations are sometimes based on theories that themselves had not been convincingly demonstrated. It is not, however, the historical ecologist's task to solve such problems. My conclusions may, as they are intended, serve as useful pieces in the great puzzle that will comprise the social history of the Árpádian period.

9. 2. Beginnings

Before 1200, we know very little of the Forests. One thing is certain: some of their territory was woodland. Secondly, they were the private property of the king. This was sometimes stressed, as in the case of Pilis, which was called the king's "very own woodland" in 1187.[25] These woods were sometimes also called *predia*. This is a special term of this period, denoting allodial, privately held estates that were managed by servants. The leader of such a unit was regularly called *procurator*.[26] Some woodlands had already been managed, as we know from the same charter of 1187, which granted five cartloads of wood daily to the hospitallers of Esztergom for "the use of their house."[27] Apart from this, the other, and certainly more substantial function of these estates was hunting. It is their characteristic feature to have contained at least one hunting lodge. Hunting, although undoubtedly the most noble sport of the time, is not to be understood as a pass-time activity for bored rulers. For most of the Árpádian period, the royal court was constantly on the move. The hunting lodges were stations on a long and oft-repeated journey on the *via regia*. Szűcs was able to pinpoint the location of most of the stations in Patak Forest.[28] The strong connection between Forests and hunting lodges was not peculiar to Hungary. In England, royal houses and Royal Forests also occurred together, and the same held true for Germany.[29] In Hungary, the royal family themselves sometimes participated in hunts, the son of St. Stephen, for example, died in a hunting accident.[30] Nonetheless, game was needed in large amounts to support the royal table, and hunting was probably carried out by professionals. We know little about them, since records of their activities do not survive in any written form in this early period.

With this scanty evidence it would be hard to say anything comprehensive about these estates, however, we should notice how similar they appear to early medieval Forests in continental Europe. One cannot, evidently, suppose that eleventh-century Hungarian kings carefully studied seventh- and eighth-century Frankish Forest administration and tried to imitate it, yet up until the point when these territories began to be more intensely involved in land management (and thus to be more influenced by the contemporary local political and administrative history) they displayed essentially similar features. This is by no means a unique phenomenon in East-Central European history. Regardless of whether one believes in theories about centre and periphery, it is unquestionable that many things that happened in Western Europe in the Early Middle Ages were, *mutatis mutandis*, repeated in the Polish-Czech-Moravian-Hungarian region in the ninth to twelfth centuries.[31] In Forests, hunting is the strongest connection. The way we get to know about these Forests is also similar. "We find out about early *foresta* ... because the kings gave them away, in part at least, to churches."[32] Bakony, as we shall see later on, is a classic example: that we know some about its early history comes from the fact that a monastery was founded right in the middle of it in the eleventh century. Thirdly, they were not necessarily woodland, or at least the contemporaries identified them much more as "marginal" (not administratively but in an economic sense) than "wooded."

9. 3. Forests of the Thirteenth Century

Like the rest of the country, wooded royal estates were fundamentally altered in the thirteenth century. This is also the first time when there are enough sources to form a coherent picture of the whole kingdom, although what to interpret as "changes," "novelties," and "remnants of an older system" is increasingly difficult to decide. Furthermore, although there is a clear connection between social and economic changes and the sudden increase in the number of sources, their

[22] The biggest Danube island just south of Budapest.

[23] For a history of Csepel, see *ÁMTF*, vol. 4, 189–206; Pesty 1880, 75–83; Zsoldos 1998, 21–23.

[24] Despite the excellent book of Jenő Szűcs 1993b.

[25] Knauz 1863, 131. "propria silva sua, que vulgo Ples nuncupatur"

[26] For examples of Zólyom, Bereg, Sáros, and Torna, see Szűcs 1993a, 23, footnotes 122–125.

[27] Knauz 1863, 131. "usus ipsius domus"

[28] Szűcs 1993a, 13–16.

[29] Steane 1984, 11–13; Steane 1993, 79–93; Bosl 1963.

[30] Waitz 1878, 36. Emeric was later canonised and his liking for the chase forgotten.

[31] For an ingenious treatment of the whole issue, see Szűcs 1985.

[32] Wickham 1994, 160.

relationship is not straightforward. If something new popped up in the sources in the 1250s, that does not mean that it was created at that time. It may have existed for decades or more, but escaped being written about.

The most obvious sign of changes in this particular case is a new title: *comes*. The word itself, of course, was not new. On the contrary, it had been the standard term for the elite of the early society, and for the heads of counties.[33] Woodland estates, as I have pointed out above, had earlier been governed by keepers (*procuratores*). Then, shortly after 1200, *comites* appeared everywhere (Bakony: 1210, Bereg: 1214, Ugocsa: 1216, Patak: 1219, Zólyom: 1222, Pilis: 1225.[34]) Alongside with this title came the new name for the estates: county (*comitatus*). Although the title of *comes* had already lost some of its eleventh-century splendour by this time,[35] a *comes* was still much more than a *procurator*.

The people in charge of Forests were "Forest-guards" (*custodes silvarum* in Latin, *erdőóvók* in medieval Hungarian). Although this position was not peculiar to Royal Forests of the thirteenth century,[36] most of the *custodes silvarum* we come across were from those places and that time. Other than single persons in charters, their existence also survives in place-names. There are at least seventeen settlements with the name Ardó/Ordó (a short from of *erdőóvó*)[37] in the Carpathian Basin.[38] One has to keep in mind, however, that Forest-guards were not confined to the villages named after them.[39] In Pilis, for example, there was never an Ardó, although guards did live in the region.[40] Forests were also characterised by the presence of settlements named after smiths (Kovácsi), huntsmen (Daróc), specialized huntsmen (Hodász – beaver hunter; Hőgyész – ferret hunter; Peszér – dog keeper; Solymár – falconer), even if only the first of these was anything like universal.[41] Chance mentions in the *Regestrum Varadiensis* preserve the memory of the royal bison hunters,[42] who were lead by their own *comes*.[43]

While these latter characteristics may be seen as the continuation and refinement of the early hunting function of royal estates, the appearance of *comites* still requires explanation. The answer, I believe, lies in the administrative system of the kingdom and in the difference between private royal lands and lands that belonged to the crown but were not private. These two had not been differentiated until recently, making it "extremely difficult to understand the essence of landed property."[44] Private lands of the royal family supported the king and the queen, and operated much the same as anybody else's private lands, albeit on a larger scale. Early wooded estates were representative examples of this type. Non-private royal lands were the "castle-lands" (*várföldek* in Hungarian), which belonged to royal castles. Every royal castle had settlements in the county it was governing,[45] which, at least in theory, meant two out of three villages in that county.[46] The *comes* was judge of free and bonded people in his county, and collected the revenues from castle-lands, one third of which he kept and two-thirds he sent to the king.[47] The castle-lands were given to the *comites* as allotment for their office. This is not unlike Western-European *feudum*, although the differences are not be underestimated. A *honor regni*, as this institution was called in the Middle Ages in Hungary, was entirely up to the king to give and take away at any point, and it never became hereditary, but remained a special form of land ownership.[48] In my view, the most fundamental difference between early royal estates and Royal Forests in this country was that Forests were no longer regarded as private land, they became similar to castle-lands, given as a *honor* to the *comes*. Where Forests still existed in the Late Middle Ages, this was undoubtedly the case: Bakony was called *honor* in 1379, and also a century later, in 1470.[49] Why my theory is rather hypothetical is because the kingdom in the thirteenth century was most certainly not the same as in the fourteenth. *Honor*, as an institution, was characteristic

[33] The word cannot be translated into English, but it has its Hungarian equivalent: *ispán*. Bak, Bónis, Domonkos and Sweeney 1989–1996, the standard reference book in such matters, uses the Hungarian word, but I decided to remain with the Latin. For different medieval meanings, see *Lexicon Latinitatis*, vol. 2, s. v. "comes."

[34] Bakony: *ÁÚO*, vol. 1, 100; Bereg: Karácsonyi and Borovszky 1903, 314; Ugocsa: Karácsonyi and Borovszky 1903, 163; Patak: Karácsonyi and Borovszky 1903, 215; Zólyom: *MES*, vol. 1, 238; Pilis: *ÁÚO*, vol. 11, 183. In Torna Forest, however, the first time we hear about the comes is 1272. *Zichy*, vol. 1, 28.

[35] Engel 1998, 96.

[36] 1181: *RA*, vol. 1, 43–44, we read about the two Forest-guards of a monastery in the village of Tur (Ro; Co. Szatmár).

[37] Karácsonyi 1902.

[38] Heckenast 1970 is a groundbreaking study to collect and examine toponyms of this type. A list of Ardó settlements is found on pages 91–93. More on servant populations is in Györffy 1983, 426–436. Such settlements were wide-spread in medieval East-Central Europe, notably in Poland and the Czech lands. For a short summary and extensive bibliography, see H. Ludat, "Dienstsiedlungen," in *Lexikon des Mittelalters*, vol. 3, 1006–1008.

[39] First pointed out in Heckenast 1970, 16–17. Heckenast stretched the theory rather far by assuming that Ardó place-names inferred villages founded to settle Forest-guards as opposed to villages of other names accommodating guards, which, he supposed, had been in existence before the Forest. This would allow an insight into the organisational process of Forests.

[40] *Custodes silvarum* in Pilis: 1285: *MES*, vol. 2, 207, 192.

[41] One should not imagine these villages as places where everybody was occupied full-time with dog-keeping or beaver hunting. Their inhabitants, just like everybody else, were primarily peasants, who had special obligations towards the king.

[42] Karácsonyi and Borovszky 1903. The *Regestrum* was compiled between 1208 and 1235. "Comes bubalorum" is usually uncritically understood as a reference to bisons, however, László Bartosiewicz tells me that up to the nineteenth century there is a general confusion in terms for bisons, aurochs, and later also buffalos.

[43] *MEO*, vol. 1, 4.

[44] Engel 2001a, 80.

[45] There also existed royal castles not governing a "traditional" county, that is a compact territorial unit. These were later phenomena inserted into the existing system in order to protect the borders of Hungary. Zsoldos 2000b.

[46] Györffy 1983, 200–203.

[47] *DRMH*, vol. 1, 34. The Golden Bull of Andrew II, 1222: "Comites iure sui comitatus tantum fruantur, cetera ad regem pertinentia ... et duas partes castrorum rex obtineat."

[48] Engel 1981; Engel 1982. Although I am convinced about the existence of *honor* as a separate institution, serious doubts were voiced right after Engel's first article. Fügedi 1982.

[49] Dl 101 919; Dl 102 568. The former is quoted in Engel 1981, 12.

of the fourteenth century. By 1500 it was almost forgotten, and very little is known of its origins. It is suspected to have existed in some form already in the Árpádian period, but to claim more than some very basic continuity would be pointless.[50] The case of Royal Forests, in fact, may be one of first examples to tackle the problem of thirteenth-century *honores*. It is worth remembering at this point that in the development of early medieval Forests in Germany, France, and Italy, land-ownership and social history were also central questions.[51] Western European Forests became *rights* over whoever's land in a complex system of property, which never happened in Hungarian Forests due to a number of reasons. Most importantly, land-ownership was becoming less and less complicated ending finally as a strikingly simple system in the fourteenth century. From that time onwards, land was always absolute property, and all landowners (meaning nobles) were subject – in theory – only to the king. There was no possibility in such a system for rights over land and ownership of land to separate.

What were Forests like in physical reality? Most importantly, they were territorial units. From the eleventh century on, Hungary consisted of countless settlements abutting on one another. Woods, just like arable or meadows, were parts of these settlements. A Royal Forest, on the contrary, was a landscape element on the level of settlements. The clearest demonstration of this is the thirteenth-century charter (better known as the Albeus conscription) that listed all the possessions of the Benedictine abbey of Pannonhalma. Vinye – a settlement in the north of the Bakony region (Co. Veszprém) – was described as having "common boundaries with these villages, that is with silva Bochon, Feneufey, Kugere, Cestuch."[52] (*Fig. 41*) There are many more cases like this in perambulations. For example, in 1274, we read about the boundary of Csaba (Co. Pilis) which is in a road "where it separates from your [the king's] Forest called Pilis."[53] In 1250, a piece of land was given to a certain Radun next to the Garam river. The territory had a common boundary first with the village of the Forest-guards, then with the Royal Forest (of Zólyom) itself.[54] This also suggests that contrary to what is universally held among historians, the villages of Forest-guards were not parts of the Forests. As territorial units, Forests could have perambulations as well. For one reason or another (perhaps because the authority of the royal house was never questioned to that extent) no Forest perambulations survive. The only example when one was commissioned was in Bakony, which will be discussed in detail later. Despite this territorial "independence," Forests were described with the same Latin word as woods: silva. In the 1274 charter of Csaba, the whole boundary started in a big wood (*silva magna*). We shall see later in examples about Bakony that

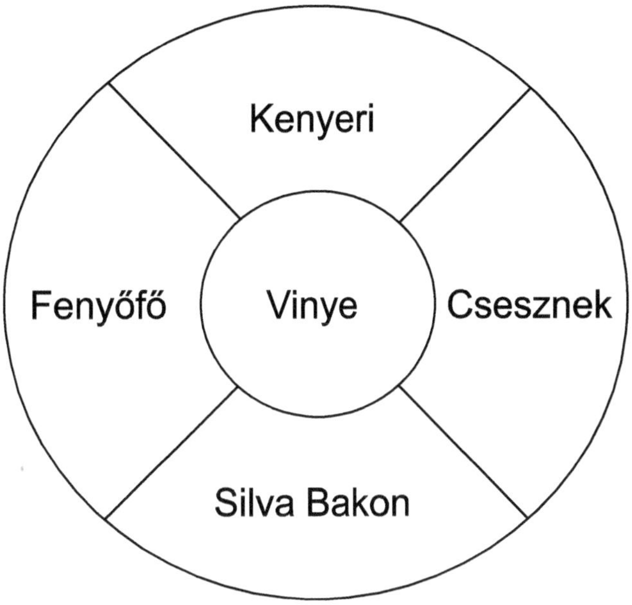

Fig. 41. Schematic representation of Vinye and neighbouring villages, as imagined in the Albeus conscription.

when the scribes meant a Forest, they tended to make their point clearer with the help of an adjective, or (as in this case) a possessive.

In connection with the previous phenomenon, we must consider how wooded Forests may have been. This is important, because ever since the pioneering study and map of Higounet in 1965, scholars have identified Forests (by whatever name they appeared in whichever period) with woods.[55] This is always a mistake, but it is not quite as grave in some periods as it is in others. Thirteenth-century English Forests were, all in all, about one-fifth wooded,[56] which is probably a low percentage when compared to earlier Forests elsewhere in Europe. Even these latter, however, were not only woodland.[57] Although they were identified – other than *foresta* – with names often derived from or associated with "woodland" (*gualdus*, *saltus*, or even *silva*), Wickham very rightly remarked that these "superficially indicate marginality or non-cultivation," and not woodland only, as apparent in such strange constructions as *silva eiusdem gualdi* in the Sabina in the eighth century.[58] *Silve* (*regales*) should not be taken at face value in Hungary, either. Royal residences were included in them from the beginning. Later, they often contained monasteries. At one extreme lies Zólyom Forest, which was mostly wooded, while at the other extreme is the island of Csepel, which appears to have been rather lightly wooded by Hungarian standards. If we call Csepel a Forest, it is almost easier to think of an English Forest than of a wood to understand what is meant.

[50] From the fourteenth century we can no longer speak about any difference between private royal lands and castle-lands. In the Late Middle Ages, the *comes* did not have to pay any money to the king from the income of his *honor*.

[51] Jarnut 1985.

[52] 1234–1240: *ÁÚO*, vol. 2, 10. "Predium Vegne ... est conterminalis cum his villis, scilicet cum silva Bochon, Feneufey, Kugere, Cestuch." The modern names of the three villages are Fenyőfő, Kenyeri, and Csesznek.

[53] *CD*, vol 5/2, 160. " ubi separat de sylua vestra Pilis vocata"

[54] *CDES*, vol. 2, 242–243.

[55] Higounet 1965.

[56] Rackham 1980, 175–188.

[57] Wickham 1994, 161 provides a most valuable outline of the problem with an awareness that is unusual among historians.

[58] Wickham 1994, 164, 166.

9. 4. The Dissolution of Forests

Many things vanished from the Kingdom of Hungary with the Árpáds. With a modernised royal court, there was no need, and probably no possibility to maintain the traditional, oversized Royal Forests. With decreasing royal power, there was great uncertainty about what would happen to them. In light of the highly variable late medieval histories of Forests, I cannot but agree with Jenő Szűcs, who claimed that "the relationships between Forests and counties display almost as many variations as the number of actual cases."[59] Zólyom Forest broke down into several counties: Zólyom, Liptó, Árva, and parts of Bars and Hont. Patak Forest was mostly integrated into Co. Zemplén, Ugocsa was made into a separate county (the smallest in the kingdom) though parts of the Forest were given to Co. Szatmár.[60] Csepel survived the changes in a largely unaltered form.[61] Other surviving Forests were transformed to fit the new state of affairs. In sum, from the fourteenth century on the histories of Forests were so different from one another that it would be pointless to discuss them together in general terms. In the following, I shall only describe the fate of Bakony and Pilis Forests. Why take *these* two Forests can be explained in many ways, some of which have to do with the histories of these Forests, others with the available sources and secondary literature. In a sense, the history of Pilis Forest is a "smaller" version of the history of Bakony. Pilis is smaller in territory, with fewer surviving sources. Many of its laconic documents become more understandable when interpreted in light of similar sources from Bakony. Also, as I have mentioned before, the characteristics of these Forests do not fit the traditional view on Forests in Hungary, therefore they force the researcher to take a fresh look at the old statements. A very important consideration is that both Pilis and Bakony are included in the *Archaeological Topography of Hungary*.[62] I shall not, however, go into details here. Throughout the next chapters I shall continuously explain my point, and it is up to the reader to finally decide whether my choices were justifiable.

---

[59] Szűcs 1993a, 23.

[60] Szűcs 1993a, 23–24.

[61] For late medieval Csepel, see Kubinyi 1965.

[62] *MRT-1–4, 5, 7.*

Chapter 10

Pilis Forest

There are many aspects to the history of a particular landscape. However, "what the historian researches and how he or she goes about it, is after all determined by the available source material."[1] If in the following I concentrate on some aspects of the history of Pilis and Bakony Forests and leave aside others, it is availability rather than preference that influenced my choices. Sources allow me to investigate the history of these Forests in four basic directions. Firstly, I shall outline their histories as Royal Forests, then I shall move on to woodlands and trees in the Forests. After this, the settlement histories of the Forests will follow, and lastly I shall deal with monastic orders, whose characteristic presence is one of the strongest links connecting the histories of Pilis and Bakony.

As described in the previous chapter, Pilis lies in what was called in the Middle Ages the *medium regni*. Geographically speaking, the biggish (ca. 400 km^2), triangularly shaped territory of Pilis is bordered on the east and north by the Danube. The longest side of the triangle (ca. 35 km) is the three sections of the elongated Pilis Mountains proper, stretching from the north-west to the south-east. North-east of this, filling up most of the area towards the Danube, is what is today called the Visegrád Mountains. Pilis and the Visegrád Mountains are geologically very different. Pilis is composed of limestone and dolomite, and is characterised by steep slopes and wonderful patches of cliff vegetation, whereas the Visegrád Mountains are composed of volcanic andesit, and are mostly uplands with oak and beech. The differentiation between the two parts, nevertheless, is modern: in the Middle Ages all this area was called Pilis. The highest peak in the region (also called Pilis) is 756 metres above sea level (*Fig. 42*).

10. 1. Early Years and the Thirteenth Century

The name "Pilis" is of Slavic origin. Some form of *pleš* exists in most Slavic languages and denotes a bare, plantless area.[2] Pilis peak, which gave its name to the whole region, has been bare for millennia, as shown by the presence of the rare, ice-age relict flower *Ferula sadleriana*, which grows on its top. This big and not very splendid looking plant does not tolerate overshading and has no ability to colonise. In other words, it did not come from some other place where taller plants overshadowed it, but has grown here continuously since at least the latest ice-age.[3] The early Slavic settlers noticed this wholly natural site and named the peak accordingly.[4]

The early history of the Forest is very obscure.[5] Other than the fact that they gave names to Pilis and Visegrád ("high castle") nothing is known about the early Slavic settlers. The territory appears as royal property in the earliest sources we possess. I have already quoted the 1187 charter that mentioned it as the king's very own Forest. That royal property meant the private possession of the Árpád family, is demonstrated in the foundation of the provostry of Dömös (ca. 1107) by Prince Álmos, brother of King Coloman (1095–1116).[6] Dömös is situated in the northern part of Pilis, and why it came into the possession of Álmos is unknown, except that he was a member of the royal family.[7] Before the provostry was founded, there had been a *regale allodium* in Dömös, as the fourteenth-century chronicle composition informs us, where King Béla I (1060–1063) died when his throne collapsed on him.[8] Archaeologists identified the ruins of the building itself.[9] This place was not the only early royal residence in Pilis. Hunting lodges lay near Pilisszentkereszt,[10] Kesztölc,[11] Pilisszentlászló,[12] and Pilisszentlélek.[13] The memory of these buildings survived in a written form because they were all transformed into monasteries. There are two other places in Pilis located by archaeologists, which may also have been hunting lodges. These, however, were never mentioned in written sources and were left to decay. What we learn from all this is that before 1200, wherever the king and his retinue stayed in Pilis, they had a lodge within a few hours' ride, and

[1] Engel 1998, 9.

[2] *FNESz*, s. v. "Pilis."

[3] Kalapos 1998.

[4] There are at least seven places containing some form of the word Pilis in their names within the Carpathian Basin. Not all of them are in mountainous regions.

[5] The most important works about the medieval history of Pilis are: *ÁMTF*, vol. 4, 581–714; Pesty 1880, vol. 1, 59–67; Zsoldos 1998; Zsoldos 2002; Tringli 1999; Kristó 1988, 252–254; Kristó 2002a; Kristó 2002b.

[6] For the foundation, see *SRH*, vol. 1, 427–428. The holdings of the provostry were enlarged and conscribed in 1138, by the son of Álmos, King Béla II (1131–1141). The conscription was published in Szabó 1936.

[7] Györffy claimed that around 1100, King Coloman donated it to Álmos. However, he only concluded this from indirect evidence. *ÁMTF*, vol. 4, 630.

[8] *SRH*, vol. 1, 360. The "fourteenth-century chronicle composition" is a by now lost text compiled at the royal court and (re)constructed from several later manuscripts.

[9] Gerevich 1983; Gerevich 1992. A famous carved stone from an excavation depicts a hunting scene. The figure of the hunter is often thought to be Prince Álmos himself. The hunt, however, is symbolic in nature, as whoever the hunter is, he is chasing a lion.

[10] Gerevich 1984; *MRT-7*, 161; Holl 2000.

[11] Gyöngyösi 1988, 48. "palatium … quod habebat in insula de Pilisio pro venationis requie" Here, archaeologist Júlia Kovalovszki identified the walls and even some carved stones from an earlier building. She made the obvious connection between the ruins and the royal residence, however, warned that the relationship between the two might not be straightforward. Kovalovszki 1992a. See also Kovalovszki 1992b.

[12] Györffy 1956, 284. "domunculum lapidea venationi regum preparata" The author put forward a convincing theory to explain the very awkward information presented in the *Inventorium* of Gregorius Gyöngyösi.

[13] Gyöngyösi 1988, 49–50. Archaeological excavations here also confirmed the existence of an earlier building, again, possibly connected to the royal residence. Lázár 1992, esp. 498, 501. In the case of Pilisszentkereszt and Pilisszentlélek, their use as hunting centres cannot be demonstrated.

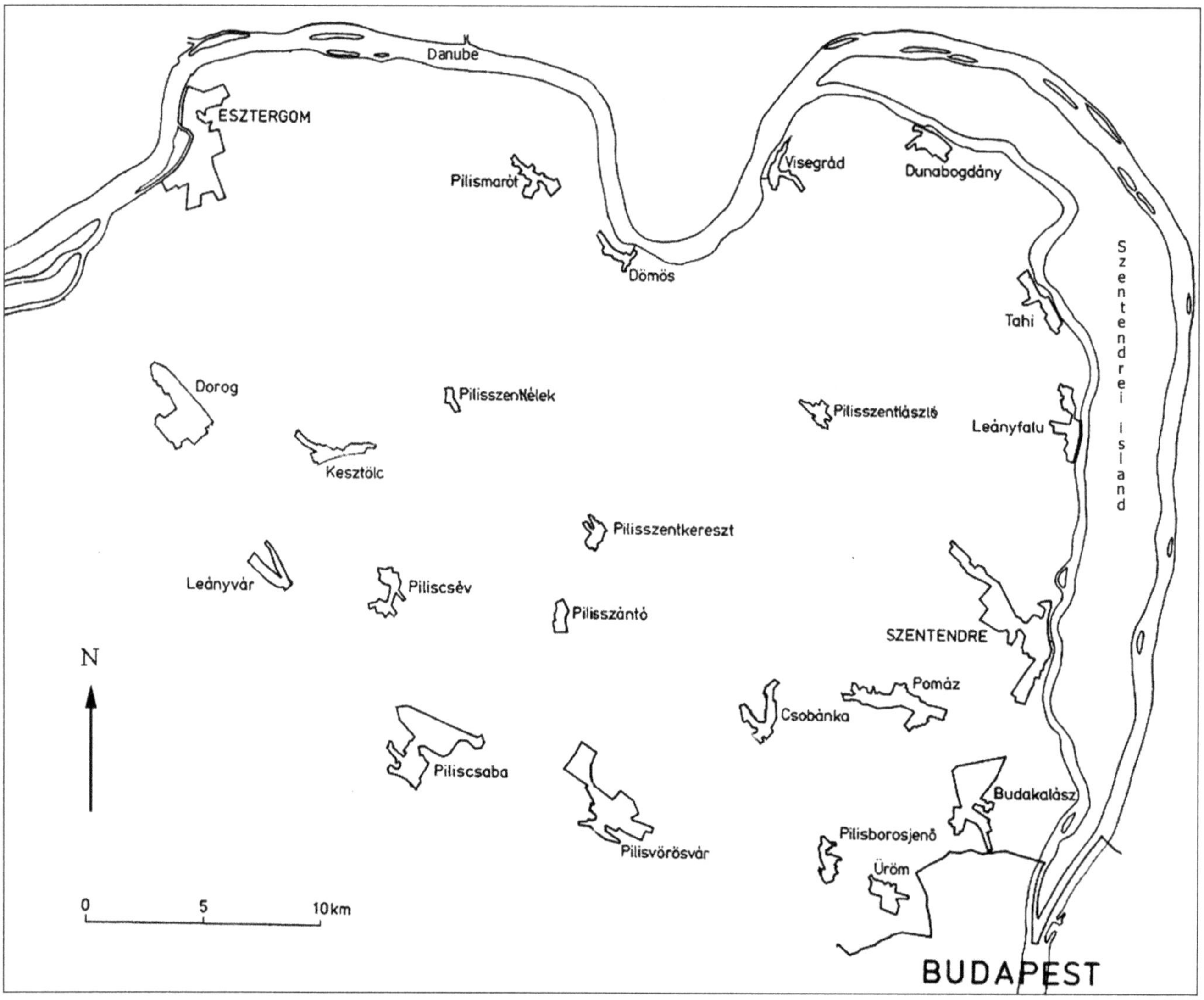

Fig. 42. The Pilis region in modern times.

the archbishop, the queen, St. Stephen tomb, and their own residence within one day's journey (*Fig. 43*).[14]

Pilis had also been transformed into a Forest by the thirteenth century. The first appearance of the *comes* of Co. Pilis is from 1225.[15] Mentioned in the year 1285, we also find Forest-guards, dwelling in Bogud.[16] The memory of other royal people was preserved in place-names. Kovácsi, the smiths' settlement stood just north of Pilis peak. Peszérd, a village that was situated a few kilometres south-east of Esztergom, was the home of the royal dog-keepers. Fedémes, at the south-eastern end of the Pilis Mountain, was named after the bee-keepers there.[17] Solymár, somewhat further to the south-east, was where the falconers lived. Not much is known about the physical extent of the Forest. The only place where its boundary is explicitly mentioned is the one quoted in the previous chapter.[18] There is another charter where I suspect we witness a mention of the Forest boundary. In 1278, Ladislaus IV donated the village of Pomáz to his daughter. Pomáz was defined as being "below Pilis Forest."[19] Although this was unusual (settlements were usually located with the "in comitatu X" formula), so far no one has noticed that the phrasing may have some extra meaning. I only became suspicious when I found a number of similar examples from Bakony. These strongly suggest that medieval charters were not tourist-books, and in the 1278 charter "below" most probably meant "next to," thus defining the boundary of Pilis Forest. In Chapter 14, I shall discuss the Bakony charters in more detail.

[14] This seemed reasonable to contemporaries, as well. Bishop Roger, in his *Carmen miserabile* written about the Mongol invasion soon after 1241, declared that Esztergom and Fehérvár were one day's journey from each other. *SRH*, vol. 2, 562.

[15] Wertner 1897, 660. For a list of all *comites* in Co. Pilis, see also Pesty 1880, vol. 1, 65–67.

[16] *MES*, vol. 2, 192, 207.

[17] Here we should note the presence of the mysterious little niches, abundant in Pilis, carved in rocks. These are called *kaptárkő* in Hungarian, an expression associated with beehives. No one really knows when these niches were created and for what purpose, but ethnographic analogies from the Balkans and a well-known early modern practice lead Andor Saád to suggest that they were connected to bee-keeping. Saád 1972. More recent research, while not negating the idea that it is possible to use some of the niches as beehives, argued that originally they had been made for some religious purpose. Baráz 2000. The authors of *MRT* opted for the first explanation. *MRT- 7*, 200. Other than this, *kaptárkövek* are surrounded by myths, and have a tremendous appeal to those involved in dreaming up a glorious past for the Hungarians. Such theories (and publications) are countless.

[18] *CD*, vol 5/2, 160. " ubi separat de sylua vestra Pilis vocata"

[19] *CD*, vol 5/2, 446. "sub sylua Pilis"

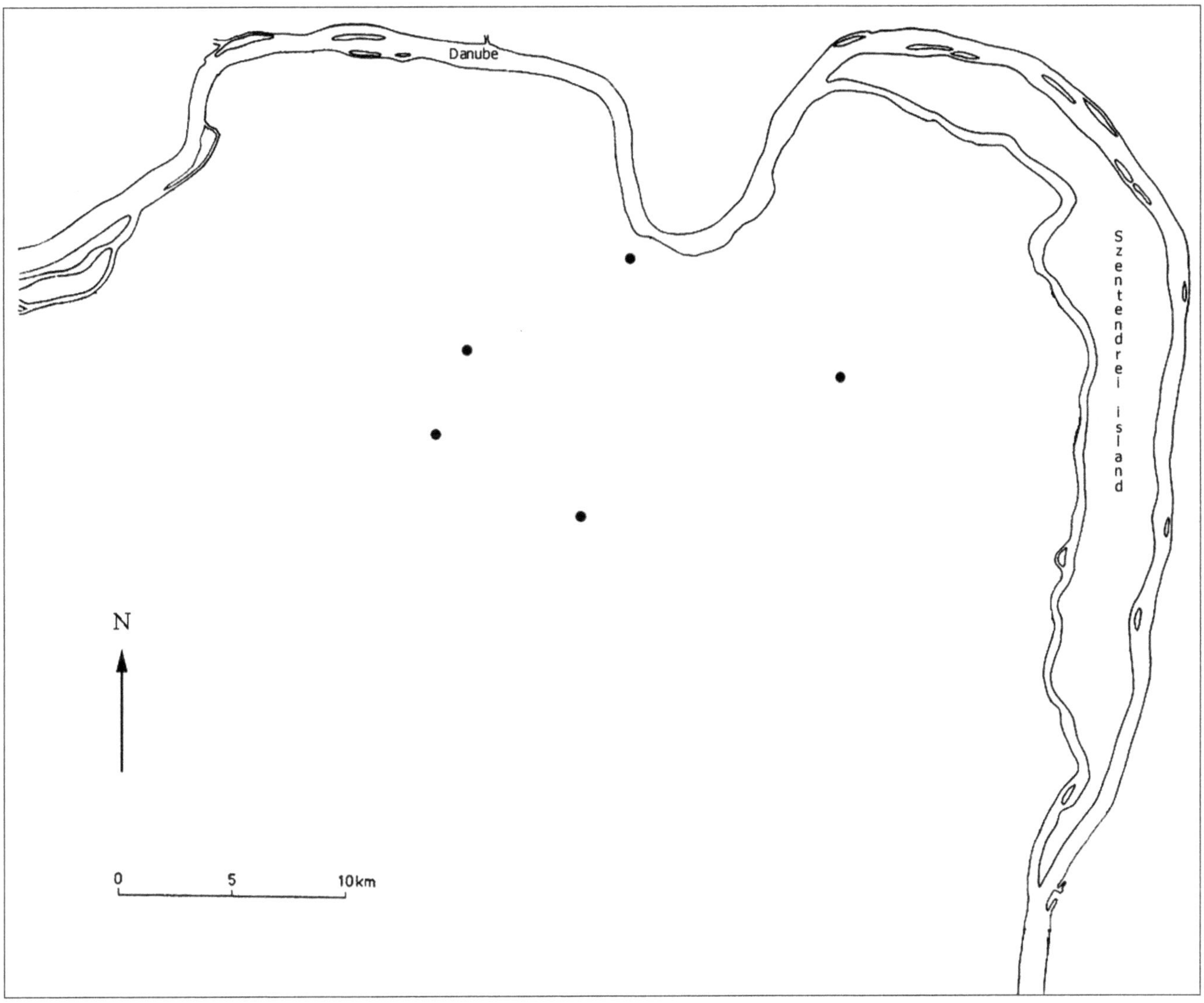

Fig. 43. Royal residences in medieval Pilis.

10. 2. Changes and Survival

Dömös and its surroundings were among the first places to separate from the royal lands. What happened here was repeated several times in the Árpádian period. In 1184, King Béla III founded a Cistercian monastery near Pilisszentkereszt,[20] which was followed by three Pauline monasteries in the second half of the thirteenth century. As demonstrated above, all these foundations were established using existing royal residences (hunting lodges or other), as a sign that by the late thirteenth century the king no longer wanted to use his Forest in the old way. He *did* visit the monasteries with his retinue,[21] but these places were more "hotels" than "residences."

Another event of the highest significance was the construction of the castle of Visegrád in the mid-thirteenth century. There had been a castle in Visegrád before, built upon the ruins of a Roman fortress on Sibrik-hill,[22] the focal place of the ancient Co. Visegrád. This fell from use by the early 1200s, for the county centre moved to Esztergom.[23] While hunting residences gradually became out-of-date and the Forest was in need of some new function, Queen Mary (wife of King Béla IV) apparently driven by her own ideas, started to build a castle above the old fortress, financed by selling her own jewels. The new castle was at least partly ready by 1251.[24] Eight years later Béla donated "the castle with the county and district of Pilis" to the queen.[25] Probably something important was revealed in two papal charters that confirmed the same donation in 1263–1264.[26] Here, it was stated that the income of the county was less than fifty golden marcs. This was not much: in the early fourteenth century, the income of the provostry of Dömös alone was estimated to be around sixty marcs, while for example the income of the chapter of Buda was two hundred.[27] With less and less desire to be ceaselessly on the move, and in an increasingly commercial environment,

[20] Békefi 1891–1892, vol. 1, 124–126. The date of foundation, set forth by Békefi, has been widely accepted, although it is based on secondary sources only.

[21] Queen Gertrudis was murdered near the Cistercian monastery in 1213. For the different sources for this event and their interpretations, see Székely 1987, 1285–7. Cardinal Gentilis, the papal legate, met Charles Robert and Matthew Csák in Szentlászló in 1308. *CD*, vol. 8/7, 62.

[22] Szőke 1986.

[23] Zsoldos 1998, 14–17.

[24] She issued a charter "in Wisegrad" that year. *ÁÚO*, vol. 7, 236.

[25] *ÁÚO*, vol. 7, 502. "castrum cum comitatu et districtu de Pelys"

[26] *ÁÚO*, vol. 8, 70–72; *ÁÚO*, vol. 3, 94–96.

[27] *ÁMTF*, vol. 4, 592.

the king cared little about a Forest that yielded such low income. He would let the queen take control of the events and the place, even appreciating her efforts. At the same time, it is true that the new castle originally had a well-defined function, also mentioned in the same charters. It was built to protect, in case of another Mongol invasion, the Dominican nuns of what is today Margaret Island (back then Rabbit Island) in Budapest.[28] Margaret, one of the nuns, was the daughter of the royal couple, which certainly explains why the queen was willing to sacrifice her jewels.

Although it is usually supposed that after the completion of the castle of Visegrád the *comes* of Pilis and the castellan of Visegrád were always the same person, there is a remarkable time-gap before this can, in fact, be demonstrated. The first to hold the two positions at the same time was Eyza, "the saracen," in 1285.[29] Before that and after 1251, we know of five *comites de Pilis*.[30] Two of them are known only by their names (Philip and Oliver) although the latter was termed *magister*.[31] Among the rest, we find one bishop (Thomas, bishop of Vác), one archbishop (Nicholas), both Chancellors of the Royal Court; and Joachim, Master of the Treasury and also *comes* of Co. Pozsony. This clearly shows that the county of Pilis was no longer simply an economic unit but had a symbolic significance. Its *comites*, very far from keepers, received their title as a sign of royal honour and cared little about the woods. Pilis was managed, in ways that are unknown to us, by lesser officers appointed by the *comites*.

After the death of the last Árpád (Andrew III, 1301) a new era commenced in the life of Visegrád and the surrounding Pilis Forest. Until the 1320s, however, Charles Robert, the new king, had more important issues to handle than Royal Forests.[32] Then in 1323, he moved his court from faraway Timişoara (Ro; Co. Temes) to Visegrád. Why he did not choose Buda (or Óbuda) we shall never know, but he probably valued Visegrád for its geographical position and strong fortress.[33] With this, the castellan of Visegrád (by this time truly unified with the position of the *comes* of Pilis) became an important position. For example in the 1340s, Töttös of Becse was castellan for a decade, being also castellan of Óbuda, and *magister ianitorum*. Benedek Himfy, the next castellan (and *comes*) later emerged to be ban of Bulgaria.[34] This "union," however, did not last long. The castellans of Visegrád used the title of *comes* of Pilis only until 1366.[35] After this time, the castellans gradually lost control over the county, or, which is equally likely, became gradually less and less interested in the county. There was most probably no need to demonstrate royal power in a county where it was so overwhelmingly present. In any case, what we do know is that the castellans ceased to call themselves *comites*, although King Sigismund addressed letters to this apparently non-existent officer as late as 1411.[36] In parallel with this, the regular institutions of the county itself began to develop. In 1333, we first hear about the "noble magistrates," this four-member judicial committee, the most important representation of the new "noble" counties, although for one reason or another there were only two magistrates in Co. Pilis.[37] By the fifteenth century, there was little difference between Pilis and any other county as far as administration was concerned. The "forest-county" disappeared altogether. What made Co. Pilis special is that it had no *comes*. As argued above, we shall probably never know the reasons behind this. Nonetheless, it is worth noting that Co. Pest (another part of the mysterious eleventh-century Co. Visegrád) shared a similar fate. Yet another inexplicable fact in the late medieval history of Co. Pilis is that it started growing in the fourteenth century, acquiring extensive territories south of its core area.[38] This is not quite as unusual as the missing *comes*, but still puzzles the modern historian, especially because the region was settled early. This *comitatus Pilisiensis* was then united with Co. Pest some time in the fifteenth century.[39]

All in all, one can detect a double structure that influenced the late-medieval history of Co. Pilis. On the one hand, there was the newly emerging "noble" county with its magistrates and territorial expansion. This line of development was entirely irrespective of the Royal Forest of the thirteenth century. On the other hand, the county was dominated by the castle of Visegrád. The castellan controlled the royal lands in Pilis, in other words what was left of Pilis Forest. The Forest, whatever the word meant in physical reality, was thought to exist as it had a hundred years before by Charles Robert, who issued a charter in 1324, in which he gave the *hospites* of Maros town (Co. Hont) the right to cut wood from Pilis to build houses or for firewood without any taxes or seal. However, if they wanted to sell the wood, so the privilege

---

[28] Or should it be Hare Island? It is usually assumed that the island received its name because of the abundance of rabbits there, but what kind of "rabbits" are meant and why they were "abundant" does not seem to be an issue. Sándor Bökönyi claimed that the island had been a reserve for the domestic rabbit, but he did not say why, other than the fact that similar things occurred elsewhere in Europe. Bökönyi 1974, 334–336.

[29] *ÁMTF*, vol. 4, 710. Györffy claimed that Philip was the first castellan of Visegrád to be the *comes* of Pilis. His argument is based upon the fact that Philip's widow made her will in Visegrád. This, in my view, does not prove that the above statement is true.

[30] *ÁMTF*, vol. 4, 692; Pesty 1880, vol. 1, 65–66. Pesty has one Peter for the year 1279, who does not appear in *ÁMTF*.

[31] *Magister*, at that time, may have meant three things: knight, officer in the chancellery, or minor canon. *KMTL*, s. v. "magister."

[32] Engel 1988.

[33] It has often been argued that Charles Robert was not very fond of Buda, which supported his rivals to the throne. This theory was discarded by Tringli 1999, 3–4. That the king may have valued the fortifications of Visegrád has been put forward by Buzás 1995, 9.

[34] Engel 1996, 34. *Banatus Bulgariae* was a shortlived (1365–1369) enterprise of King Louis I. The ban, however, counted among the barons, the highest officials of the kingdom.

[35] With the exception of Leusták of Jolsva in the late 1380s and early 1390s. The following description of the development of Co. Pilis is based mostly on Tringli 1999, 165–169.

[36] *BTOE*, vol. 3, 295.

[37] Conventionally, we speak about two basic types of county organisation in medieval Hungary. First, there was the so-called "royal" county lead by a *comes* appointed by the king. Then, along with the formation of nobility, and the overall changes in the thirteenth century, counties attained a form of self-governing independence (the best indicators of this are the presence of the above-mentioned noble magistrates and the regular assemblies of the nobles) although the *comes* was still there and still essentially a royal officer. This model, however lumpy, outlines a basically valid line of development.

[38] Tringli 1999, 137–138; Horváth 1995, 50.

[39] Horváth 1995, 64.

went, they could do it only with the consent of the *comes* of Co. Pilis.[40] This charter reveals the legal system of woodland management in Pilis. The *comes* controlled woodcutting. It was most probably his seal that was needed in usual cases, and it was also this *comes* who collected the taxes for timber and wood. The restriction that the *hospites* may not cut any wood for sale is commonplace: most of the privileges concerning woodland rights contained it. Few other such charters survive. In 1388, Sigismund ordered the *comes* to let the nuns of Óbuda cut wood if they needed it for building or fire. The form "silva nostra Pilissiensis" still reflects thirteenth-century royal attitudes.[41] Something similar was repeated in 1468, that is approximately a century later.[42] This mandate was not dedicated to the *comes* any more, but to the castellan of Visegrád. The woodland in question was no longer called Pilis, instead we read about the woods of Visegrád. From these woods, the nuns acquired the right to take wood "for their kitchen" for a year. That the *comes* (or castellan) was probably involved in the sale of wood and timber, and made good use of the closeness of major cities can be deduced from the 1411 charter of Sigismund mentioned a paragraph above.[43] The nuns *de Insula leporum* complained that when their people of the village of Csaba carried wood towards Buda and Esztergom to sell it there,[44] the people of the *comes* held them up. Were the peasants from Csaba interfering with the *comes*' wood business? Even if his actions certainly did not lead to the desertion of Csaba, as the nuns envisaged,[45] he undoubtedly did not fancy other people taking advantage of the Buda and Esztergom markets.

10. 3. Parks

There was another landscape element to appear in Pilis in the Late Middle Ages: parks. A medieval park (or deer-park) was an enclosed area to keep wild animals in for hunting. In Hungary it would be very difficult to estimate how well-developed the network of parks was in the Middle Ages, but for example in England, Rackham estimated that there were about 3200 parks in 1300.[46] The most famous Hungarian parks included for example Nyék (Co. Pilis), near Buda.[47] Among these medieval parks we find another one in Pilis, near the castle called Ákospalota, which belonged to the archbishop of Esztergom at the time when works may have started to set up the park.[48] Archaeologists did not find anything definitely related to this construction, but Zolnay suggested that certain ruined walls near Esztergom might have belonged to this very park.[49] This place is famous in Hungarian historiography because Lodovico da Bagno, an Italian at the court of Tamás Bakócz, archbishop of Esztergom (1497–1521) reported having seen a gigantic animal here, which may have been a bison or an auroch.[50] Most probably the same kind of animal escaped from this park in December 1526, as related by Herberstein, diplomat from the court of King Ferdinand I. Herberstein wrote that the king went hunting for the animal which was on the loose, and did not return to Esztergom until midnight, by which time everybody feared the worst.[51] The park was also described by the humanist Nicolaus Olahus in 1536.[52]

10. 4. Conclusions

As a Royal Forest, Pilis had a special position in medieval Hungary, being in the *medium regni*. Its early history was dominated by a dense network of royal residences for the itinerant court. In the thirteenth century, Pilis acquired the characteristic features of Royal Forests: it had a *comes*, Forest-guards, and royal servant population. However, physical royal presence in the Forest was gradually declining, and the residences were transformed into monasteries. Spiritually, nonetheless, the kings kept their control. As we shall see later on, all monasteries were royal foundations. Still in the thirteenth century, the construction of the castle of Visegrád further strengthened royal control. In the Late Middle Ages, the overwhelming royal influence was somewhat balanced by the local nobility. In their understanding, Co. Pilis had nothing to do with the Royal Forest any more, and became a regular "noble" county. As some charters of Sigismund attested, this was not necessarily how the kings understood the situation. They tried to keep whatever was useful for them from the Forest system. The missing *comes* and magistrates certainly testify that the county was not quite as "regular" as the nobles would have wished.

---

[40] *AnjOkl*, vol. 8, 10. Maros (today Nagymaros) is close to Visegrád but on the opposite side of the Danube. It has rights in Pilis (and not in Börzsöny) because Eyza (*comes* of Pilis and castellan of Visegrád in 1285) acquired the rights to settle *hospites* on its territory. *ÁMTF*, vol. 4, 710.

[41] *BTOE*, vol. 3, 18.

[42] Dl 16 631.

[43] *BTOE*, vol. 3, 295.

[44] "wood cut in their own woods and transported in their own carts" "in propriis curribus ipsorum ligna in silvis ipsarum dominarum succisa" The nuns seem to have been much irritated.

[45] "propter quod dicta possessio Chabya … desolationis pati opptineret ruinam"

[46] Rackham 1986, 123.

[47] Zolnay 1971, 129–132. The walls of Nyék (today within Budapest) were still visible in 1931. Garády 1931.

[48] For a description of Ákospalota, see *MRT*-5, 219–220.

[49] Zolnay 1961, 220.

[50] For the letter, see Óvári 1890, 253–254. Zolnay 1971, 111–112; Csőre 1980, 148.

[51] Szamota 1891, 148–149.

[52] Olahus 1938, 13.

Chapter 11

Trees and Woodland in Pilis

11. 1. Trees in Perambulations

I know of eighteen perambulations with respect to the medieval Pilis region. The first one dates to 1249, the last one was carried out in 1479, but most of them come from the late thirteenth – early fourteenth centuries.[1] The various trees mentioned in these documents reflect a patterning not unlike the ones we have seen in the charters of the chapter of Veszprém and those from the years 1417–1420 (*Fig. 44*). Out of nine different types of trees, walnut is by far the most common. The only worthy rival to this tree is pear, while the remainder are referred to a maximum of five times. *Figure 45* illustrates the approximate whereabouts of the trees. Two smaller regions are the best documented: one of them is east of Esztergom, the other is in the south-east, the square-shaped territory formed by today's Pomáz, Budakalász, Pilisborosjenő, and Üröm villages.

type of tree	*occurrences*	*type of tree*	*occurrences*
walnut	17	plum	2
wild pear	14	elm	2
oak	5	willow	1
crab-apple	4	*rakatyafa* (*Genista* sp.?)	1
"fruit-tree"	3		

Fig. 44. The frequency of trees mentioned in perambulations from the Pilis region.

The frequent mentions of wild pear are not at all surprising. It is also frequently referred to in the other two datasets. The high relative proportion of walnut is unusual, but not outstandingly so. Oak, however, faired rather poorly in these perambulations. In general, the list is dominated by fruit trees (82 %). Even without the reference lists of Vkpmlt and the years 1417–1420, it is clear that the trees mentioned in the perambulations of Pilis give us no clue as to the composition of woodland in the Middle Ages. These trees were mostly free-standing, and found around settlements. All we can say is that in this region of the kingdom people preferred to have their boundaries marked by fruit trees. One wonders, however, whether the abundance of fruit trees might reflect the existence of a well-developed system of horticulture. Archaeological finds unquestionably prove that the medievals consumed many fruits and by far not only wild ones: they were active in breeding various kinds of domesticated fruits.[2] There are differing opinions about how this was done. Other than gardens in the modern sense of the word, which undoubtedly existed, some claim that medieval fruit gardens were parts of woods,[3] although at present we do not exactly know how to imagine this. Ethnography may be of help. Bertalan Andrásfalvy described a method used until recent times. After a wild fruit tree had been found, people made a clearing around it in order to provide the tree with enough space. Then, the tree was grafted. This process was repeated in large numbers, so that hundreds of wild fruit trees were often treated in such a way around settlements. A very interesting aspect of such trees, argued Andrásfalvy, was that they had been in common possession up until the eighteenth century.[4] Here again, one has to ask oneself about the validity of ethnographic analogies. Why should medieval peasants have done the same things as early modern peasants? There is, nonetheless, in the Pilis region a reference to a wood (*nemus*) of plums, walnuts, and other fruits.[5] In another case, a *nemus* of plum trees was mentioned.[6] In Chapter 6, I have already touched upon the problem of the different meanings of *nemus*, and the reader may remember that the word apparently has such a "garden" meaning in other places as well, although at present that is all we know.

11. 2. Trees in Archaeological Sources

Bálint Zólyomi and István Précsényi published the pollen analysis of the sediment from a small pond near the ruins of the Cistercian abbey near Pilisszentkereszt.[7] A comparison with pollen analyses in the Balaton region demonstrated that the Pilisszentkereszt pollens reflected a reasonably local vegetation. The results of this survey were – in short – as follows: The largest amounts of pollen came from pine trees although this was due to selective preservation, and cannot, under any circumstances, be interpreted as a proof of the presence of extensive pine woods in Pilis. (This is reasonable, even to the non-specialist, because the nearest native pine-woods lie many kilometres away.) A large amount of *non-arbor* pollens were found, which appears to reflect gardening in the neighbourhood of the pond, around the monastery. Walnut produced especially high levels of pollen. This coincides almost too well with the written evidence. However, Zólyomi and Précsényi (quite rightly) remarked

[1] The full list: 1249: *CD*, vol. 4/2, 39–41; 1274: *CD*, vol. 5/2, 159–161; 1285: *MES*, vol. 2, 192–193, 207–208; 1288: *MES*, vol. 2, 243–245; 1292: *MES*, vol. 2, 322–324; 1299: Békefi 1891–1892, vol. 1, 417–420; 1322: *AnjOkm*, vol. 2, 15–17; 1324: *AnjOkm*, vol. 2, 120; 1326: *MES*, vol. 3, 94–97; 1326: *MES*, vol. 3, 82–84; 1333: *MES*, vol. 3, 228–230; 1353: *AnjOkm*, vol. 6, 147–149; 1355: *AnjOkm*, vol. 6, 279–284; 1367: Dl 5 623; 1367: *CD*, vol. 9/4, 77–81; 1393: Bártfai Szabó 1938, 105–106; 1457: Dl 15 186; 1479: Knauz 1863, 124–130.

[2] See for example Gyulai 2001, 186–192; Torma 1996.

[3] Csőre 1980, 121.

[4] Andrásfalvy 1989, 45–47.

[5] *AnjOkm*, vol. 2, 16. "nemus prunorum, nucum et aliorum fructuum"

[6] *HO*, vol. 8, 411.

[7] Zólyomi and Précsényi 1985.

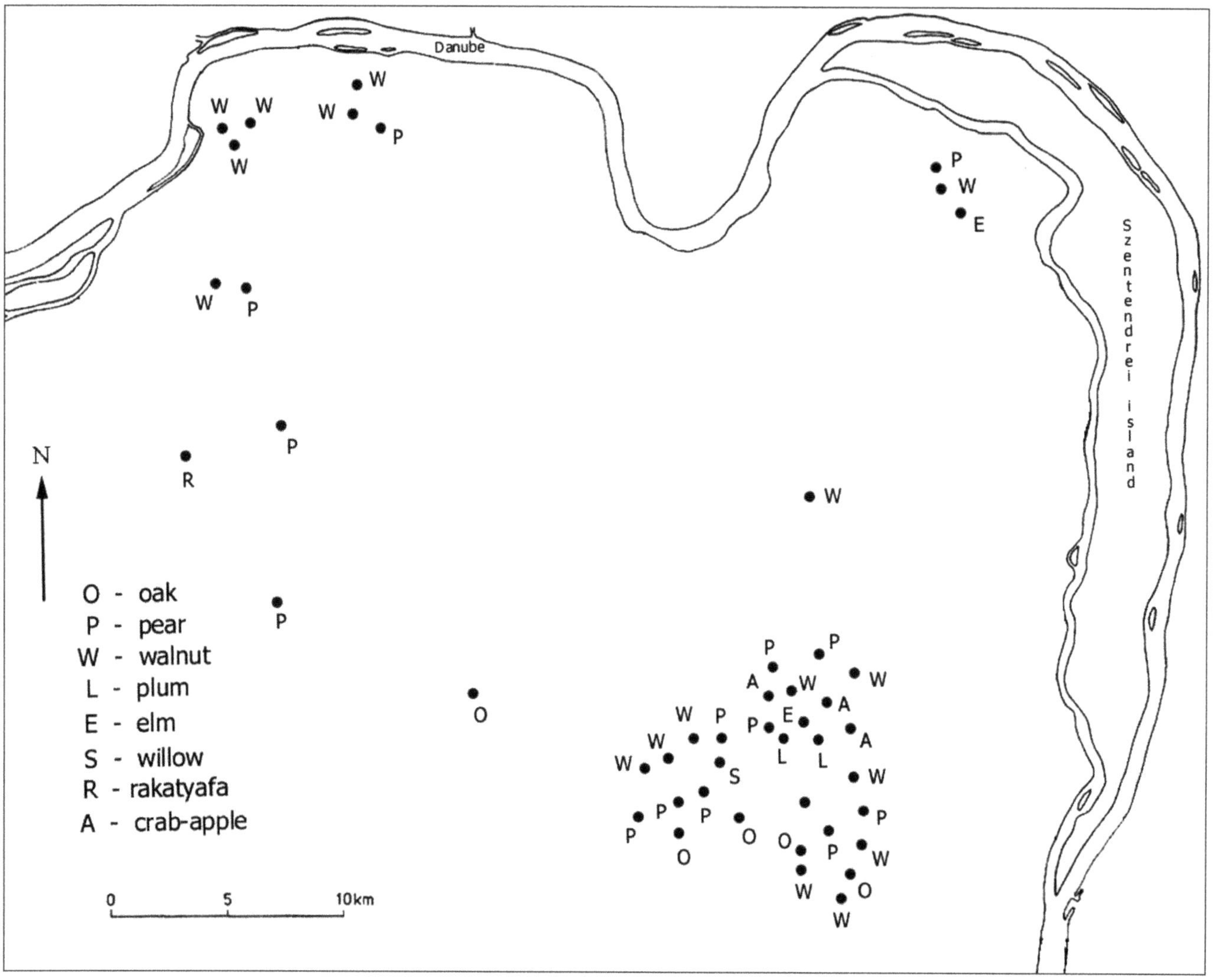

Fig. 45. The spatial distribution of trees mentioned in perambulations from the Pilis region.

that walnut produces much more pollen than most other trees, thus one needs to be careful not to overestimate its presence. Nevertheless, as the same phenomenon is suggested by both written and archaeological sources, I propose that walnut was a characteristic tree in the medieval Pilis region.

As for woodland trees proper, Zólyomi and Précsényi established the medieval proportions of beech, hornbeam, and oak as 9, 16, and 75 %, respectively and relative to each other.[8] They came up with similar proportions for lime, alder, and hazel. It is not my task to criticise their methodology here but it needs to be pointed out how they worked out the "real" proportions of various tree species from the number of pollen grains in the core. This methodology was based on comparison to the Balaton samples, while the same Balaton samples were used as reference points when discussing the results thus acquired. This, to my mind, is circular argumentation.

Environmental archaeological research was also carried out in the garden of the Royal Palace in Visegrád from 1993 to 2001.[9] The material preserved in the medieval wells of the garden was analysed, both pollens and wooden remains including plant macro-rests (fragments of trees and their seeds) (*Fig. 46*). From the latter, an impressive fifty-nine species identified include for example oak, hornbeam, small-leaved lime, beech, wild-cherry, hazel, and walnut (this last one again in high numbers). The pollen analysis, unfortunately, did not provide quantitative data, but only a stratified floral list.[10] The analyst (Enikő Félegyházi) nonetheless, proposed that the lowest strata of the sample from the fourteenth-century well represented the original vegetation of the neighbourhood. This was rather similar to what botanists imagine as "potential" vegetation: pedunculate and sessile oaks, hazel, blackthorn, and hawthorn. Ash and willow were growing on the banks of the Danube. Walnut was already present at this early stage, and thus, here as well as basically everywhere else in the country, appears to be a pre-medieval import. Despite the low quality of the finds, the gradual appearance of more and more plants specific to the garden is easy to follow. All this was then destroyed in a great fire.

Similar to the analysis of Zólyomi and Précsényi, pine is well-represented in these samples, as well. Furthermore, we know that the shingles of the cover on one the wells were also made of this tree,[11] which might imply that, after all, pines

[8] Zólyomi and Précsényi 1985, 158.

[9] Pálóczi Horváth 1997; Pálóczi Horváth and Torma 1999; Pálóczi Horváth 1998; Pálóczi Horváth 2000b. I should like to thank András Pálóczi Horváth for providing me with published and unpublished material about the excavations.

[10] Félegyházi 1998.

[11] Pálóczi Horváth and Szőke 1995–1997, 48.

Fig. 46. The medieval garden at Visegrád in an early phase of reconstruction, hence the lack of lawn and plants. Note the well where some of the material discussed here came from. July 2002.

may have grown naturally somewhere not very far.[12] The presence of pine is further strengthened by the sequence that Félegyházi established, where one can differentiate between the original pines and the newer, cultivated species of the trees that decorated the medieval garden.

What do we know about the present vegetation of Pilis? Several works are available on the topic, the most detailed one published by András Horánszky in 1964 about the Visegrád Mountains, that is the northern, volcanic part of Pilis.[13] Horánszky distinguished seven associations. The natural vegetation in most of the region, he argued, is dominated the sessile and turkey oaks accompanied by hornbeam. This is reinforced by a map in the National Atlas of Hungary, compiled by Bálint Zólyomi.[14] The soil of the Pilis region is also most suitable for oak, being a brown woodland soil with clay illuviation.[15] The altitude of the Pilis and Visegrád Mountains also puts them into the oak – oak-hornbeam region. In Hungary, the dividing line between oak, which prefers lower regions and beech, which tends to populate higher altitudes, is about six hundred metres.[16] This, however, does not apply to steep northern hillsides where beech grows in Pilis as well. To avoid contradictions in their theories, botanists call these beechwoods "extrazonal." Does this not suggest that beech should somehow not grow here? Trees, fortunately, care little about precisely conforming to zones.

Forestry has had a very heavy impact on the region. Although even in the eighteenth century the whole of Pilis was a coppice-wood supplying huge amounts of firewood for Buda and Pest,[17] little of this is visible today. Areas capable of fair wood production (that is, by modern forestry standards) have all been transformed into *hochwald*. Coppice stools, where surviving, have been singled out and are a common feature of the woods (*Fig. 47*). The most interesting areas, archaeologically speaking, are those on steep slopes, with bad soil, or with any other feature that has made them marginal for foresters. These woods have not been cut for a century or more. They are, almost without exception, ancient coppice woods.[18] The individual stools can be of considerable size, and thus of great age (*Fig. 48*). The average stool, however, is not more than a metre across.[19] Nonetheless, in the places examined,

Fig. 47. Singled-out coppice stool near Visegrád. April 2000.

Fig. 48. Ancient coppice stool in Pilis. October 2001.

[12] Although rafting pines down on the Danube was always an option. It should be noted that it is hardly possible to make such shingles from any other tree than pine.

[13] Horánszky 1964. The Hungarian version of the same was the author's doctoral dissertation, submitted in 1957.

[14] Pécsi 1989, 89.

[15] Stefanovits 1963, 155, 317.

[16] Soó 1965, 107.

[17] Magyar 1998, 102–108.

[18] Let me remind the reader that the same holds true for all "natural" remnants of woodland steppe.

[19] This is my own impression rather than anything else.

trees grow rather slowly.[20] The tree composition is much more varied than in "forestry woods" and includes lime, ash (*Fraxinus ornus*), service, *Quercus pubescens*, wild-cherry, as well as many other species together with the common oaks and hornbeam. Although foresters claim that oak "does not like to grow" in these places, but is overwhelmingly present at its preferred locations, this is not necessarily so. Here is how the present tree composition of Pilis has been achieved:

> "All woods ... are managed by foresters. In the zones of the turkey and sessile oaks it is not the mixed woods of these species which grow, but *Quercus petraea* monocultures. The shrub layer is also subject to heavy "unnatural selection." In practice, this means that from time to time foresters visit each area, and cut down all trees and shrubs other than sessile oak. Therefore natural vegetation-dynamical processes, interactions between species and groups of species ... in the tree layer cannot be studied at all."[21]

What trees would be growing where foresters now operate and in what proportion is difficult to tell. "Potential" vegetation maps are there to help, but woodland is more unpredictable than mid-twentieth-century botanists liked to imagine. What is certain is that at least ten types of trees would appear in any one wood in Hungary, if left alone.[22]

11. 3. Woodland

Establishing the medieval wooded area in Pilis is an even more complicated task and provides fewer results than that concerning types of trees. Estimations do not exist for the region, and the eighteen perambulations offer very little in terms of woodland description. Is this because woods were a commonplace, and therefore useless in boundary definitions? Pilis, in the Middle Ages but also today, has been characteristically different from the Great Plain in having larger *continuous* wooded areas. We are not concerned with "islands of woodland in a sea of arable," but rather with a tree-dominated area where the boundaries of woods were loose, and mostly determined by settlement expansion. Pilis was outside the region of thirteenth-century *melioratio terre*,[23] there were neither *hospites* nor *locatores*. Thus, the traditional medieval vocabulary of woodland clearance is missing here.[24] The reasons behind this are twofold. Firstly, there was no land to donate (the main driving force behind *melioratio terre*), or rather no wish to donate land. Secondly, Pilis, as we shall see in the next chapter, had a well-developed settlement system by the thirteenth century.

The lack of information in medieval written sources means that one has to consult other types of evidence in hope of finding something relevant about the amount of woodland in the Middle Ages. Maps would be an obvious choice, however, the first map that provides information accurate enough for our purposes dates to 1782–1784 (the first Ordinance Survey). Prior to this, detailed maps were only made for small areas, mostly in the immediate neighbourhood of castles.[25] The data on the first OS maps can be compared with the maps of the second and third surveys. Although it would be tempting to propose that the late eighteenth century must have been quite similar to medieval times, this hypothesis is almost certainly incorrect because of the intervening Ottoman occupation. As is clear from *Figures 51* and *52*, the whole settlement system of Pilis was destroyed during and after the sixteenth century, together with all four monasteries. This would have brought about woodland regeneration on a large scale within some decades. The effects of the continuous wars in the region are hard to estimate. Several works ventured to demonstrate that the Ottoman wars were accompanied by extensive woodland clearance in order to obtain enough wood for maintaining castles and fortification systems in the country.[26] These studies, however, are based on the misconception that once a tree is cut, it dies. After the Ottomans were finally driven out of the country at the end of the seventeenth century, the area was repopulated by Germans, Slovaks, and Serbs.[27]

Even this rather basic outline of the early modern history of Pilis demonstrates that woodland was in constant flux from the sixteenth to the eighteenth centuries. Thus, it would be misleading to identify an eighteenth-century map with the medieval situation. It is not, nevertheless, entirely unavailing to study the OS maps in detail. The most heavily wooded land on the first OS map is the middle of the Visegrád Mountains, which was already this way in the Middle Ages. On the other extreme, the least wooded part of Pilis in the late eighteenth century was undoubtedly its south-eastern part, which, once again, is likely to have been the same in medieval times, since this area was then densely populated. Most of the fruit trees in medieval perambulations were recorded in this region. The Ordinance Survey shows very little woodland, but extensive vineyards here. (Is not walnut even today remarkably common in vineyards?) Another possible similarity is the case of Szentkereszt and the surroundings of the former Cistercian monastery (*Fig. 49*). Here, the immediate neighbourhood of

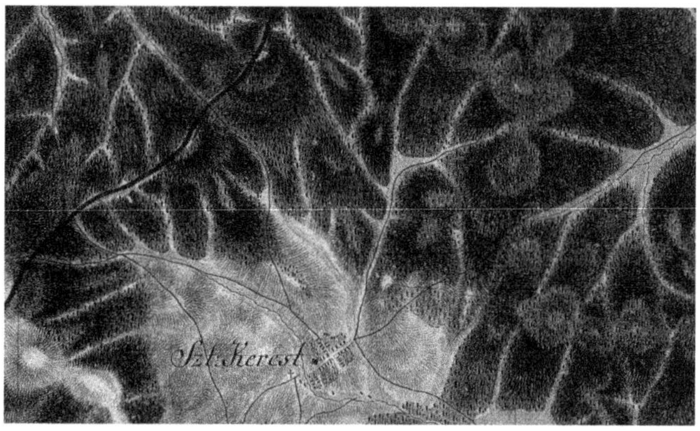

Fig. 49. The surroundings of the Cistercian monastery on the first Ordinance Survey.

[20] Csontos, Tamás and Kalapos 2001.

[21] Csontos 1996, 11–12.

[22] Bartha 2001, 25.

[23] Körmendy 1995. See also Chapter 4.

[24] For what was conventionally used, see Csőre 1980, 133–137.

[25] Jankó 1995, 35.

[26] Sugár 1995; Pálffy 1995. The latter work contains an intriguing and detailed description of the huge amount of wood used in Košice (Sk; Co. Abaúj).

[27] For Serbs, see Dóka 1981, 15–16.

the eighteenth-century settlement was not wooded and even a vineyard may be observed, although the whole clearing was rather small, in much the same way as suggested by the medieval pollen record.

11. 4. Conclusions

As far as individual trees are concerned, the species that would have caught the eye of the medieval traveller as unusually abundant appears to have been the walnut. This tree was well-represented both in the written and the archaeological sources. Fruit trees, in general, were a speciality of the region. How intensely fruit gardens were managed, however, remains an open question. Most probably there were many versions from the formal gardens of the Royal Palace and the Cistercian monastery to the extensive treatment of fruit trees in woods envisaged by ethnographers. Information on woodland composition is missing from the written sources, but archaeological material suggests that we should imagine it as similar to the "potential" vegetation of botanists. This, however, is rather vague and disregards local variations, changes in time, the effects of coppicing, etc. Even less data is available on the extent of woodland in Pilis in the Middle Ages. Written evidence does not exist, and quality maps are too late for our purposes. *Faute de mieux*, we have to suppose that there was woodland where there were no settlements, and involve another set of data in the investigation: topographical information.

Chapter 12

Settlements in Pilis

12. 1. Problems of Methodology

The settlement system around any Forest provides relevant information on the history of that Forest. As I have argued in the concluding remarks of the previous chapter, a number of Hungarian Forests, including Pilis, are exceptional in the sense that changes in the settlement system are the only real possibility to get a – however vague – idea of how wooded these Forests were in the Middle Ages.

Before we analyse the archaeological evidence concerning settlements, we have to take a look at the method with which it was collected. All that is argued in this chapter is based on two volumes of a mighty undertaking: *Magyarország Régészeti Topográfiája* (*Archaeological Topography of Hungary*).[1] The *MRT* was designed to gather all known archaeological details (sites and finds) of a territory from prehistory to the early modern times, but the authors did more than that. They used a method called field survey or fieldwalking to establish the settlement system of the areas under study.[2] In Chapter 2, I have already briefly outlined what fieldwalking is: trained archaeologists walk through a certain territory in a certain pattern and collect everything that can be considered an archaeological find, which in practice means mostly pieces of broken pottery. This work can be anything from very *intensive* (covering a given unit of land on multiple occasions regardless of whether it promises to be a fruitful enterprise or not) to *extensive* (only parts of an area are searched systematically rather than the whole). Extensive field surveys are quicker and less time consuming, and thus are suitable for larger areas as well. The writers of *MRT* used this method to establish the locations of as many settlements as possible. The surveys were also *targeted*. Places to be investigated were chosen based on documentary evidence and geographical features. The first one was especially important. Despite being archaeologists, the authors did excellent archival research, allowing them to know what to look for and approximately where.

It is not my task here to reconsider the work in *MRT-5* and *MRT-7*, yet a few remarks seem necessary for orientation. Jankovich noted that there was not one single well-designed concept and practice in the individual volumes.[3] The results of the field surveys therefore differ from volume to volume, which makes using them increasingly complicated. The editor of both *MRT-5* and *MRT-7* was István Torma. Although regarding *MRT-7* he stated that the authors "almost entirely managed to reconstruct the medieval settlement system and follow the history of the individual settlements,"[4] the book hardly ever provides information concerning the nature of possible settlements. The same holds true for *MRT-5*, although in this volume one can sometimes find conclusions drawn from the archaeological evidence. In other words, the writers (in *MRT-5* mostly István Horváth) tried to interpret the findings of fieldwalking with respect to the settlement type, although this interpretation was by no means systematically carried out.

The quality and quantity of the field surveys in the Pilis region should also be considered. Torma claimed that "the Dorog district is one of the areas in Hungary that have been most thoroughly studied through field surveys,"[5] and we may assume an at least workable quality in the case MRT-7, as well. One aspect still remains problematic: woodland is entirely unsuitable for fieldwalking. Archaeologists look for their potsherds in arable land aided by the plough; in woods the ground is covered with leaves, and the only help is the occasional badger.[6] It is not said explicitly in either volume that woods were avoided in the field surveys, but the opposite would have been unusual. The sites that were discovered are conspicuously in areas that presently have no wood cover.[7]

The dating of most of the settlements was done with the help of the potsherds. Archaeologists agree, however, that the dating of pottery is vague and imperfect.[8] This is especially true for the tenth and eleventh centuries.[9] Although the dating of late medieval pottery is more secure, any conclusions about a possible settlement system based upon potsherds alone must be used with caution. Another problem is the relationship between archaeological sites and possible settlements. While talking about topography, Jankovich stressed that "we are speaking about sites and not settlements."[10] Although this sentence is meant to warn archaeologists not to draw direct conclusions about the population from the number of sites, it also means that a site is not necessarily a settlement. Some finds are called in archaeological literature *background noise*. This means that it is not known how they got to the place at which they were found, but are presumably not related to any sort of underground cultural feature. With settled populations, these finds are usually spread around settlements.[11] I tried to exclude

---

[1] *MRT*-5 and *MRT*-7. Thus far ten volumes have been published in the series, covering approximately one tenth of (present-day) Hungary.

[2] For a detailed description of fieldwalking, see Jankovich 1993.

[3] Jankovich 1993, 5, 46–47. Jankovich 1985, 284.

[4] István Torma, Introduction to *MRT-7*, 9.

[5] István Torma, Introduction to *MRT-5*, 9.

[6] To the best of my knowledge, archaeologists do not use plants as indicators. They should: the stinging-nettle, for example, indicates phosphate in the soil, which usually accumulates through human agency in or around settlements. Compare Rackham 1996, 135–136.

[7] Compare Aston 1985, 99.

[8] János Makkay, Introduction to *MRT-8*, 31; Jankovich 1985, 285.

[9] Jankovich 1985, 284.

[10] Jankovich 1985, 287.

[11] Jankovich 1993, 10.

this "background noise" from my analysis. Nevertheless, it is probably not the task of the historical ecologist to draw conclusions where archaeologists feared to tread. The typology of medieval Hungarian settlements, which may be of use in resolving this problem, will be discussed later.

12. 2. Settlement System

Medieval documents mentioned thirty-seven settlements in Pilis. They were all connected to archaeological sites, with the exception of Dersülése.[12] *Figure 50* illustrates their first and last appearances in the charters, supposing that most of them existed in the period between the two dates. The shape of this chart would suggest that the number of settlements underwent constant growth in the Pilis area: there were three settlements in 1100 and twenty-seven in 1500. However, the value of such a chart is highly questionable, considering the fact that the number of surviving written documents also displays the same numerical development. In other words, this chart may well be taken as an illustration of the growing production and survival of charters during the Middle Ages in Hungary.

The archaeological finds in many cases reveal that a particular settlement existed long before its first appearance in written documents. Out of the thirty-seven settlements in the Pilis region, eighteen can be predated this way. In six settlements, no information is available because they lie in an area inaccessible to archaeology (Bajon, Nyír, Bogdán, Kande, Fedémes, and Káloz) and the above-mentioned Dersülése cannot be located, which leaves eighteen changes out of thirty cases. If we compare their distribution in time to that illustrated in *Figure 50*, we cannot fail to see the changes. One can still find twenty-seven settlements in 1500 while only twelve were noted in 1100, but the number of villages for 1100 quadrupled if we include the archaeological evidence. The overall picture also undergoes modification. *Figure 51* indicates that from a small number of settlements there was a sudden "boom" in the thirteenth century, and then the number of settlements remained basically the same until the Ottoman invasion of the sixteenth century.

In addition to settlements that we know of from documents, there were many others never mentioned in writing, and discovered in field surveys. When we take a look at the chart that depicts these settlements (*Fig. 52*) we see a rather different picture from the previous ones. This chart, although it can display tendencies only, is clearly contrary to that seen in Figure 51. There are many settlements in the eleventh century, another peak in the thirteenth century, and then the number *decreases*. At this point, one must ask two basic questions: Is this discordance between the charts specific to the Pilis region and if so, why?

All this is related to the general problem of medieval Hungarian settlement system and the changes it underwent through time. The most influential historian in this field was István Szabó, who published two major books: *The Formation of the Settlement System in Hungary* and *The Medieval Hungarian Village*.[13] In the former, he presented

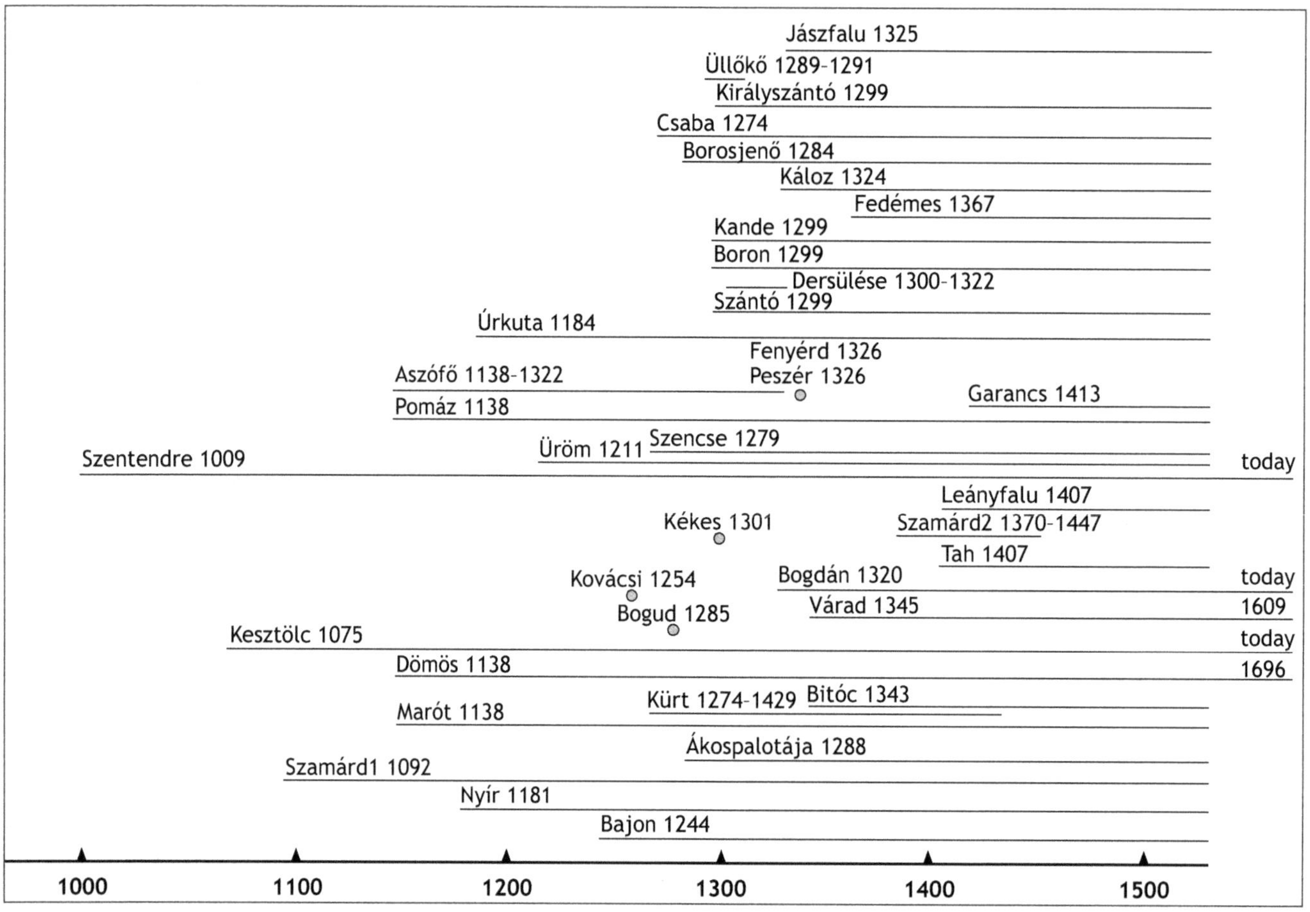

Fig. 50. Settlements mentioned in written documents.
The settlements existed until the Ottoman invasion, unless otherwise indicated.

[12] *MRT-7*, 205.

[13] Szabó 1966; Szabó 1969.

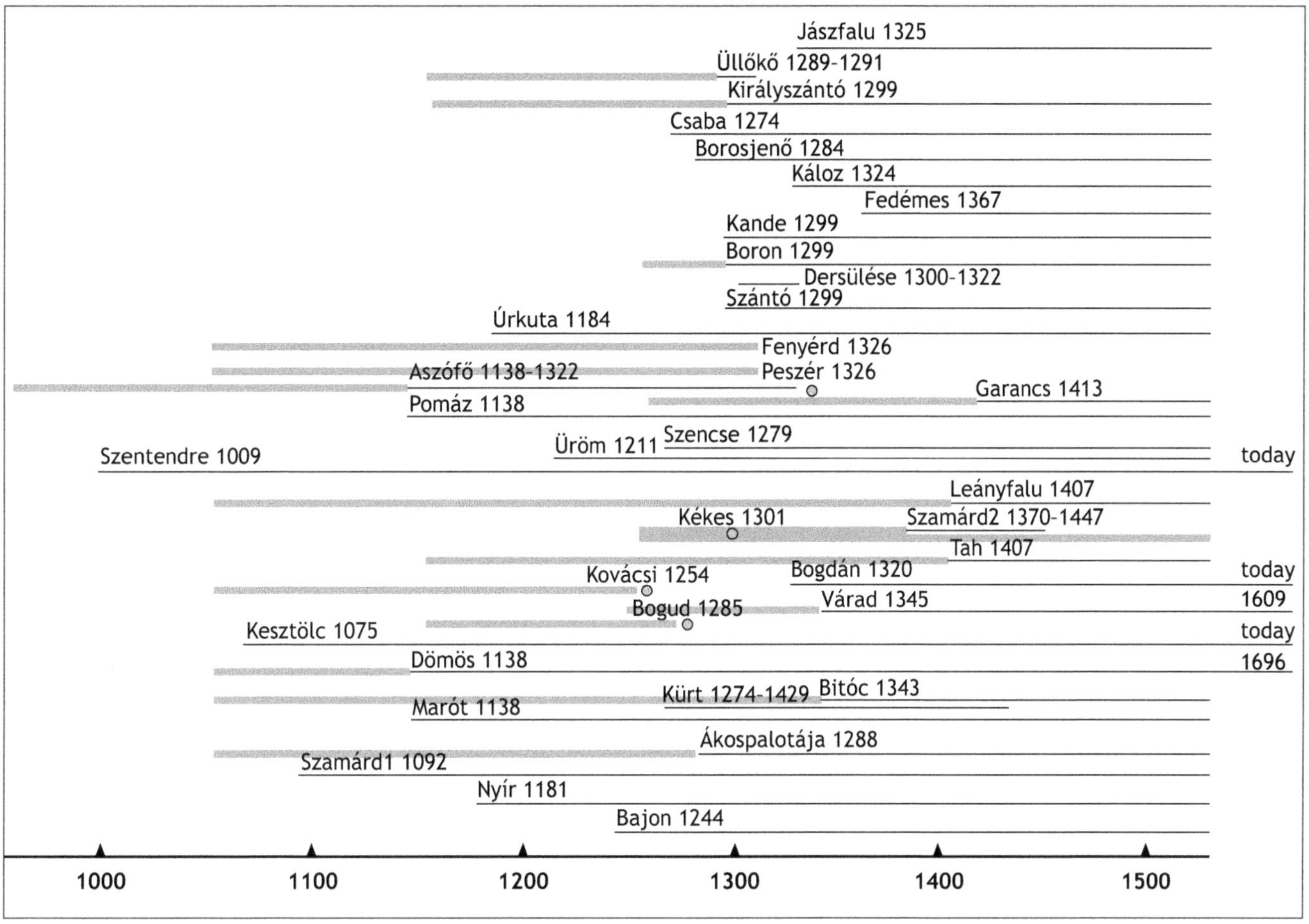

Fig. 51. Settlements mentioned in written documents supplemented by archaeological dates. The grey lines indicate a date using archaeological information.

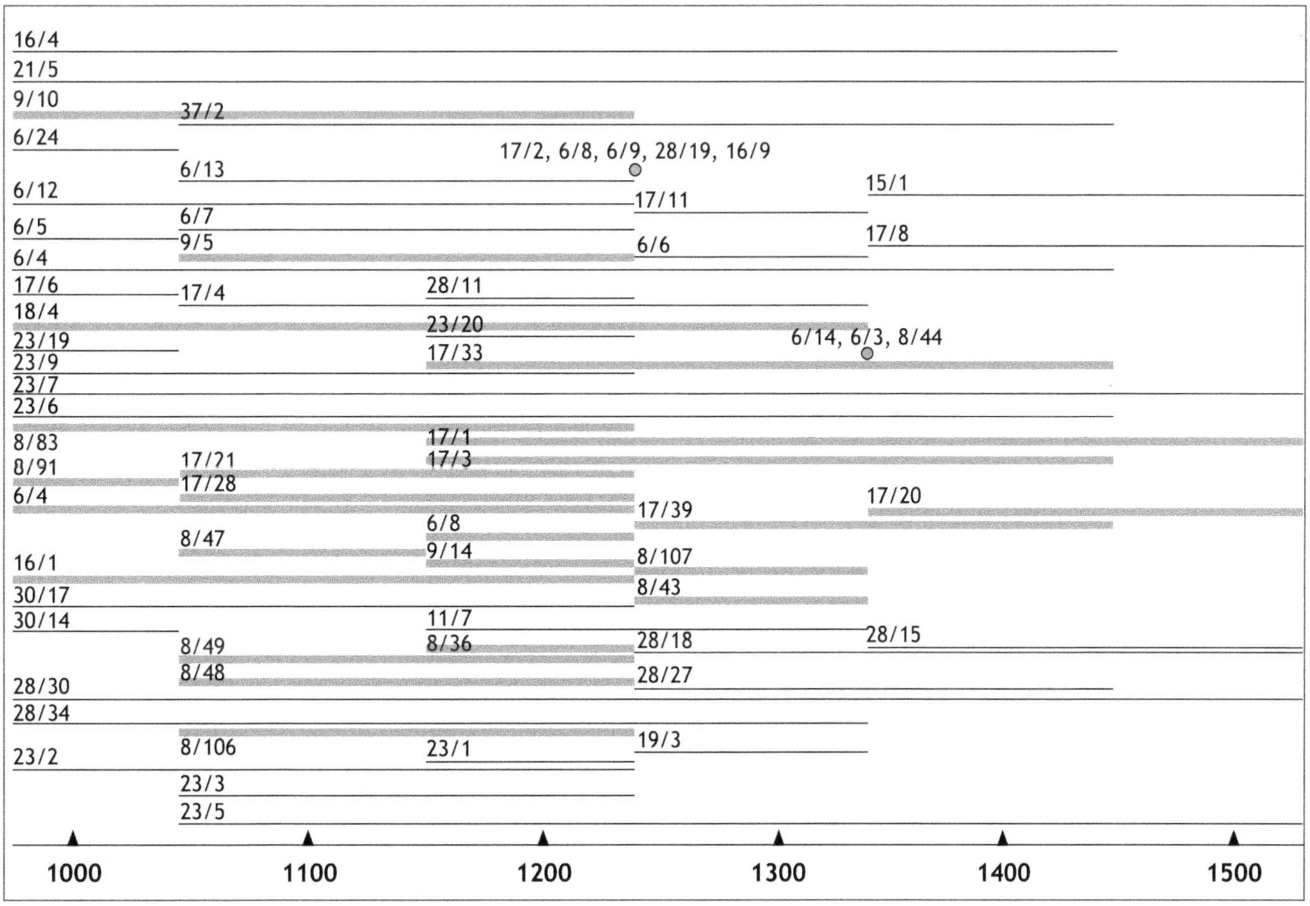

Fig. 52. Settlements not mentioned in written sources. Grey lines – MRT-5; Black lines – MRT-7

his view concerning the changes in the settlement system of medieval Hungary, a view now commonly accepted. His argument was as follows: after the Conquest of the country, the Hungarians settled down and a fairly firm settlement system[14] was established during the eleventh century. The twelfth, thirteenth, and early fourteenth centuries were the period that Szabó called the time of "excessive disintegration and spread," that is, the settlements became smaller and more numerous. In the late fourteenth and fifteenth centuries this process was reversed, and many settlements disappeared, either by being integrated into a bigger village or deserted. The reasons behind their disappearance were manifold: economic, social, agrarian structural, wars, the poor living conditions of the peasantry, natural catastrophes, and epidemics. This process was a general European phenomenon, Szabó claimed, in accordance with the population decrease experienced in the continent during the same period.[15]

As compared to Szabó's theory, the settlements of the Pilis region (especially those without written documents) are not an exception to the general medieval Hungarian trend. However, the settlements with written documentation increase in number, which does not fit Szabó's pattern. One cannot assume that every single settlement that was recorded in writing was necessarily bigger or more significant than those that survive only in archaeological finds, but in general this seems to be the case. These were the settlements that formed the backbone of the settlement system of the Pilis region; they represented solid ground within the turmoil of the appearance and disappearance of smaller settlements. A linguistic analysis of the settlement names may be of importance here. Twelve out of thirty-seven names in the Pilis region originate from personal names.[16] It would be tempting to suggest that these names came into being in the early period of the Hungarian Middle Ages and did not change.[17] In fact, out of these twelve settlements, Bogdán is documented the latest (1320), but in this case we can only rely upon written documents. The majority of the twelve settlements were born at least as early as the eleventh-twelfth centuries. Many settlements recorded in writing can be dated to the thirteenth century, and after that the number of settlements with written documents remains approximately the same. The year 1300 shows that thirty of those settlements with written documents already existed, there were also thirty settlements mentioned from 1400, and twenty-seven from 1500. If we compare this with the information available about the settlements without written documents (a decreasing number in the fourteenth and fifteenth centuries), it can again be suggested that the settlements with written documents were more stable than the ones we know of from archaeological evidence alone.

The settlements that do not appear in charters are more problematic. First, as I have already mentioned, their dating must necessarily be imprecise. Secondly, one always has to be cautious about their size and significance. It is perhaps understood that not all of them were nucleated villages, but in many cases even the term *settlement* needs further explanation. As illustrated in *Figure 52*, the highest number of these archaeological sites date to the Árpádian period. What János Makkay wrote about *late medieval* Co. Békés applies to the Pilis region as well: "The stray finds and the sites related to smaller settlements are almost entirely missing."[18] Among the Árpádian period settlements, we can distinguish two major types.[19] "Village settlements" are those with rich find materials, often with the remains of a church, covering a larger territory. In other words, these settlements resemble the majority of those with written documents. The other type comprises "habitation-like settlements," usually situated in a smaller area, represented by a few sherds. At present it is difficult to decide whether they were parts of a bigger community or independent entities. Jankovich described the following possibilities: they were "1. part of a spacious village with a loose structure, 2. straggling bits of villages, 3. abodes (a. summer abodes, b. abodes of herdsmen, c. abodes of nomads.)"[20] Similar options were presented by József Laszlovszky in 1986, who also described a characteristic feature of these smaller sites, namely that they were very often situated in chains along rivers and streams.[21] This was typical of the Pilis region as well. All three options are possible in every case, and one has to investigate and decide the question for each individual site. But let me stress again that the Pilis region does not present an exception to the general rule that these settlements were characteristic of the Árpádian period.

Figure 53 shows the location of all possible settlements. Along the Danube, their distribution seems even, and they appear not to have entered the depths of Pilis Forest. However, as we have already seen, this may well be the result of the technique of fieldwalking. This drawback would not affect the settlements that have written documents, since charters and their survival do not depend upon geographical features, woodland in this case. Nevertheless, the settlements that have charter evidence basically surround the Forest. It appears that the people living in more stable settlements preferred to stay outside of the woodland, or vice versa, that those settlements that managed to survive longer were the ones that were not within the woods.

---

[14] One cannot be careful enough with the word "village." Although Szabó used it freely in all his books, I mostly prefer to write "settlement."

[15] Szabó 1966, 186.

[16] As introduced in Chapter 2, place-names based on personal names are very problematic. "Puszta személynév" (literary "pure personal name") – the technical term used to describe them – is hard to translate into any language. "Pure" does not carry nationalistic connotations, it simply means that – apparently – the name of the given settlement originates from an actual person's name. In many cases it does not change any further (the best known example of this is Buda). At present, linguists do not know what to do with this type of toponym, which, as the Laws of Murphy dictate, is the most common in the whole corpus. Even the usual high-precision dating ("early") was seriously questioned by Kristó 1976, 38. This leaves us with a frustratingly huge pile of indecipherable information. In Pilis, the categories with the second and third most settlements are the ones referring to geographical features and those referring to the occupation of the inhabitants, with six and five settlements, respectively. In addition, three more villages are combinations of pure personal names and common nouns.

[17] Contradicting what is written in the previous footnote, and following mostly Kázmér 1970.

[18] János Makkay, introduction to *MRT-8*, 31.

[19] János Makkay, introduction to *MRT-8*, 31; Jankovich 1985, 285–287.

[20] Jankovich 1985, 286.

[21] Laszlovszky 1986b, 137–138.

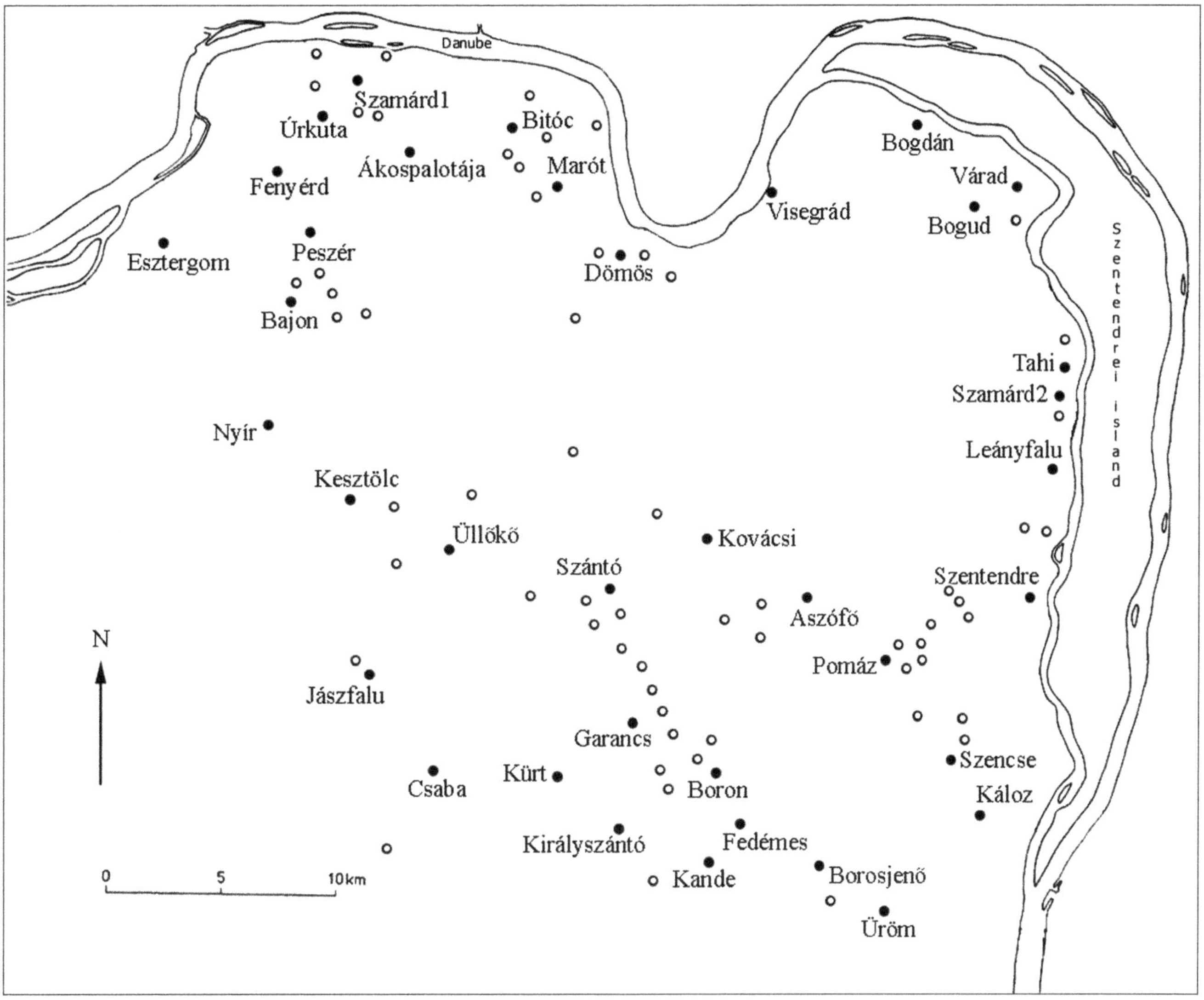

Fig. 53. Medieval settlements in Pilis. Dots: settlements mentioned in written documents. Circles: settlements located through fieldwalking, not mentioned in written documents.

There were two areas where the "purely-archaeological" settlements largely outnumbered the documented ones. One of them is the region of today's Pomáz, the other is the valley between the middle part of the Pilis Mountains (Hosszúhegy) and the south-eastern end of it (Kevélyek) together with the northern part of the valley of Pilisvörösvár, which is the south-western side of Hosszúhegy. This is partly explained by the geography of the region. The longish limestone Pilis Mountains rise steeply from the neighbouring flat areas. Around here, the dividing line between habitable and non-habitable places is sharp. Many smaller settlements were established here in the Árpádian period, in accordance with general Hungarian trends. Early settlers inhabited all possible places, but did not venture the impossible. Late medieval desertion in the Pomáz and southern Pilis regions resulted in two possible models. One of them was created when a settlement, established previously along with many others in the twelfth and thirteenth centuries, "took the lead" and became the most important village of the territory, as in the case of Szántó (*MRT-7*, 17/9) or Boron (*MRT-7*, 6/10). The other model is represented by Szencse (*MRT-7*, 23/8) or *MRT-7* sites 6/4 and 23/5, where an early settlement survived the "latecomers."

The northern part of the region offered more possibilities for settlement expansion towards woodland. The (more) gently rolling slopes would have been manageable for medieval settlers, as some sites within the hills amply demonstrate. However, very few sites have been found in these places. Providing that all this is not the result of survey techniques, it seems probable that whoever owned the woods preferred them uninhabited.

12. 3. Conclusions

It is reasonable to propose at this point that the Pilis region with its hills and woods was not an exception to the general patterns of change in the medieval Hungarian settlement system. The fact that many settlements disappeared in the fourteenth and fifteenth centuries was an overall trend in Hungary – and also in Europe. Medieval people, in general, did not inhabit the depths of the Pilis Forest, but stayed mainly in the valleys. Although this might seem obvious, the reasons behind it may be very complex. The most influential of these reasons was probably the very existence of the Royal Forest.

Chapter 13

Monasteries in Pilis

A typical feature of medieval Hungarian Forests (not exclusively but especially of Pilis and Bakony) was the presence of two monastic orders: the Cistercians and the Paulines. It is also true that their settlements were the only units deliberately placed *within* the Forest. It would be all too easy to claim that monks were looking for uninhabited places in order to be able to dedicate themselves to heavenly matters, however, let us remember that Pilis was right in the centre of everything that was mundane. The Cistercians had one monastery in the region, while the Paulines possessed three: Szentkereszt (Keresztúr), Szentlászló, and Szentlélek.[1] A look at the map (*Fig. 54*) immediately makes it clear that these four monasteries were situated – perhaps not by accident – so as to form an almost continuous line of holy places in Pilis Forest. One was able to reach any one of them starting from any other within one day, even on foot.[2] Their density was the highest around the north-western end of the Pilis Mountains. Szentkereszt and Szentlélek, together with the Cistercian monastery formed a small-scale version of the triangle constituted by Esztergom, Visegrád, and Buda.

There were obvious differences between these two orders. The Paulines were followers of Paul the hermit; the Cistercians had less of an eremitic devotion. The Cistercians were an international, early twelfth-century organisation, whereas Paulines (officially called *Ordo Fratrum Heremitarum Sancti Pauli Primi Heremitae*) were the only Hungarian order, established in the thirteenth century, precisely in Pilis. The White Monks are known to have invented and used new agricultural methods and are also remembered as the order that cleared and cultivated much previously "pristine" land; the Paulines had no such reputation. This list goes on endlessly. Most of these assumptions are nowadays seen as topoi, however, that does not necessarily make them untrue. One has to compare each case against the official ideology. It therefore seems reasonable to examine the two orders separately.

13. 1. The Cistercians

The Cistercians, in a sense among the prime products of the Middle Ages, do not need much introduction. The original idea behind their foundation was to return to the pure life of the rule of St. Benedict, from which – so the Cistercians said – the Benedictine order had greatly deviated. The Cistercian way of life, in theory, was puritan and simple.[3] Shortly after its foundation, the order became immensely popular, and by the middle of the twelfth century the number of Cistercian monasteries exceeded three hundred.[4] Historians generally agree about the strange fact that this popularity must have been very surprising for the Cistercians themselves.[5] There are many explanations for the success, but what is most relevant for us is the Cistercians' new economic structure and their unusual behaviour towards nature in general. The Cistercians typically cultivated their lands in *granges*. A grange was "a property wholly made up and run as a demesne and cultivated by conversi, i.e. lay brethren of the order."[6] This meant a huge mass of free labour for the Cistercians. Southern argued that the Cistercians – led by the *Regula* – developed an economic system, which, by pure accident, coincided almost perfectly with the needs of the twelfth century.[7] In the beginning, at least to the middle of the twelfth century, the White Monks refused to accept rents, tithes, and tenants.[8]

A much more complicated issue is their role in internal colonisation. For decades, if not centuries, scholars were mislead by what the Cistercians had said about themselves. Their *Regula*, as they continuously stressed, enjoined them to settle in the wilderness. The traditional view of the Cistercian monk is that of a brother who settles in the middle of nothing, then clears and cultivates the land.

> "Cistercian monasteries almost invariably were situated in forests; in hilly or rocky places ... ; or in politically contested border areas ... Through colonisation and with enormous energy the White Monks drained swamps and erected dikes and dams, cut down forests, and brought the land under cultivation. The monks acted as pioneers in undeveloped areas, and they acquired the reputation of being successful agricultural innovators ..."

wrote Bennet D. Hill in the *Dictionary of the Middle Ages* as late as 1983. As landscape historians tell us more and more about the medieval countryside, we have to realise that this traditional view should be replaced by a more balanced picture. Twelfth-century Europe was rather densely settled, and, although naturally there were places where the monks could fulfil the requirements of their *Regula* to the letter, many of the foundations were established in areas which were not especially deserted.[9] As Janet Burton wrote in 1994: "not all their sites were remote, and the researches of historical

[1] Other monasteries belonging to the Paulines existed nearby. One was in Visegrád from 1493. Hervay 1984a, 169. According to *MRT-7*, 68, Dömös was also transformed into a Pauline monastery for some time in the 1440s. However, these monasteries were late foundations and outside the Forest. Thus, I shall not include them in my study.

[2] As monks were supposed to travel.

[3] Among the numerous books and articles about the Cistercian order, a good starting point is the monograph of Lékai 1977, with an extensive (although by now somewhat outdated) bibliography. For a short summary it is still best to consult Southern 1970, 250–272.

[4] Southern 1970, 254.

[5] Constable 1996, 217.

[6] Postan 1975, 102.

[7] Southern 1970, 225.

[8] Duby 1962, 199.

[9] For a bibliography of this debate, see Lékai 1977, 431–432.

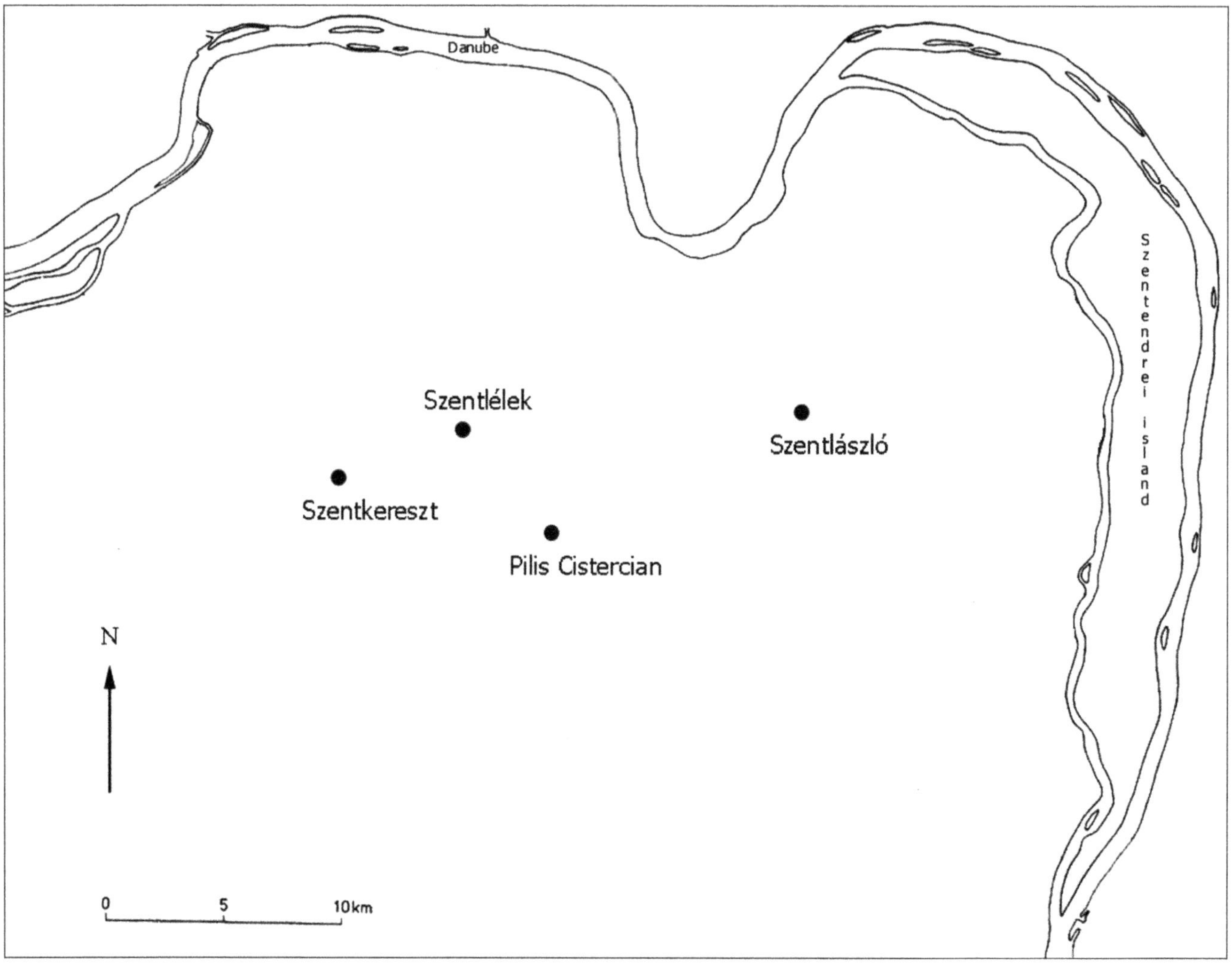

Fig. 54. Monasteries in Pilis Forest.

geography have indicated many cases where lands given as Cistercian endowments were far from marginal."[10] What is now gradually becoming *communis opinio* was summarised by Giles Constable:

> "The wilderness and solitude in which the new houses were often said to be located was a topos based on the description of the desert as a 'place of horror and vast wilderness' in Deuteronomy 32.10. They were usually much closer than the sources suggest to settled areas and to roads and rivers, as the contradictory topos of hospitality shows... Even in Eastern Europe and Poland the Cistercians usually settled on or near settled lands and did not play a major role in clearing new land."[11]

It was certainly true, however, that the Cistercians introduced new agricultural techniques, and it was probably in this sense that their influence on the landscape was the most significant. As Michael Aston wrote about Cistercian colonisation:

> "Even though such activities by Cistercians may have been overemphasised by researchers in the past, and such landscape changes were in any case widespread in the twelfth century, it was nevertheless a major aspect of early Cistercian monasteries to modify dramatically the landscape in which they were built."[12]

Part of this colonisation problem was the clearance of woodland. In this respect, one thing cannot be denied: a great change in attitudes. As Georges Duby wrote about France: "The image of a monk clearing woods definitely does not fit the French landscape of 1075," but "Cîteaux restored each monk's physical contact with the land. The initials of the *Moralia in Job*, ... actually depict monks using axes to clear woods."[13] At this point I may even quote the well-known lines of Gerald of Wales, the late-twelfth century writer: "Settle the Cistercians in some barren retreat which is hidden away in an overgrown forest; a year or two later you will find splendid churches there with fine monastic buildings."[14] Opinions differ among historians, but it is obvious that once a monastery was situated within a wood, land simply had to be cleared in order to produce food. There is another, much neglected aspect in establishing Cistercian attitudes towards woodland. Both in England and on the Continent, the White Monks were associated with animal husbandry. Duby claimed that "Cistercians lived the pastoral life and developed

[10] Burton 1994, 72.

[11] Constable 1996, 120.

[12] Aston 1993b, 74.

[13] Duby 1978, 43.

[14] Gerald of Wales 1978, 106.

primarily by stockbreeding."[15] It is a well-known fact that medieval pasturing took place to some extent within woods. Even special laws existed to regulate this process, although, at least in England and the Netherlands, we know that they must not be interpreted at face value.[16]

13. 1. 1. Cistercian Industry

Another, equally important aspect that needs consideration is the industrial activity of the monks. In recent years, more and more attention has been paid to for example iron production or glass making, with special attention to the possible interactions with the landscape.[17] I will bring some examples to illustrate the growing significance of this phenomenon. Grenville Astill wrote about no less than a "medieval industrial complex and its landscape" in connection with Bordesley abbey, England, alongside an exhaustive description of the abbey's mills, workshops, and resource supplies.[18] Astill also claimed that "such industrial complexes were relatively common to judge from documentary accounts and surveys."[19] Paul Benoît, when describing *l'espace industriel cistercien* around Cîteaux, distinguished between three spaces: the first one was the immediate neighbourhood of the monastery with the inevitable water-system, the second one a circle of a few kilometres with the granges and most industrial activities, whereas the third one was a loose structure of mines and forges in the region.[20] All these industrial activities demanded huge amounts of wood, thus encouraging planned woodland management.[21] In Normandy, to bring another example, a wood to supply a forge was planned to operate on an eleven-year coppice cycle.[22]

13. 1. 2. Cistercian Economy in Hungary

As opposed to the field of international scholarship, little is known about the Cistercian economy in Hungary.[23] As a starting point, Ferenc Hervay compared the incomes of the different monasteries in a study in 1984.[24] In 1994, Beatrix Romhányi provided an excellent overview of the Hungarian Cistercian monasteries in light of new international research.[25] She stressed that the Hungarian monasteries received huge incomes from tolls and the salt trade, and that "most of the Hungarian abbeys were located in already populated regions, close to the main trading routes."[26] Although there existed some exceptions to this rule, evidence for internal colonising activity is present in one case only.[27] As for the Cistercian industrial activity, Romhányi reported that it "is nearly unknown even now." However, she mentioned Pilis, as a clear example of metallurgy.[28] In general, she concluded that "the Cistercians had an economic structure completely different from the ideal... The Cistercians in Hungary were primarily active in the secondary economy, ... their life was based, from the moment of their foundation, on toll incomes and trade."[29] Furthermore, she even suggested that the monasteries had problems with their work-force (that is, lack of *conversi*) which may have been one of the reasons for the lack of internal colonisation. Another observation of hers on the same topic was that in Hungary the Benedictines played the pioneering role in colonisation in the eleventh century, while in the late twelfth century – when many Cistercian monasteries were founded – the kings preferred to have collectively privileged populations under direct royal control, rather than the largely independent people of the Cistercians, in the outskirts of the kingdom, where there was still land to donate.[30]

13. 1. 3. Pilis Cistercian Monastery

I am in a fairly lucky position concerning the "Pilis monastery,"[31] because it was studied extensively from both sides, that is, from the historical and from the archaeological one, by Remig Békefi and László Gerevich, respectively. Békefi was a nineteenth-century scholar, who collected almost all written data and analysed the historical development of the monastery, while Gerevich worked for decades (1967–1981) on the archaeological site, the results of which were, alas only partly, published in 1984.[32] Later this century, another historian, Ferenc L. Hervay continued working on the history of the monastery, which ended in slight modifications of Békefi's results.[33] In 2000, the finds from the archaeological excavations were, at long last, published in a separate volume.[34]

---

[15] Duby 1962, 142. For England, see Burton 1994, 236.

[16] Rackham 1980, 155; Dirkx 1998.

[17] For an overview, see Aston 1993b, 128–132.

[18] Astill 1993.

[19] Astill 1993, 303.

[20] Benoît 1994.

[21] Rackham 1980, 153–154.

[22] Arnoux 1990. The whole book is a good collection of essays to demonstrate the connection between metal industry and woods, albeit mostly on early modern material.

[23] The only study in this regard up until the 1980s was Elek Kalász's essay on the monastery of Szentgotthárd (Co. Vas). Kalász 1932. This work was politely described as an "excellent overview of Cistercian economic ideas," in Romhányi 1993–1994, 186. In practice, this means that the author described Cistercian economy in Hungary "as it must have been," without using a single local source. It is almost ironic that any study on the subject starts with a reference to Kalász' work, although, as far as its ideas about Cistercian economy in Hungary are concerned, it would be better to forget it as such. It is only useful as a study on the estates of the Abbey of Szentgotthárd.

[24] Hervay 1984b, 26–27.

[25] Romhányi 1993–1994, 180–199.

[26] Romhányi 1993–1994, 184–185.

[27] Romhányi 1993–1994, 190. The monastery of Cîrţa (Ro; Szászföld – not a regular county but a more or less autonomous territory of the Saxons) was notorious for being so far off to the east that its abbot was required to go to only every third of the annual meetings of all Cistercian abbots in France.

[28] Romhányi 1993–1994, 192.

[29] Romhányi 1993–1994, 197.

[30] Romhányi 1993–1994, 196, 199.

[31] This (*monasterium de Pylis*) is what it was called in the Middle Ages.

[32] Békefi 1891–1892; Gerevich 1984.

[33] Hervay 1984b; Hervay 1985.

[34] Holl 2000. Some more architectural remains were published in Takács 1994b. Remains of the well were described in Takács 1992.

The short history of the monastery is as follows: It was founded by Béla III, king of Hungary on 27 May 1184.[35] This was the fifth Cistercian foundation in Hungary, after Cikádor (Co. Tolna), Igriş (Ro; Co. Csanád), Zirc (Co. Veszprém), and Szentgotthárd (Co. Vas).[36] Soon it became one of the most powerful monasteries in the kingdom. The building of the monastery must have been well-developed by 1225,[37] and Gertrudis, the wife of Andrew II, was buried here as attested by a charter issued by his son, Béla IV, and by the shrine itself.[38] The rather complicated story behind the foundations of the early monasteries has some relevance to our perspective as well.[39] The first monastery, Cikádor, was founded in 1142, forming part of the first great European expansion of the order. However, in contrast with what was happening elsewhere in Europe, no foundations followed in Hungary in that period. It was King Béla III who revitalised the issue, undoubtedly as part of his overall western-oriented foreign policy. The above four monasteries, and a fifth one to follow immediately (Pásztó (Co. Heves), 1191) were all his foundations. The kind of role royal power played in the process at this time is demonstrated by the fact that afterwards there were no more direct royal Cistercian foundations.[40] We can also identify the group of people that possibly initiated all this. Starting with Lucas around 1150, a number of students studied in France, mostly at the University of Paris. These learned men later acquired important positions both at the royal court and in church administration.[41] Lucas, for example, became archbishop of Esztergom, and the same career awaited Job, a member of the next generation of "French" students. They must have had a significant role in arranging the marriage between King Béla III and Margaret of France, daughter of King Louis VII, in 1186. Knowing that Béla's Cistercian monasteries were all directly affiliated from France one cannot fail to see the connection. The importance of the Pilis monastery, which is partly explained by its proximity to the royal court, is presumably also the result of the efforts of the above-mentioned archbishops of Esztergom.

The monastery soon grew strong enough to affiliate and populate three new monasteries: Pásztó (1191), Bélháromkút (Co. Borsod, 1232), and Ábrahám (Co. Tolna, 1270).[42] The Mongol invasion of the mid-thirteenth century reached the monastery in its prime, but did not destroy it completely, because as early as 1244, the abbot was acting on behalf of the pope.[43] Nevertheless, the archives were destroyed, and Béla IV had to produce a charter in 1254 in order to grant all former privileges to the monastery.[44] After 1356, when a German monk was promoted to be abbot of Pilis, German influence was strengthened in the monastery, which was once again destroyed in 1526, this time by the Ottomans. Its books and charters (what had been left and produced since the thirteenth century) did not survive. The monks themselves fled to Heiligenkreuz.[45] After this, the monastery was so much in ruins and beyond memory that later the Pauline order claimed its territory, backed up by the misleading place-name of a nearby village (Pilisszentkereszt). It was Békefi's book in 1891 that clarified that the ruins that had then still been visible belonged to the Cistercian monastery of Pilis.[46]

Since its documents were destroyed twice, examining the history of this monastery is not easy. Almost two hundred charters mention the monastery or rather the abbot in some way, but only two of these come from its original archives. We can have some knowledge about its general political history, but as far as written sources are concerned, not much may be found about its economy, let alone its woodland management practices. As a result, I had to rely more upon archaeological evidence. The two problems I shall concentrate on are identical with those connected with the Cistercian order in general. First, where and how the Cistercians settled in Pilis and second whether they cleared woodland in Pilis or exploited it any other way.

The Pilis monastery is the perfect example of how a Cistercian monastery can be secluded and in a central position at the same time. Situated on the eastern slopes of Pilis Mountain proper, it lay only eighteen kilometres from Esztergom, approximately the same distance from Buda, and even closer to Visegrád (*Fig. 55*). That is to say, for the White Monks to feel far from the "mundane world" they only had to look out the window and imagine that there was nothing beyond the trees, but if they wanted to be actors in Hungarian or international politics they could travel to one of the above places in less than one day. Indeed, the abbot of Pilis had many important commissions both from the pope and the king of Hungary.[47] Thus, the position of the monastery seems to confirm those new ideas about Cistercian settlements being not secluded at all, and strengthens the image one has about the special geographical position of Pilis Forest. Another interesting aspect of the monastery of Pilis is that it was founded in a location that had some kind of an earlier settlement on it.[48] It is not known whether this settlement was already deserted when the White Monks started their own

[35] As I mentioned in connection with the Royal Forest, this was set forth in Békefi 1891–1892, vol. 1, 124–126.

[36] Hervay 1991, 479. This book is the Hungarian translation of Lékai 1977, with an extension of the history of the Cistercians in Hungary, written by Hervay.

[37] Gerevich 1984, 16.

[38] Békefi 1891–1892, vol. 1, 317. "... in qua eciam corpus matris nostre honorabiliter requiescit." Takács 1994a.

[39] The following is largely based on Laszlovszky 1994 and Laszlovszky 1997.

[40] Romhányi 1993–1994, 182–183. There were two other unsuccessful foundations.

[41] A famous example of their influence is that which they had on the documents issued by the royal chancellery. Kubinyi 1975.

[42] Hervay 1985, 598.

[43] Békefi 1891–1892, vol. 1, 244, the charter is on pages 140–141.

[44] Békefi 1891–1892, vol. 1, 316–319.

[45] Hervay 1985, 598–599.

[46] Békefi 1891–1892, vol. 1, 122–124.

[47] Békefi 1891–1892, vol. 1, 128–168.

[48] This settlement may have been a Benedictine abbey. Gerevich 1984, 3–4. The existence of this Benedictine abbey was explicitly denied in Hervay 1984b, 147, however, he revised his own views the following year and wrote that it was not as yet possible to take sides: Hervay 1985, 597. What is relevant here is that there *was* some kind of an earlier settlement.

buildings or whether the locals were driven away.[49] In any case, it seems that the land that was given to the Cistercians had not been totally uncultivated, but the question still remains as to how much time passed (if at all) between the abandoning of the early settlement and the coming of the Cistercians. If it was more than a few decades, there must have been some regeneration of woodland, and consequently, the White Monks may have undertaken their own "recolonising" activity. As for other evidence of woodland clearance, no written sources exist which testify that the Cistercians were involved in such an activity. The only proof that the White Monks (that is their people) ever touched a tree is from 1468, when Abbot Hermann was summoned in one of those common cases of illegal woodcutting that I have analysed in Chapter 6.[50] That there was no more arable land around the monastery than what was strictly necessary is demonstrated by Zólyomi and Précsényi's pollen analysis, which described cereal pollens in very low numbers.[51] The Pilis Cistercian monastery had many different sources of income, mostly from tolls and from their estates. The most important of these was one third of the toll from one of Hungary's most significant ports in Bratislava. Another third of this toll went to Pannonhalma, the leading Benedictine abbey in Hungary, which strengthens the peculiar dependence of the Hungarian Cistercians on incomes from the secondary economy, in other words living very much like Benedictines. To the Bratislava toll belonged the toll from the town of Spišský Štvrtok (Sk; Co. Szepes). The monastery also had toll-incomes from Győr and Óvár (Co. Moson), and it possessed houses in Bratislava, which were let to tenants.[52] All this suggests that they did not need to live on the incomes produced by their immediate environment. But they still had to live somehow, that is they had to cultivate the land surrounding the monastery. Indeed, during the excavations of the monastery, several agricultural tools were found, some of them capable of woodland-clearance.[53] However, let us remember Romhányi's proposition that there may have been a shortage of lay brethren, without whom agricultural development was impossible on the one hand and pointless on the other. The monks also needed wine, which they obtained from the vineyards on their estates in Bratislava, Devín (Sk; Co. Pozsony), Bitóc (Co. Pilis), Buda, and Pest.[54] Nevertheless, they also produced wine around the monastery, because a pair of vine-pruning shears was also found during the excavations.[55] Ethnographers claim a connection between early medieval wine production and woodland, based on – yet again – mostly recent analogies, where a common way of growing grapevines was to use trees.[56]

Fig. 55. The ruins of the Pilis monastery, covered by bushes. To the right is Pilis peak, in the foreground lies the present village. May 2000

Pilis is one of the few Hungarian monasteries where we have clear evidence of industrial activity. The outer courtyard of the monastery was covered with a thick layer of slag, and the furnace itself was excavated. Ores were smelted into copper, iron, lead, and probably precious metals, supported by a sophisticated water system.[57] In addition to this, three sites were discovered a few kilometres south-east of the monastery which were most probably medieval ore mines.[58] We have no written evidence that these mines provided the monastery's resources, but it seems likely to have been so. The village of Kovácsi, which has a reference to blacksmiths in its name, was in the White Monks' possession in the mid-thirteenth century.[59] Kovácsi lay very close to the mines, thus, it cannot be excluded that this settlement was some kind of an extension of Cistercian metallurgy in the region. In addition to metallurgy, a highly productive brick and floor-tile workshop functioned in the monastery, which also must have required much firewood.[60] If the Cistercians were in fact involved in such industrial activities, then the most probable place for them to get wood was the woodlands in Pilis, and the most probable method for producing sufficient quantities was coppicing.

13. 2. The Paulines

"Humanae fugiens laudis Waacz, eminus auras
deserti petiit solus ad antra Pilis"

These are the opening lines of Chapter 2 of Gergely Gyöngyösi's *Vitae Fratrum*, written in the first half of the

[49] One of the myths about the Cistercians is that they liked to get rid of the locals to create deserted places where these were not available. As with many others stories, this has also been questioned. See for example Constable 1996, 118–119.

[50] Békefi 1891–1892, vol. 1, 421.

[51] Zólyomi and Précsényi 1985, 154.

[52] Békefi 1891–1892, vol. 1, 228–237.

[53] Gerevich 1984, figure 92.

[54] Hervay 1984b, 151–152.

[55] Gerevich 1984, figure 92; Holl 2000, 39.

[56] Égető 1980, 61, 68. For modern analogies: Égető 1983. There is, in fact, medieval pictorial evidence of this in Esztergom. On a surviving piece of the now destroyed twelfth-century throne from the cathedral of Esztergom we may observe a man standing on a huge tree with a sickle in his hand. The tree is overgrown with grapevines, and it is harvest-time, because the grapes are fully ripe. Takács 1994b, 235–236.

[57] Holl 2000, 46; *MRT-7*, 162.

[58] *MRT-7*, 23/14, 23/17, 23/18.

[59] *MRT-7*, 23/26.

[60] Holl 2000, 47–66.

sixteenth century. This distich sets the tone for the way the Paulines related to Pilis Forest.[61] The Pauline monks needed the Forest in order to escape the world much more than the Cistercians; for them, Pilis was a phenomenon described by Jacques Le Goff as the "desert-forest."[62] This term means that the role that had been played by the desert in early monasticism in the east was later adopted by woodland, since woods represented a different, unknown, and secluded world in medieval Europe. This approach was present in Gyöngyösi's language, the word *desertum* appears several times in the text, always as a reference to the woods of Pilis.[63]

In the Pauline tradition, not Waacz, but Özséb (Eusebius), a canon of Esztergom was venerated as the founder of the order.[64] But, as described in the *Vitae Fratrum*, the Pauline order was not born from nothing since there had been hermits living in the caves of the Pilis Mountains.[65] This tradition is supported by the results of archaeology. Many smaller caves can be found in the Pilis region and at least three of them – in the immediate vicinity of the monastery of Szentlélek – can be fairly safely associated with the Pauline order.[66] Özséb organised the so far sporadic eremitic life into an order, and founded a monastery in Szentkereszt.[67] The monks asked for the *regula* of Augustine, which they were refused in 1263, but granted in 1308 by Gentilis, the papal legate.[68] Around 1300, a new monastery was founded in Szentlőrinc (Co. Pilis), which assumed the leading position from Szentkereszt in the early fourteenth century.[69] In 1327, Pope John XXII ordered a census of the Pauline order, the result of which showed that in those years thirty monasteries existed with twenty, fifteen, or twelve monks, each.[70] By 1470, the Paulines possessed 51 monasteries. This number grew to 63 by 1526.[71] In spite of being a Hungarian order, the Paulines had foundations in Germany, Poland, Austria, Portugal, and Italy.

Szentkereszt was of crucial importance in the early history of the order. As we have already seen, this monastery was founded by Özséb around 1250. The Paulines themselves were called *fratres sancte crucis de heremo* even in the early fourteenth century.[72] The monastic buildings were situated east of today's Kesztölc at an altitude of about 280 metres, halfway between the Cistercian monastery and Esztergom. If we consider that Özséb was a priest in Esztergom, the location seems reasonably close to the town while still secluded enough.[73] In 1285, the royal army burnt down the building, which damage was compensated in 1289 by the donation of a deserted territory near the monastery by Ladislaus IV.[74] The fact that the leading position among the Pauline monasteries was later adopted by Szentlőrinc, which is today within the boundaries of Budapest, signified a distinct change. The Paulines preferred to have their administrative centre closer to Buda, one of the centres of the whole medieval Hungarian kingdom. The monastery of Szentkereszt was most probably destroyed by the Ottomans in 1526, and when searched for later, it was for some time identified with the ruins of the Cistercian monastery near Pilisszentkereszt.

Szentlélek is a few kilometres north-east of Szentkereszt. According to the *Vitae Fratrum*, Béla IV donated his hunting lodge to the monks in the "Pilis island," and Ladislaus IV repeated the donation in 1287: "in insula Pilis quendam locum Benedwelgye vocatum cum quodam palatio ibidem existenti."[75] Different interpretations were suggested concerning this mysterious "island." The expression appears four times in the *Vitae Fratrum*, in connection with different locations.[76] Thus, associating it with one particular place is not possible. Most probably Sándor V. Kovács has the right answer. He wrote in the appendix of the Hungarian translation of the *Vitae Fratrum* that *insula* is likely to refer to its isolation from the world.[77] This monastery was also destroyed by the Ottomans some time between 1526 and 1543.[78]

The third Pauline monastery in Pilis Forest was Szentlászló, the only one which is not visible today in any form, because a Baroque church was built upon its ruins. It was situated ca. six kilometres south of Visegrád, at an altitude of about 350 metres. The time of its foundation is given in a rather confused manner in the later sources. According to the *Inventarium*, Andrew I gave his hunting lodge to the Paulines, which is obviously impossible, since this king reigned between 1046

[61] Gyöngyösi 1988, caput 2. Naturally, this image, close to *devotio moderna*, was largely influenced by Gyöngyösi himself. Mályusz 1971, 264–266.

[62] Le Goff 1985.

[63] Besides the lines quoted above, for example "De transitu Waacz heremitae ad interiora deserti Pilisiensis" Gyöngyösi 1988, caput 2.

[64] There was an early settlement in Ürög (Co. Baranya), but this first settlement and its *regula* are somewhat marginal in the whole story of the order. Bencze and Szekér 1993, 7. For a general description of the development of the Pauline order and for a catalogue of their monasteries, see Romhányi 1996, 57–63.

[65] Gyöngyösi 1988, caput 5.

[66] *MRT-5* 18/5; *MRT-5* 18/6; *MRT-5* 18/7. The tradition of these three caves was still present in the fifteenth century, as attested by the words "prope speluncam triplicem" Gyöngyösi 1988, caput 8.

[67] Gyöngyösi 1988, caput 8.

[68] Bencze and Szekér 1993, 7–8.

[69] Bencze and Szekér 1993.

[70] Gyöngyösi 1988, caput 21.

[71] Hervay 1984a, 163.

[72] Mályusz 1971, 258.

[73] On the whole problem of the identification of the location of the monastery, see *MRT-5*, 235–236; Kovalovszki 1992a.

[74] Gyöngyösi 1988, caput 15.

[75] Gyöngyösi 1988, caput 14, 15. However, Béla IV's donation has several interpretations in the secondary literature. Györffy listed it under Pilisszentkereszt. Györffy 1956, 283. The *Documenta Artis Paulinorum* connected it with Szentlászló. *DAP*, vol. 2, 408. The *MRT* connected it, probably as a result of miscommunication among the writers, both to Szentlélek and Szentlászló. *MRT-5*, 297; *MRT-7*, 167. In fact, it is difficult to decide whether it can be connected to Szentlászló or Szentlélek. I am inclined to agree with Hervay, who, in the footnotes of the critical edition of the *Vitae Fratrum* argued that this donation is most likely to refer to Szentlélek. Gyöngyösi 1988, 209. (For an explanation also see here.)

[76] Gyöngyösi 1988, caput 10, 14, 15.

[77] Gyöngyösi 1983, 355. Despite the title, this book is the translation of the *Vitae Fratrum* and was, curiously enough, published before the publication of the original Latin version, with several mistranslations.

[78] *MRT-5*, 299; Lázár 1992.

and 1060.[79] The monastery first appeared in documents in 1291, when it was registered in a letter of Benedict, bishop of Veszprém.[80] Strangely enough, Gyöngyösi considered King Charles Robert to have been the founder of this monastery.[81] Györffy suggested that Charles Robert might have had the monastery rebuilt.[82] The monastery was the scene of a very important political meeting in 1308, between Cardinal Gentilis, papal legate, and Matthew Csák.[83] Szentlélek, as all the other monasteries in the Pilis region, was destroyed by the Ottomans around 1526.[84]

There were some common features in all the Pauline monasteries. First, they were all royal foundations,[85] even though the eremitic movement itself seems to have been spontaneous. Probably the kings recognised a trend "that sprang from real material conditions,"[86] and utilised the fact that the hermits already lived in Pilis Forest. The Forest, as I have argued above, was undergoing changes, and basically lacked a proper new function, which the monastic orders were then to provide successfully. Secondly, the locations of these monasteries were similar. All of them were situated within the woods, in the valleys between the mountains. This was a general trend among all Pauline settlements up until the Late Middle Ages.[87] This would suggest that the monks were looking for secluded places to hide from the world, but – just as in the case of the Cistercian monastery – in Pilis Forest one could never be *very* far from the most important towns of the kingdom. The monasteries avoided main trading routes – but not much. They typically lay along such routes (like the one that crossed the Forest) a few kilometres off, connected to it by a minor road. Lastly, an important feature, and a speciality of the Paulines, was to form small groups of monasteries. Such groups were not only present in Pilis, but also in Zemplén and in the southern regions of Bakony. Through the research of Károly Belényesi, we are now aware that members of these groups were closely connected, not only geographically but also economically and spiritually.[88]

Little is known about the Pauline economy in general, and hardly anything about the Pilis region. As all three monasteries were destroyed in the sixteenth century, the original archives do not survive. Indeed, what we know about the Paulines is almost solely dependent upon a humanist prior: Gyöngyösi. He compiled the chronicle of the order, and, as preparation for this, he did the "fieldwork" of visiting individual monasteries and copying charters found there. *His* picture of the Paulines is that of an order that did not care much for economic success apart from the needs of everyday life. Literacy was not important, either. It is true, regardless of Gyöngyösi, that the monasteries only kept the absolutely necessary written material: charters proving their rights to land.[89] We can read in the *Vitae Fratrum* that in the fourteenth century 10–20 brethren inhabited a monastery,[90] and although we know that they were accompanied by lay brethren,[91] the overall number of Pauline monks in the Pilis forest is unlikely to have exceeded one hundred and fifty in the Middle Ages.[92] These people, reasonable followers of Paul the Hermit, appear – unlike many other monastic orders – to really have lived a "self-sustaining" way of life.[93] Éva Knapp, when describing Pauline economy in Co. Baranya, arrived at the same conclusion.[94] The only special feature we can detect, and which will reappear in Bakony, is the Paulines' interest in mills. By the end of the fifteenth century, the monastery of Szentlászló had five mills in the region, two of which the monks probably built themselves.[95] Still, the success of the order in Pilis was ensured by the fact that the Pauline way of life coincided with what the royal house preferred to see in their Forest.

13. 3. Conclusions

Although founded on different bases, the Cistercians and the Paulines had a somewhat similar relation to Pilis Forest. The geographical position of the Forest made it possible for the two orders to achieve a status peculiar to this region, namely that Pilis was a place secluded enough to be appropriate for monastic orders, even for the followers of Paul the Hermit, but at the same time the monasteries were within walking distance of the most important lay and ecclesiastical centres of the kingdom. The Cistercian monastery was founded by Béla III near Buda and Visegrád and most probably supported by several archbishops of Esztergom with special French connections; in the case of the Pauline order, the kings only fostered a spontaneous process. Nevertheless, the fact that all four monasteries were royal foundations demonstrates the kings' interest in maintaining control over the monastic orders in the Forest.

---

[79] *DAP*, vol. 2, 408.

[80] Gyöngyösi 1988, caput 10.

[81] Gyöngyösi 1988, caput 24.

[82] Györffy 1956, 284.

[83] *CD*, vol. 8/7, 62.

[84] *MRT*-5, 167.

[85] In this respect similar to the early Cistercians.

[86] Mályusz 1971, 257.

[87] Hervay 1984a, 170–171.

[88] Belényesi 2004. My sincere thanks to the author and his wife for providing me with a copy of the unpublished text.

[89] Without these, anyone, especially landowners outside the local nobility, could lose his or her lands.

[90] Gyöngyösi 1988, caput 21.

[91] Mályusz 1971, 272.

[92] These numbers are reinforced by Romhányi's calculations, which were based upon the architectural analysis of the monastic buildings. Romhányi 1996, 61.

[93] Nevertheless, it is interesting that the Paulines had important incomes from the secondary economy, just like the Cistercians. Mills were of special importance to the monastery of Toronyalja (Co. Nógrád), in the Börzsöny Mountains. Miklós 1997. The same held true for some (but not all) monasteries in the Zemplén region, supplemented by possible trade in wine. Belényesi 2003. Mills can also be observed in the case of Szentlászló, *AnjOkm*, vol. 7, 70; *MRT-7*, 28/3; and most probably Szentlélek, *MRT-5*, 18/2.

[94] Knapp 1994.

[95] Dóka 1981, 10.

The Cistercian monastery in Pilis does not conform with the traditional image of the order as taking an active part in internal colonisation. However, fruit gardens and workshops for metal, brick, and floortile production can be supposed to have existed around the monastic complex. The Pauline order exploited the woods less than the Cistercians. What is true for the Cistercians, but especially in the case of the Paulines, is that further archaeological studies around the monasteries would largely extend and probably alter our perception of their relations towards woodland.

Chapter 14

Bakony Forest

> "The prince [Álmos, son of St. Ladislaus] constructed a monastery at Dömös, and asked the king [Coloman, his brother] to come for the consecration, and so it happened. After this the palatines accused the prince of preparing a trap to kill the king ... The indignant king wanted to capture him, but the most reverend bishops and other good magnates knowing this was false ... made peace between them ... And the king sent the prince in peace to hunt in Bakony."

Extract from the fourteenth-century chronicle composition, the year was ca. 1107.[1]

Bakony is much harder to define than Pilis. All Hungarians know that "it is those mountains north of Balaton," but anything more precise is rather challenging. Geographically speaking, only its southern boundary, Lake Balaton, can be easily identified. Even this, nonetheless, is questionable. Today Co. Veszprém extends down to the lake, but until the twentieth century a thick line north of Balaton belonged to Co. Zala. Bakony has two basic parts. The higher, northern part is separated from the lower, southern region by a wide valley, in which a major road – of possibly Roman origin – has carried passengers from Fehérvár to the west since at least the eleventh century. Veszprém, the largest city in the region guards the eastern end of this valley. Bakony is essentially limestone and dolomite, and the southern region merges seamlessly into spectacular basalt hills just north of Balaton. The highest peak in the northern part is Kőris Mountain (709 m) while in the south there is Kab Mountain (599 m). The whole territory is large: some sixty kilometres lie between the village of Dörgicse next to Balaton and the castle of Ugod in the north. One has to travel about half of this distance from Ajka in the west to Tés in the east. The name "Bakony" derives from somebody's personal name, and thus provides little information.[2]

Every discipline has its own definition of Bakony. In plant geography, and this is to be observed by the historical ecologist, *Vespremiense* and *Balatonicum* (that is, the upper and lower parts of Bakony as opposed to the basalt mountains north of Lake Balaton) are separated because their vegetations differ. The former, on the other hand, includes the Vértes Mountains, east of Bakony.[3] Economic geography defines Bakony basically as those lands lying above 200 metres, including all the northern shore of Lake Balaton.[4] Foresters differentiate between "Upper Bakony" and "Lower Bakony."[5] In historical studies, Bakony has not been considered a separate entity. Historians have mostly worked within Co. Veszprém if there was need for a unit larger than individual settlements.[6]

My definition cannot be any less arbitrary than the others. I include the region on the northern shore of Balaton, despite the vegetational differences, because historically it is inseparable from the rest of Bakony (except on a county basis). The eastern boundary is the valley of Mór (Co. Fejér), following the railway from Fehérvár to Komárom (Co. Komárom). On the north and west, I also conform to the 200-metre contour line.

Before we embark on the investigation of Bakony as a Forest, we should note that this region has been a favourite subject of Hungarian scholars for more than a century. The first, and in some ways still the best book about it was the polyhistor Flóris Rómer's work in 1860. This was followed shortly by Károly Eötvös' two-part description.[7] It was in the nineteenth century that Bakony gained its reputation as a hiding place for outlaws, a romantic Hungarian version of Sherwood Forest. Although Rómer disappointed all believers by stating that not a single bandit was to be seen in Bakony, the image of a godforsaken place proved to be stimulating to modern ethnographers.[8] Both general and specialist studies are abundant,[9] but they rarely have direct relevance upon the subject of this book.

14. 1. Early Years

As ever, little is known about the eleventh and twelfth centuries. The story of Álmos, the treacherous prince, tells us that in the early twelfth century the royal family was in charge both in Pilis (where Dömös is located) and Bakony, and that going from one to another in a "hunting tour" was normal

---

[1] *SRH*, vol. 1, 427–428. "Dux autem construxit monasterium de Demes rogavitque regem, ut ad consecrationem veniret, et sicut factum est. Postquam palatini accusabant ducem, quod insidias preparasset, ut interficeret regem . . . Rex autem iratus voluit capere eum, sed reverentissimi episcopi et alii boni principes scientes falsum esse . . . pacificaverunt eos . . . Dimisitque ducem in pace, ut in Bokon venaretur."

[2] *FNESz*, s. v. "Bakony." There was a settlement in 1193 with the name Bakony, which disappeared or changed its name by the fifteenth century. This Bakony was in the eastern part of the region, so much so that it lay in Co. Fejér, not in Co. Veszprém. Whether this place gave its name to the whole Forest cannot be decided. Mikos 1935, 161. The name, however, was not unusual in those times. The same charter included, as part of the perambulation of Újudvar (Co. Zala, south-west of Balaton), references to a "sylua Bakon."

[3] Soó 1965, 110.

[4] Wallner 1937, 49–52.

[5] Danszky 1973, vol. 1, 290.

[6] Éri 1969.

[7] Rómer 1860b; Eötvös 1909.

[8] And filmmakers. There have been a number of films about its outlaws, the latest (*Sobri*) hit the big screen in late 2002.

[9] See for example Vajkai 1959; Tálasi 1939. Specific works on woodland include Herkely 1941; Hegyi 1978. Presently ethnobotanical research is being conducted in the region. Kindly told me by Kata Frendl.

practice. (Although we know that for Prince Álmos hunting was some sort of a drug to take his mind off his misfortunes – he was originally intended as king instead of his slightly disabled elder brother, Coloman.)

An important difference between the early histories of Pilis and Bakony is that monasticism was present in Bakony from the eleventh century. First, probably St. Günther lived here as a hermit,[10] and legend has it that later St. Gerald, a Benedictine monk of Venetian origin, having met King Stephen on his way to the Holy Land, decided to stay in Hungary and live in Bakony. He went to a place called Bél, and stayed there for seven years.[11] It was at this place that St. Stephen founded a Benedictine monastery in the eleventh century.[12] This early monastery was a solitary venture in the woods. Lands in the immediate surroundings of Bél were conspicuously missing from its holdings. This becomes interesting when compared to a nearby early Benedictine monastery, Pannonhalma, which was given all the land adjacent to it in a rather wide circle. In Bakonybél, the king could have created a similar "estate" – he owned the land – but he preferred not to do so. One can only suspect the connection between this and the Forest, and it kindles the imagination why the same thing (lack of donations in the immediate neighbourhood) should have happened to Fulda in the eighth century in a certain *silva Bochonia*.[13]

Just like Pilis, Bakony was entirely part of a regular county (Veszprém) in the early Árpádian period. This county, as I have pointed out in Chapter 9, was mentioned in the 1009 charter of St. Stephen, in which the king defined the territories of the bishopric of Veszprém.[14] That there may still have been a sense of physical difference between Co. Veszprém and Bakony is demonstrated by the 1086 conscription of the holdings of the monastery of Bél, in which a certain village was defined as situated "next to" Bakony.[15] Being "next to" a Forest, as argued in connection with Pilis, may have carried a meaning of delimitation, of which I shall quote more examples in the following paragraphs.

14. 2. The Thirteenth Century

As opposed to earlier times, there is much more material for the thirteenth century.[16] In the first charter that described the Forest, we are already witness to a fully developed system. The year was 1210, and the abbot of Pannonhalma was complaining that the Forest-guards of Kenyeri had been troubling his people in three villages. The guards had demanded money (*pensio*) and the *comes' descensus* (the right for hospitality) in the villages.[17] The first *comes* we know by name was Lucas (1232).[18] Although from this very document it is clear that Lucas himself was acting in local matters – he had to assign the territory to be donated to the chapel of Szentkereszt[19] – two years later, Ladislaus of the powerful Kán kindred was *comes* of Bakony and judge royal at the same time.[20] The reader may remember from Pilis that it was also from around the mid-thirteenth century that people of higher status started to acquire the title of *comes*, and the position separated from everyday duties. In fact, the first *vicecomes* of Bakony appeared in 1240.[21] In 1258, Dénes, probably of the Tomaj kindred, son of a count palatine, was *comes* of Bakony and Szolnok.[22] Dénes also had his *vicecomes*, Saul.

Bakony Forest had its characteristic population. There was one settlement named after its Forest-guards (Ardó), and documents testify that guards also lived in Kenyeri, Berend, Németi, and Lövöld.[23] Hőgyész, west of Pápa, was the dwelling place of ferret-hunters. There were four settlements of blacksmiths (Kovácsi) in Bakony.[24] Other than these, hare-hunters (*leporariferi*) were recorded in Németi, and the keepers of hounds (*liciscarii*) in Nevegy.[25] The dog-keepers (*caniferi*) of Szentgál, although their fame began in the fourteenth century, were already present in the thirteenth.[26] The Forest-guards, despite the name, were mostly hunters. In the above charter of 1240, we can see them defending their position in the Forest.[27] It so happened that in 1232, King Andrew II donated to the chapel of Szentkereszt in Bakony as much land as "can be reached from the cemetery by two bowshots" in all directions.[28] For eight years apparently nothing happened. Then Béla IV, not exactly a follower of his father's will in other regards, ordered the *comes* of Bakony to act on the same orders as his predecessor. The rather reluctant *comes* Donát reported that he could not do this because on the southern side of the territory the royal cooks contradicted the decision, while "in the east we were no less hindered because of the rebellion of the Forest-guards, who claimed that the place was good for hunting

[10] *SRH*, vol. 2, 388.

[11] *SRH*, vol. 2, 472. Later Gerald became the bishop of Csanád.

[12] *SRH*, vol. 2, 422.

[13] Wickham 1994, 157.

[14] *DHA*, 52–53.

[15] *DHA*, 247–260. "iuxta Bokon" This charter has long been regarded a forgery, however, recently György Györffy (in *DHA*) convincingly argued that its first section – in which the settlement in question was described – was in fact genuine.

[16] Pesty 1880, vol. 1, 201–212; Kristó 1988, 260–262.

[17] *ÁÚO*, vol. 1, 100.

[18] *ÁÚO*, vol. 1, 293.

[19] Not to be confused with the settlement of Szentkereszt, in the north-eastern part of Bakony – *MRT-4*, 79/9.

[20] *ÁÚO*, vol. 6, 549.

[21] *ÁÚO*, vol. 2, 111.

[22] *ÁÚO*, vol. 2, 314–315; Wertner 1902. Szolnok was one of the most important counties of the age, unusual in having two geographically separate parts. The *comes* was (from 1263 to 1441) also the voivode of Transylvania. Kristó 1988, 434–440.

[23] *ÁÚO*, vol. 2, 112; *ÁÚO*, vol. 1, 100; *ÁÚO*, vol. 2, 314–315; *ÁÚO*, vol. 4, 48. In the Late Middle Ages, Lövöld broke down into two settlements, one of which was called *Ordó*lövöld. *Csánki*, vol. 3, 213.

[24] For these settlements, see Heckenast 1970, 89–132.

[25] *RA*, vol. 2/2–3, 130, 231.

[26] *ÁÚO*, vol. 4, 111.

[27] *ÁÚO*, vol. 2, 111–113.

[28] *ÁÚO*, vol. 1, 293. "a cimyterio potest bis saggita attingere"

because it was a gathering place for game."[29] Donát had to call all parties together to finally reach a solution. The guards succeeded in having the lands of the chapel assigned elsewhere. Besides hunting game, it was also their responsibility that no one grubbed out trees in the Forest. This in practice meant that they had to watch the boundary of the Forest and make sure that the neighbouring settlements did not expand over it, as was the case in 1258, when at the end of a court-case the people of Vinye had to swear to the Forest-guards of Kenyeri they would not touch any living tree in Bakony Forest and they had to pay some fine for having done so in the past.[30] In 1244, King Béla IV ordered the *comes* of Bakony (thus the guards) to let the nuns of Veszprémvölgy have trees "that were necessary for the reparation of their house or the rebuilding of their church."[31]

The Forest was very much visible on the ground. As opposed to Pilis, where we could see the boundary of the Forest in only one perambulation, there survive a number of such cases from Bakony. In 1233, for example, an intriguing court-case was related by Bartholomeus, Bishop of Veszprém: Lucas, *comes* of Bakony, went to the king and reported that the people of Billege – a neighbouring village – "were not satisfied with their own rights, crossed the boundaries …, extended themselves into the royal rights, that is, they abused royal lands and woods, above the donation of his majesty, the King." Then followed a perambulation, in which part of the boundary was between Billege and – very boldly – the King.[32] Obviously they were talking about Bakony Forest. We also learn that there were "lands and woods" in Bakony, which description visualises a landscape of different land-uses with larger chunks of woodland. In other words, although Bakony was termed *silva* it was not merely an oversized wood. What made it what it was was its ownership. Once the king gave away a piece of land, it no longer formed part of *silva Bakon*, regardless of its wooded or non-wooded nature.

In the Albeus conscription of the properties of the abbey of Pannonhalma, quoted in Chapter 9, we could already see that the village of Vinye had common boundaries with Fenyőfő, Kenyeri, Csesznek, and Bakony Forest.[33] In 1240, when the lands of the chapel of Szentkereszt were defined after the rebellion of the Forest-guards, part of the boundary was held in common with "royal land."[34] Similarly to the case of Billege, this "royal land" must have referred to Bakony Forest. In 1239, a perambulation took place in Ajka (western side of Bakony), in which the eastern boundary of the given territory was "next to Bakony."[35] Fifty-three years later, the same parties (albeit the next generation) divided their holdings in Ajka. Several boundary signs were mentioned that separated Ajka from *silva Bakon*.[36] Furthermore, when walking through the countryside, the officers recorded a wood that they crossed. Although both Bakony and this anonymous wood were termed *silva*, they did not represent the same level of abstraction. The latter was just a simple wood belonging to a settlement. Bakony was more similar to the settlements themselves in the respect that they had common boundaries. The writer of the charter may have shared my thoughts, as he, when mentioning *silva Bakon* again, added a little explanatory note: "silva Bokon, silva videlicet regalis" – "Bakony Forest, that is a Royal Forest." There was no separate name to differentiate between a simple wood and a Royal Forest, however, the two *silue* did not even resemble each other.

In Pilis, there was a settlement (Pomáz) that was once defined as being "below" the Forest. I wrote at that time that this might be another way of catching a glimpse of the physical boundaries of Royal Forests and that there were more such cases in Bakony. The earliest of these is the one I have quoted from the 1086 conscription from Bél. Probably the most interesting is the interpolated fourteenth-century copy of the 1055 foundation charter from the Tihany Abbey.[37] This included a certain Tósok "over" (*ultra*) Bakony, and one Csősz "in" Bakony. There is not much difference in the geographical position of the two settlements, a simple look at the map would not suggest one being within as opposed to the other being over; in fact, they are both in a marginal position. Whoever produced this charter had a clear idea of what was to be understood by Bakony. Other settlements "over" Bakony were Apáti and Szőlős[38] (both in the fourteenth century). A village could also be "under" (*sub*, *subter*) Bakony as was the case with Vinye, Szentkereszt and Uzsol.[39]

Fundamental changes came with the 1270s. First and foremost, the mighty Csák kindred got hold of the title of *comes*. They were certainly not keepers, and acquired the title as another position that brought dignity and money. Why they were interested in Bakony is partly because the original possessions of the kindred were in neighbouring Vértes.[40] The first Csák to appear in the history of Bakony was ban Csák (he had the name of the forefather of the kindred, a usual feature) who was called no less than "perpetual *comes* of Bakony" in 1270.[41] The position of the *comes* of Bakony was, as it seems,

---

[29] "ex parte vero orientali nichilominus impediti fuimus propter rebellionem custodum silvarum asserencium locum illum aptum ad venacionem, eo quod esset ibi status ferarum" Note the resemblance to the types of woodland in estimation guidelines.

[30] *PRT*, vol. 2, 300–301. Wenzel (*ÁÚO*, vol. 2, 314–315) misread the name Cugeri (later Kenyeri) as Churgi, which lead to the assumption that there was a place called Csurgi (the alleged pronunciation of the misread name) in the Middle Ages. This has been countlessly repeated since, mostly because it got into the (otherwise high quality) historical geography of Dezső Csánki.

[31] *CD*, vol. 7/1, 362. "ligna in silva Bokon quae pro reparatione curiae vel ad reaedificationem ecclesiae necessaria fuerint"

[32] *CD*, vol. 7/1, 235. ". . . populi de villa Billique nolentes esse contenti iure suo metas . . . transeuntes, ad iura regalia se extenderunt, usurpantes scilicet terras et silvas regales, ultra donationem regiae maiestatis."

[33] *ÁÚO*, vol. 2, 10.

[34] *ÁÚO*, vol. 2, 112. "terra regia"

[35] *HO*, vol. 3, 4. "iuxta Bochon"

[36] *HO*, vol. 1, 88–90. For example "due mete que separant a silva Bokon prenotata" or "venit usque ad terminos silue Bokon."

[37] *DHA*, 153–156.

[38] Df 200 784; *Zichy*, vol. 2, 190.

[39] *ÁÚO*, vol. 2, 314–315; *ÁÚO*, vol. 7, 371; *CD*, vol. 7/1, 333.

[40] Kristó 1986, 21–24.

[41] *HO*, vol. 3, 17–18. "Chak banus filius magistri Demetrii perpetuus comes de Bokon" This rather enigmatic "perpetuus comes" could be connected to the fact that three members of the Csák kindred were appointed *comes* of Bakony, however, since neither of the other two was called "perpetuus," it may simply have meant that Csák was *comes* of Bakony for his lifetime.

"inherited" within the Csák kindred, although not necessarily upon the death of the previous title-holder.[42] In 1280, a certain Stephen, from another branch of the Csák kindred, was the *comes*.[43] The above ban Csák probably did not have any children, and we do not know when he died, therefore Stephen may have "inherited" the title from him.[44] However, already in the next year, Demeter, the nephew of the above-mentioned Csák, was the *comes*.[45] The solution to this apparently abrupt, although not at all unusual change may lie in the nature of what Bakony Forest meant at that time. In the 1280 charter, Stephen and his brother Peter were donated the castle of Kobersdorf (Au; Co. Sopron) with all its incomes. Although the charter does not describe an exchange – otherwise common – it might appear as though Stephen had given away a *honor* – Bakony Forest, for the castle in Co. Sopron.

The Csák did not merely become *comites*. Beside the position, they were also donated land in Bakony. In 1270, King Stephen V donated four villages: Lövöld, Rendek, Esztergár, and Fenyőfő to ban Csák. Lövöld, as we have seen above, was the home of the Forest-guards. Four years after the donation, in 1274, the guards were still living there, as we know from a court-case that was administered in front of the chapter of Veszprém, in which three guards from Lövöld accused a certain Pous and Laurence of killing a relative of theirs.[46] Whom did these guards belong to? Other *custodes silvarum* ceased to be guards as such, when in 1273 they were given noble status by Ladislaus IV. Their privilege charter also relates that they were elevated to noble status because they had been fighting for the king below the castle of Győr, just north of Bakony.[47] The hunters of Németi were also ennobled in 1275.[48] In 1279, one of these hunters was called *comes leporariferorum* and was donated some land.[49] We know that in those days everything was uncertain in the country, and this particular case is a fair demonstration of that. If Heym, as the hunter was called, was elevated "from the status and condition of hare-hunters with all his lands"[50] to the ranks of noblemen, why was he still *comes* of the same hunters four years later? What was, in any case, a *comes leporariferorum*, never heard of before or after?

Perhaps it is useful to take a closer look at the reign of Ladislaus IV to put the events of Bakony Forest into a larger context.[51] The king, nicknamed "the Cuman" after his mother, was born in 1262. When his father died and he was crowned, he was only ten years old. For almost a decade, but truly for most of his (short) life he was a puppet of the powerful and power-hungry magnates of the kingdom – the Csák versus the Kőszeg and Gutkeled. Ladislaus was not necessarily weak; it is more that he was not strong enough. It is interesting to note that the most representative member of the Gutkeled kindred at that time was Joachim, several times *comes* of Pilis. This is perhaps another reason why the Csák wanted to be *comites* of Bakony: to achieve equal positions at court. Although there is evidence that Ladislaus ennobled Forest-guards elsewhere in the country as well,[52] Bakony seems to have played a special role in his troubled life. In 1278, in a donation charter to the nunnery of Veszprémvölgy, the king described how he had been going "through Veszprém to Bakony Forest for the consolation of hunting."[53] The aforementioned Heym received his lands from Ladislaus in 1279 because he had been performing faithful service in the same "hunting of consolation."[54] Ladislaus IV resembled his twelfth-century relative, prince Álmos; hunting in Bakony was probably a real consolation to him, and he did not hesitate to raise his favourite Forest-guards and hunters to more prestigious social positions. Irrespective of how it may have been with Ladislaus, Heym began one of the success stories of late medieval Hungary – the Himfi kindred.

There was another important element in the history of these decades, and in fact in the whole history of late medieval Bakony and Hungary alike: castles.[55] Not much have been said about them thus far, because in Pilis they were absent all through the Middle Ages. Bakony, however, was more like the rest of the kingdom. In Hungary, only the king had the right to build castles until the middle of the thirteenth century. Then, after the Mongol invasion of 1241, the king encouraged the construction of castles, so that people could protect themselves in case of another attack. At the same time, and also earlier, huge tracts of royal land were given away to laymen and the church. Landowners tried to concentrate their holdings around their – already existing or would-be – castles. This way, compact units of land of formerly unknown size came into the possession of one person. The king ceased to be the greatest landowner, the previous servant-based manorial economic system collapsed, new social layers emerged, and castles took over the role of *politicum*.[56] Owning a castle meant much more than the ability to defend one's self. It was

[42] There were never any hereditary titles (feudalism, if you like) in Hungary in the Middle Ages.

[43] *ÁÚO*, vol. 12, 288–289.

[44] We know that Csák was dead by 1309, when his widow was mentioned. *AnjOkl*, vol. 2, 242. On the Csák kindred, see Karácsonyi 1900–1901, vol. 1, 313–365; Engel 2001b. Stephen apparently did not follow ban Csák straight into the position of *comes*. There was a certain Gule in 1279, of whom we know nothing but his name. *RA*, vol. 2/3, 231. In accordance with this, Stephen was *not* called *comes* in a donation charter of the same year. *RA*, vol. 2/3, 253. In this charter it was reported that Stephen had asked the king for a deserted settlement in Bakony.

[45] *ÁÚO*, vol. 9, 295.

[46] *ÁÚO*, vol. 4, 48–49.

[47] *HO*, vol. 8, 151. On the problem of the origins of the Hungarian nobility, and what it meant to be a *serviens regis*, which the hunters became, see Fügedi 1998, 35–42.

[48] *CD*, vol. 5/2, 252–254.

[49] *RA*, vol. 2/2–3, 231.

[50] "de statu et conditione leporariferorum cum omnibus terris suis"

[51] Kristó 1994a, 5–43.

[52] Podhorany (Sk; Co. Sáros). 1283: *HOkl*, 99–100.

[53] *CD*, vol. 5/2, 256–259. "per Wesprimium ad syluam Bokon ad solatia venationis . . . transitum faceremus"

[54] *HOkl*, 80. "in solacionis venacionibus"

[55] For general information, see Fügedi 1977; Engel 1977; Engel 1982.

[56] Engel 1982, 881.

political capital. In fact, the political history of late-medieval Hungary is the story of who owned or held which castles at what times. I shall return to this problem when discussing the fourteenth century.

The Csák possessed the first castle in Bakony. Ugod (northern Bakony) was an ancient hereditary settlement of the kindred, where they built a castle some time between 1287 and 1301,[57] that is, after they had acquired the four villages mentioned above. With this, the royal house started to lose control over Bakony Forest and a process was launched in which the Royal Forest was to be transformed into a private Forest. In 1281, King Ladislaus IV repeated the 1270 donation of the four villages, this time to Demeter, the nephew of ban Csák.[58]

14. 3. The Late Middle Ages

Starting from the years following 1281, we lose track of the *comites* of Bakony for more than three decades. Demeter, the last *comes* we know of from the thirteenth century was reported dead in 1287.[59] He had two sons: Maurice and Csák. The former, after an unsuccessful marriage, decided to become a monk, whereas the latter died without heir in 1309. Ugod castle thus came into the possession of their closest relative, Peter, who had a great-grandfather in common with Demeter. He sold the castle to a powerful nobleman of the time, John of Kőszeg.[60] What is interesting here is the absence of the *comes* from the sources from 1281 to 1316. The easiest solution would be to attribute this to a lack of sources, which, of course, is always an option. It is, nonetheless, unlikely to be the case, as there is no comparable gap in the three-hundred-year history of Bakony Forest. It is also possible that the *comitatus*, as such, ceased to exist at the end of the thirteenth century. Although the Ugod branch of the Csák kindred was not quite as authoritative and powerful as the Trencsén, we still know of one marriage with a Kőszeg daughter, and one with an Aba, that is, there was at least some orientation towards the "petty kings."[61] We also suspect that the Csák had another castle beside Ugod, Hölgykő, built around 1300. This castle appeared in the sources as late as 1310, when it was royal property, yet it stood in the territory of Lövöld, a village the reader may remember from the 1270 donation charter. Thus, although we have no direct evidence, Hölgykő may have been built by the Csák and taken over by the king later.[62] In sum, once the Ugod Csák showed disinterest in co-operation with the royal house, which we can merely suppose, it would have made little sense for the king to appoint the *comes* of Bakony.

In 1301, Charles Robert was elected king of Hungary, but it was not until 1323 that he controlled the kingdom.[63] About half of the castles became royal possessions again, including Ugod, which the king confiscated from the Kőszeg ("notorious infidels" of his, as he called them)[64] and donated to a faithful Czech knight, Čenik, in 1328. After this, Ugod was never royal property again, and lost connection to the position of the *comes* of Bakony. The other royal castle in Bakony in the early fourteenth century was Hölgykő. From 1310 to 1325 we know of three castellans (*castellanus* – the leader of a castle appointed by the owner) by their names. None of them, however, were *comites* of Bakony. The only *comes* in the sources was a certain Dominic, who was the man of one of the castellans of Ugod in 1316.[65] This sounds rather complicated. Why would the *comes* of Bakony be subjugated to a castellan? On the other hand, the Forest – as a piece of land – was less sensitive to whatever was going on in royal circles. I have already quoted the perambulation of Ajka from 1292, where we read about the "silva regalis," which acquires extra meaning in this context: most probably there was no Co. Bakony at that time, yet people knew where Bakony Forest started. The locals still remembered what was theirs and what was the king's.

A charter testifies that Ladislaus Himfi was the *comes* of Bakony in 1321. Ladislaus was the son of Heym, the ennobled hunter from a few paragraphs before. Appointing him turned out to be the beginning of an ambitious programme of Charles Robert to settle the situation around Bakony Forest. As a next step, the king decided to resolve the problem of royal populations in the region. There had been, in the thirteenth century, a complicated network of Forest-guards, hunters, smiths, and more. We have also seen that by the end of that century this network was no longer functioning, but there was nothing to replace it. Therefore Charles Robert chose three villages where there were already guards and hunters (Németi, Csepel, and Szentgál) to be the dwelling place of all such people in Bakony, and exempted them from all jurisdiction but his own.[66] Indirect evidence supports the idea that all remaining hunters and guards moved to these places: in 1351, Ordólövöld was reported deserted.[67] The king's actions, of course, caused trouble. These three villages, like most others in Bakony, were not his. In 1319, they belonged to Thomas, son of Lőrinte, as we know from a case of ordinary *acta potentiaria*.[68] In 1328, Charles Robert

[57] Engel 1996, 449–450, with references to earlier literature. There were two other castles in the immediate region built before 1300: Csesznek and Essegvár. They do confirm the general image of the weakening of royal power, but had little to do with Bakony county as an adminisitrative unit, at least not in the thirteenth century.

[58] *ÁÚO*, vol. 9, 295.

[59] *CD*, vol. 7/2, 117.

[60] *CD*, vol. 8/3, 579–590; Fügedi 1977, 207.

[61] "Petty king" is a modern expression created to describe the most powerful magnates of around 1300. It is misleading in the sense that the magnates themselves never wanted to be kings.

[62] The idea is Pál Engel's 1996, 328.

[63] Engel 1988.

[64] *CD*, vol. 8/3, 583.

[65] Dl 49 899. For the *comites* of Bakony from the fourteenth century on, see Engel 1996, 101–102.

[66] Dl 24 891. The hunters, in theory, were exempt from all taxation, but this was not kept to for example in 1488, when they had to pay one florin per household to King Matthias' warfare expenses like all other inhabitants of the county. Solymosi 1984. For the history of the hunters of Szentgál, Németi and Csepel, see Magyar 1984. Note, however, that although the study is a useful tool to decipher the chain of events, I disagree with the author on most major issues of interpretation.

[67] *MEO*, vol. 1, no. 72.

[68] *AnjOkm*, vol. 1, 525–526.

had to issue another charter for the people of Németi, Csepel, and Szentgál, this time obviously against the above Thomas. In this document the king was a bit more verbose than in the first one. He related that these hunters had originally been *butharii*,[69] and were "graciously translated from this condition to the service of royal hunting" in some distant past.[70] He found them still necessary for royal hunting, so he withdrew all donations that he had ever made from their lands and possessions. This, he added, especially applied to Thomas, son of Lőrinte. With all the rhetoric, the message was clear. In 1328, the king was already in the position to simply inform Thomas: "I have changed my mind."

Having clarified who was who and with what responsibilities in the Forest, the king wanted to know what exactly the Forest was. In 1325, he ordered the chapter of Fehérvár to carry out a perambulation of Bakony Forest.[71] This – as far as I know – was the only case when a perambulation was mentioned in direct connection with a Royal Forest: not the perambulation of a neighbouring settlement, but of the Forest itself. It is, in fact, our only proof that such a thing was, at least theoretically, possible. I do not need to stress how much it would mean to be able to read this perambulation. However, not only is the original mandate unavailable, but we do not know whether the perambulation itself ever took place or not.[72] In the very same year the castellan of Hölgykő and the *comes* of Bakony were the same person: the aforementioned Ladislaus Himfi, as if Hölgykő had replaced Ugod. Next year, Charles Robert managed to take over another castle in the neighbourhood: Csesznek. This had been the property of the Trencsén branch of the Csák kindred since 1309, and after the death of the powerful Matthew in 1321, the king presumably had little trouble in making Matthew's relatives exchange four castles – including Csesznek – for two in Co. Tolna, "exchange" being a rather polite term for this shamelessly aggressive royal transaction.[73] Csesznek castle was also governed by Ladislaus Himfi. Thus, from 1326 we witness a situation far clearer than at any time since the mid-thirteenth century. Administratively, Bakony Forest included two castles: Hölgykő and Csesznek with their settlements, ruled by the *comes* of Bakony, and three special settlements for royal hunters. The boundaries of the Forest were most probably clarified through a perambulation. We can only regret not knowing which way the officials and neighbours were going when walking on the boundary of the Forest.

At this point we have to return to the problem of castles in order to understand what it really meant that the *comes* of Bakony was castellan of Hölgykő and Csesznek. As I have mentioned before, in the fourteenth century, castles were the most important elements in the political history of Hungary. A "castle" equalled not only the fortress, but also the several settlements that belonged to it (*pertinentie*, as they were called). There were two types: private and royal castles. A private castle was a nobleman's own possession, with which he could do whatever he pleased. These, quite naturally, were governed by castellans appointed by the owner. For a long time royal castles were imagined the same way, save in the person of the owner. However, the research of Erik Fügedi and Pál Engel has demonstrated that this was not the case.[74] Now we know that the majority of the royal castles were not managed by the king himself (that is by his castellans) but belonged to a *honor regni*. For example the ban of Slavonia (a major *honor*) held some fifteen castles in the middle of the fifteenth century. "Holding" these castles meant that he appointed the castellans and all the income from the castles was his. Medieval people interpreted this not as an office but as a form of ownership. The only – albeit major – practical difference between one's own castle and a royal castle held as *honor* was that the former could be inherited while the latter was given only for as long as the king saw fit.[75] This was not just a theory: the Angevins very often took away and gave out *honores*, and the reasons are rarely obvious to the modern researcher. For as long as the system was functioning, the king did not have to give reasons: his will was a good enough reason. Besides major *honores*, like the above ban of Slavonia, there were many lesser ones, mostly connected to counties. Almost every *comitatus* (in this sense *comes*-ness) was a *honor*, which included judicial power over the inhabitants and control over the royal castles of that county. It was also customary that high officials of the kingdom held – *pro honore* – several counties together. Some offices tended to be connected always to the same counties (judge royal), others always to different ones (count palatine).[76]

It is then obvious that Bakony Forest could not exist in these new circumstances without royal castles. The first of these was Hölgykő, joined by Csesznek in 1326.[77] From 1335, the *comes* of Bakony was that of Co. Győr as well. This person was rarely the castellan of the castles in person, more frequently he appointed some of his men. The counties themselves were often managed by his *vicecomites*. This system was stable until around 1364, when King Louis the Great, son of Charles Robert, decided to demolish Hölgykő and donate its lands to a newly founded monastery of the Carthusian order.[78] The foundation charter of the monastery

[69] Most probably a kind of cooper.

[70] Dl 24 891. "de dicta conditione . . . ad servitia regali venationi placide translatos" This, by the way, is the only evidence we have as to how certain people became royal hunters in the Árpádian period.

[71] *MEO*, vol. 1, no. 65.

[72] In *MEO*, only a Hungarian summary of the document was included. From this, it is clear that the original was not a perambulation, but only the king's mandate to carry out one. Tagányi reported having seen the charter in the Himfi archives. Most of the material from these archives are now kept in the National Archives of Hungary, but the first few dozen documents are not there. Pál Engel told me that these charters were (most probably) in a certain person's private collection in Vienna, and she would not let anybody have a look at them. In other words, we cannot even tell whether the original mandate is still extant. An unlikely document as it may be, one must note that the Himfi archives are special in the sense that they preserved many rare document types. While at some point in time in most collections the owners got rid of such useless bits as account books or mandates, the Himfi, for some reason, apparently preserved every scrap of paper.

[73] *VeszpReg*, 76.

[74] Fügedi 1977; Engel 1981; Engel 1982.

[75] "durante beneplacito nostro"

[76] Engel 1996, 1, 6.

[77] Engel 1996, 101–102.

[78] *CD*, vol. 9/5, 239–244.

also revealed that only three settlements belonged to Hölgykő: Lövöld, Ordólövöld, and Nyúl (in Co. Győr). That Bakony Forest was a *honor* is clear from the sources. It was thus termed by King Louis in 1379, and by Queen Elisabeth five years later.[79] The queen's wording was so characteristic that it is worth quoting: "to our faithful man, master Akus, among other *honores* the *comes* of Bakony."[80] In 1392, King Sigismund donated Csesznek to John and Nicholas of Gara,[81] thus Co. Bakony would have ceased to exist. We hear about it for a short period at the beginning of the fifteenth century, when Sigismund got hold of the castle of Essegvár (west of Veszprém) upon the pretext that its previous owner (John, son of the same Thomas that we have already met in conflict with the hunters in the three privileged villages) died without heirs (although this was not exactly true).[82] Even Essegvár was a private castle again by 1439 (*Fig. 56*).

Beside the administrative version of the story, there was always Bakony Forest as a physical entity. Sources suggest that castles and *honor* apart, the everyday functioning of the Forest remained unchanged. First, it still had visible boundaries, and sometimes in the same places as a hundred years before. In 1349, thirty *iugera* of arable were perambulated in Ajka, and the boundary ended next to Bakony.[83] In 1372, a rather complicated perambulation took place in Haraszti.[84] Here, quite unlike in most other cases, it was almost impossible to carry out the actual walk because the parties constantly disagreed with each other. In an almost parody-like manner, where one said there was an earthen boundary, the other claimed to see a mere pile of earth; where one tried to make the officials believe that a valley was a ditch, the other insisted that a valley could never be a ditch, and that this particular valley was in any case an ancient road, and so on. Importantly for us, at one place they found three earthen boundaries, one of which was supposed to be that of *silva Bokon*.

The *comes* and his people – the powerful *comes*, of course, substituted by his *vicecomites* – were actors in much the same business as in the thirteenth century. In 1337, Charles Robert ordered the *comes* not to claim any money from the people of the abbot of Bél, providing that they remained within the boundaries of their own settlements.[85] In the fifteenth century, a multiple structure may be observed. On the one hand, the kings were still distributing privileges in

Fig. 56. Castles in Bakony.

[79] Dl 101 919; *PRT*, vol. 8, 416.

[80] "fideli suo magistro Akus, inter alios honores comiti de Bakon"

[81] *HO*, vol. 7, 428.

[82] Engel 1996, 310; Engel 2001b, s. v. "Lőrinte nem Essegvári" The interpretation of the events, as always, depended on who was considered an "heir."

[83] *HO*, vol. 3, 160.

[84] *PRT*, vol. 8, 387–389.

[85] *AnjOkm*, vol. 3, 338–339.

Bakony Forest, for example in 1406 or 1453.[86] In the latter case, King Ladislaus V acted as if nothing had changed since the mid-thirteenth century. He called Bakony "our Royal Forest," where all "Forest-guards so far appointed and to be appointed in the future" were to follow his orders.[87] On the other hand, the Gara family, owners of Csesznek castle since 1392, supposed that with the last royal castle of Bakony, they inherited the Forest as well. In 1401 and 1417, Nicholas of Gara was issuing charters which were very similar to those we have already encountered, except that it was not the king but Nicholas who told the *comes* and the guards what to do.[88] The most representative example of what had changed since the time of the Árpáds is the 1450 foundation charter of the Pauline monastery of Porva.[89] Ladislaus of Gara, son of the above Nicholas, and count palatine like his father and grandfather before him, described why and how he decided to establish a monastery. There had already been, Ladislaus explained, a little chapel at the place where he and his ancestors had hunted, together with a stone house and a *curia* that had been built to serve the hunters. It was his father's idea to build a monastery there, but he died before he could carry out his plan. For the salvation of his father's soul and that of all his relatives, Ladislaus at last founded the monastery. He donated to the future monks a village, three mills, a house in the town of Pápa, two ploughs of land around Porva, and the right to use Bakony Forest. This charter, *mutatis mutandis*, could be a thirteenth century royal document. Every element is there. Hunting, the hunting lodge that was transformed into a monastery, the donation of two ploughs of land in the immediate neighbourhood, and a settlement further away (that is, providing the monastery with arable and income), and finally extensive rights to use the Forest. It appears as though the ability to do such things had "descended" from a royal level to that of the magnates, even if the Gara were one of the most highly ranked families of the age.[90] Ladislaus of Gara, when talking about Bakony, added that he had "perpetuum comitatum" of the Forest. In his mind, then, it was clear that the Forest belonged to him. However, as we have seen above, this was probably not what the king thought. Furthermore, there was one other party with similar claims on the position. In 1470, John of Essegvár came before the magistrate of Co. Veszprém and accused the royal hunters of Szentgál, Horhi, and Németi of abusing the "lands and places, meadows, woods, and incomes of the *honor* or *comitatus* that is commonly called Bakonispansagh," which belonged to him together with the castle of Essegvár.[91] This castle was the one connected to the "refounding" of Bakony Forest, which, nonetheless, was already in private hands in 1439. The words are unmistakable again. Bakony was a *honor*, connected to Essegvár, so John argued. What Ladislaus of Gara, or the king for that matter, would have had to say about this we cannot tell. What is for certain is that all three parties could not be right at the same time.

Finally, we must return to the royal hunters of the three privileged villages. Many think that it was only the territory of their villages – that is the southern side of Bakony – that was Bakony Forest proper in the Late Middle Ages. However, I have already discussed a charter of 1372 in which the boundary of Bakony Forest was reported to be next to Haraszti, which is in the north of the region.[92] The social position of these hunters was so unusual and archaic that they had to have it confirmed by almost every king, and they were in constant conflict with neighbouring people who were reluctant to accept it. In the fourteenth and fifteenth centuries regular inhabitants of the kingdom (not townsfolk or churchmen) were either nobles or peasants. The hunters of Bakony were somewhere in between, and it was mostly the point *from where* they were observed that determined to which side they seemed closer. For John of Essegvár, they were peasants, but King Sigismund addressed them as his "faithful hunting people or servants," and King Ladislaus V as his "faithful, circumspect, and prudent men, hunters, and conditionals."[93] The chapter of Győr tried to remain neutral, and referred to the "people" ("populi") of Horhi and Németi.[94] The territory of the hunters' competence was at least as big as the actual territory of their villages, that is, the central part of southern Bakony. There is some evidence to suggest that they operated in northern Bakony, as well. In 1439, the abbot of Bakonybél complained that his men killed two deer around the monastery, but the hunters took the animals from them.[95] It is worth noting that the problem of the abbot was not that the royal hunters were where they should not have been. Instead, he was referring to a privilege given to the monastery by St. Stephen, which assured the monks of half of the animals killed by anyone between four rocks: Kerteskő, Hegyeskő, Fehérkő, and Odvaskő (all of which are still on the map). The deer, continued the abbot, were brought down near Fehérkő, thus taking them away infringed on the monastery's rights.[96] At this point, one might speculate about hunting rights in the Forest. As I have outlined in connection with *silva sub venatione* in Chapter 6, hunting is supposed to have been restricted only on

[86] *HO*, vol. 2, 315–316.

[87] "silva nostra regalis" "silue nostre custodibus constitutis et constituendis"

[88] *ZsOkl*, vol. 2/1, 124–125; *ZsOkl*, vol. 6, 157.

[89] Dl 14 424. There is no sign that the charter ever had a seal, and the handwriting itself is of low quality, therefore the document appears more like a preliminary version than the final *privilegium*. This was observed by Péter Banyó.

[90] This was observed in Britain as well, where "a Forest was the supreme status symbol of the king and was aspired to by earls and a few princes of the Church." Rackham 1986, 131.

[91] Dl 102 568. "terrarum ac locorum, fenilium et pratorum necnon silvarum et proventuum honoris seu comitatus quod communiter Bakonispansagh nuncuparetur"

[92] *PRT*, vol. 8, 387–389.

[93] Dl 102 568. "providos . . . in dicta Zenthgaal, ac Horhy et Nemethy commorantes" *Providus* was the usual term in this period to denote peasants; Köblös 1997, 51–52; Dl 66 273. "fidelibus nostris, circumspectis et prudentibus viris venatoribus et conditionariis nostris" *Circumspectus* and *prudens* were the usual titles given to burghers.

[94] *PRT*, vol. 8, 508–509. Some time in the late fourteenth century, Horhi replaced Csepel. Magyar 1984, 109. That the chapter was talking about – at least partly – the hunters is attested by the mention in it of George of Hymhaz, direct descendant of Heym, the thirteenth century ennobled Forest-guard.

[95] *PRT*, vol. 8, 508–509.

[96] There is no proof that this right was in fact given to Bakonybél by St. Stephen. The earliest document to contain it is dated to 1037, but that is a mid-thirteenth-century forgery. *DHA*, 113–119. Abbot Michael, nonetheless, believed that the right had originated in the eleventh century.

a territorial basis in medieval Hungary, so that it was in close connection with the ownership of the land. With this in mind, it is not easy to interpret Ladislaus V's charter to four nobles of Ajka, which gave them the rights to use Bakony Forest freely, "except for the hunting of game."[97] Would it be possible that the hunters of the three villages had the exclusive right to hunt in what was imagined as the physical territory of Bakony Forest? Or else, could it be that in some vague form the physical Forest and the legal Forest already separated?

14. 4. Conclusions

We do not know much about Bakony before the thirteenth century. Most of the otherwise sparse data available come from documents connected to the foundation and functioning of the Benedictine abbey in Bél. Two things make the early history of Bakony comparable to that of Pilis. First, hunting was an important function at both places, a feature that was also characteristic of early medieval European Royal Forests. Secondly, the territory of Bakony was also part of a regular county, therefore the formation of Co. Bakony around 1200 cannot be explained in terms of administrative marginality. In the thirteenth century, Bakony Forest was similar to other Royal Forests in having a *comes*, Forest-guards, servitor populations, and a boundary detectable on the ground. Changes started in the 1270s. For some time, the *comites* were exclusively members of the powerful Csák kindred, who were also donated villages in the region. Their influence was further strengthened when they built a castle in Ugod and possibly another one in Hölgykő. From the 1280s to the 1320s, the situation was rather confused in Bakony. Many Forest-guards were ennobled, and the *comes* was missing for more than three decades. At the same time, the physical boundary of the Forest seems to have existed as before.

The reign of Charles Robert was a turning point in the history of both Bakony and Pilis. The king, having secured his position in the country, took his time to handle such minor issues as Forests. In Pilis, this was done somewhat unintentionally when he moved his court to Visegrád, locating a powerful centre in the Forest whose influence no element in the later history of the Forest could avoid. On the other hand, Charles Robert's choice was not accidental. Pilis was, after all, in the *medium regni*, with or without Visegrád. The new royal centre only reassured the stability of this position. When in Chapter 10 I discussed the double structure that was characteristic of late medieval Pilis (Visegrád and the Royal Forest on one side and the "noble" county on the other), I did not mention the role of castles at all. This was because there were no castles in Pilis (other than Visegrád, that is). There existed the basically non-fortified *castellum* of the Cyko kindred in Pomáz,[98] however, the residence of this middling noble family was very far from having the political significance of a proper *castrum*. All this was typical of the kind of control the royal house was eager to exercise over Pilis Forest. In Bakony, there were four castles (Ugod, Hölgykő, Csesznek, and Essegvár) that were connected to the Forest in different periods. The double structure was present here as well. Charles Robert grouped the remaining Forest populations into three privileged villages in an ambitious relocation programme, and made them royal hunters. In the fourteenth and fifteenth centuries, there were ongoing conflicts between these people and the holders of the *honor* of Bakony Forest (especially when the powerful Gara appeared on the scene) concerning the territory and the administration of the Forest. No one was really right in these debates: *what* one was considered to be was largely influenced by *who* was looking at the person. With the appearance of the Gara, a remarkable process commenced, in which the Royal Forest was transformed – at least in the eyes of the Gara – into a private Forest. A symbolic expression of these changes was the foundation of the Pauline monastery of Porva in 1450, to which we shall return in Chapter 17.

---

[97] *HO*, vol. 2, 315. "demptis venacionibus ferarum"

[98] Virágos 1997. For the Cyko kindred, see Virágos 1999.

Chapter 15

Trees and Woodland in Bakony

15. 1. Trees in Perambulations

I have found forty-six medieval perambulations that concern the Bakony region. Trees, as one would expect, occur in them in large numbers. Out of the 1474 boundary signs, 251 comprised trees and shrubs (*Fig. 57*).

type of tree	*occurrences*	*type of tree*	*occurrences*
oak	123	*gyümölcsény*	3
wild-pear	44	maple	2
willow	13	juniper	2
"tree"	10	thorn	2
crab-apple	10	hazel	2
sorb	7	plum	2
beech	4	*frutex*	1
shrub	4	hornbeam	1
walnut	4	*arbustum*	1
poplar	3	"fruit-tree"	1
elm	3	*barkóca*	1
lime	3	hawthorn	1
troncus	3	*ornus*	1

Fig. 57. Trees mentioned in the perambulations from the Bakony region.

In comparison to the reference lists in Chapter 7 and the list from Pilis, nothing distinctive may be observed. The first three places, as usual, are occupied by oak, wild pear, and willow. Crab-apple is well-represented, while walnut lags somewhat behind. This reinforces the impression that the high percentage of walnut in Pilis has a special significance in that area and should not be expected elsewhere. The Vkpmlt perambulations, however, feature walnut in considerably higher numbers than the documents related to Bakony. The two datasets partly overlap, but the Vkpmlt corpus focuses more heavily on the northern shore of Lake Balaton, whereas the Bakony charters exclude those documents from Vkpmlt that refer to the north-west of Balaton and include many perambulations from northern Bakony. Walnut, then, appears again as a tree that could be found in more densely settled regions (usually with many vineyards), but rarely outside them. As for oak, one should by no means interpret the dominance of the tree as representative of its dominance in the vegetation of medieval Bakony, because oak overwhelms almost any such list.

15. 2. Medieval Woodland

Sources allow us to assess the role that woodland had in the medieval countryside of Bakony from two directions. We can study the occurrences of individual woods (without knowing their sizes), and the data from estimations.

Perambulations mentioned many woods. These data, supplemented by terriers, conscriptions, and estimations contain information about one hundred and ten woods in Bakony,[1] sometimes referred to as simply "silva" but often specified in more detail and with an individual name (*Fig. 58*). Most of these woods never counted among the woods of Bakony Forest, which, like Pilis, can only be approached indirectly, supposing that in the inner regions of Bakony there were woods where there were no settlements.

When interpreting *Figure 58*, one must keep in mind that there are far more sources for the southern region than for the northern. In other words, the fact that more woods appear in the south has no relevance concerning the actual density of woodland in Bakony. Moreover, because logic would prescribe that perambulations mention smaller woods, fewer woods could infer more woodland. The perambulations in the southern area visualise a landscape of numerous but discontinuous woods, while in the north whole boundary walks took place within woods, thus the word "silva" would have been meaningless in many documents.

The map also tells us that different types of woodland were fairly evenly distributed in Bakony. The more one progresses to the north, however, the less one finds *nemora* and *permissoria*; and the more *silue glandinose* appear. If this is not due to a lack of sources, one possible explanation could be that with larger tracts of woodland in the north, there was less specialisation towards different types of woodland and more intensive management. The many acorn-bearing woods all come from one single source, and reflect extensive management. Coppice-woods were not at all recorded in the north.[2]

Estimations confirm the general picture derived from perambulations, namely that woodland was dominant but not overwhelming. Ganna, in the west of Bakony, was estimated in 1366 to contain sixty percent woodland.[3] An all important source is the 1478 Gara estimation, which I discussed in detail in Chapter 5. As we have seen, the Gara possessed the castle of Csesznek, which had previously been a royal castle and belonged to Bakony Forest for some time, but they also understood that this connection did not cease when the castle became theirs. Eleven settlements belonged to Csesznek and here also, sixty percent of their territory was occupied by woods. The same source contains remarkable information about the castle of Somló and its fifteen villages. This castle was on the western fringes of Bakony, while the *pertinentie* were located mostly in the low-lying lands west of it. This, although close to Bakony, was a different world. One would expect to see less woodland recorded, but in fact, the contrary is true. Almost eighty percent of the territory of the fifteen villages was woodland, which signals a warning to the modern

[1] For a detailed list, see Appendix 8.

[2] Although the estimation Df 18 145 covered several villages where all types of woodland were recorded.

[3] Solymosi 1998, 268–269.

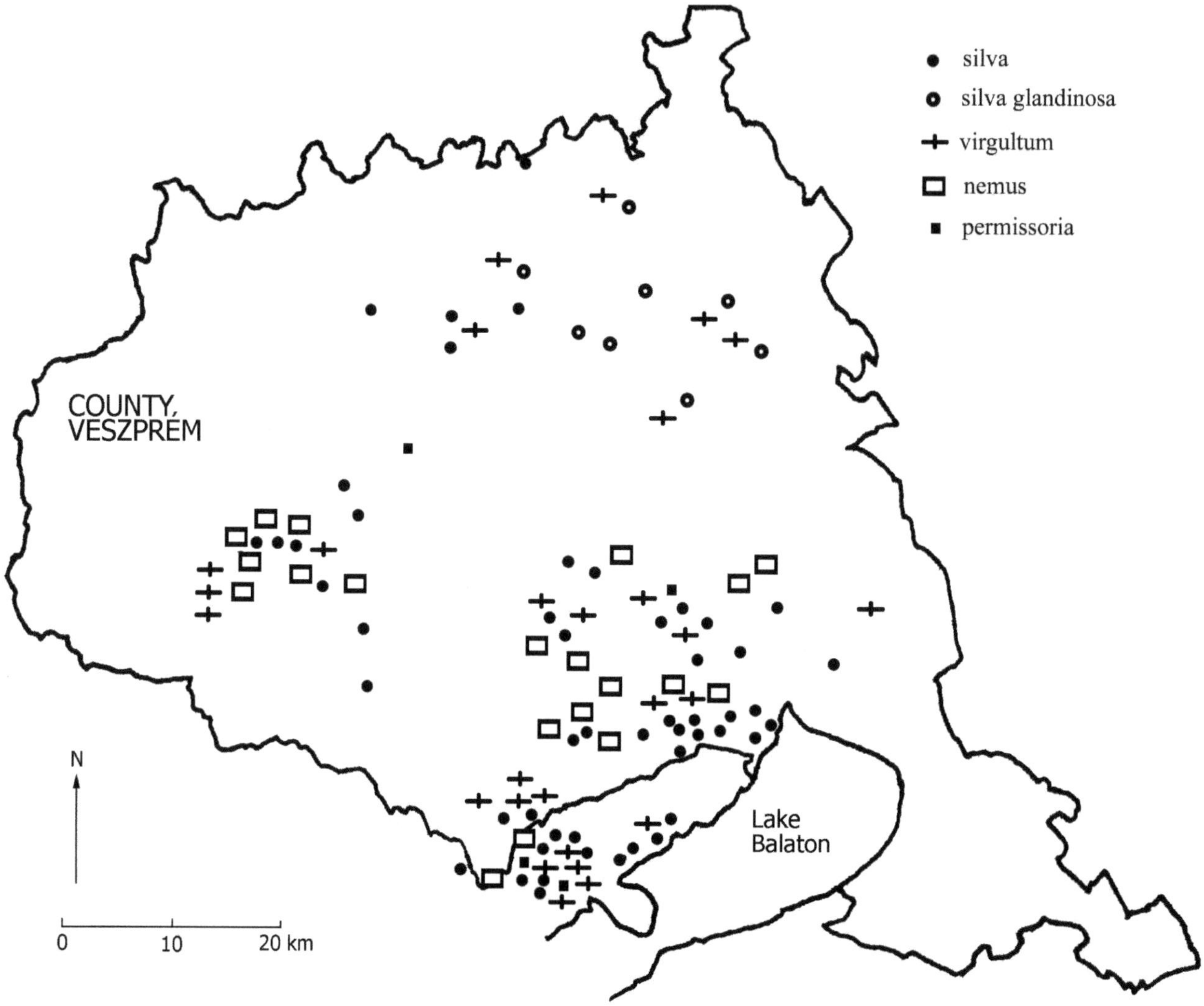

Fig. 58. Woods in Bakony, mentioned in written sources. The symbols do not represent the size of the individual woods.

researcher that analogies with the modern countryside (or is it the modern way of thinking?) can produce entirely misleading results. Today, Bakony is more wooded than any other region in this neighbourhood; in the Middle Ages this was not so.

15. 3. Woodland since the Middle Ages

By the Late Middle Ages, the Bakony region was approximately sixty percent woodland, which was way above the average of the time, but certainly not the kind of wildwood historians relish imagining. In a sense, the landscape was similar to that of today: woods were continuous in the inner regions (especially in the north) and discontinuous elsewhere. This situation was drastically altered by the Ottoman wars of the sixteenth and seventeenth centuries, when people abandoned their villages and fled northwards. In Pilis, I assumed woodland regeneration to have taken place in this period; the same was probably true also in Bakony. Ernő Wallner, when describing the changes of woodland in Bakony, rightly remarked that at the time of the first Ordinance Survey, woods were probably more extensive than at the end of the fifteenth century, especially in the north.[4] From his articles, which were based on a careful analysis of the OS maps, it is easy to follow the disappearance of woodland since the end of the eighteenth century.[5] The present vegetation of Bakony was described by Gábor Fekete.[6] Specialities, such as the yew-wood of Szentgál and the autochthonous pinewoods of Fenyőfő, were treated by Antal Majer.[7]

15. 4. Wood-Pasture

Bakony has had a longstanding reputation for pannage. Once, it even boasted its own local swine variety.[8] This would involve the presence of wood-pasture in addition to woodland. At present, Bakony is one of the few regions of Hungary where wood-pasture is preserved in the landscape, although almost without exception it is not used and thus threatened by infilling. I have located eleven wood-pastures during fieldwork (*Fig. 59*).[9]

[4] Wallner 1941, 29.

[5] Wallner 1942; Wallner 1943.

[6] Fekete 1964.

[7] Majer 1980; Majer 1988.

[8] Matolcsi 1975–1977.

[9] I would like to thank Imre Sonnevend of the Balatonfelvidéki National Park for information about the possible sites, and Tamás Koller of the Pangea Environmental Organisation for guidance during fieldwork.

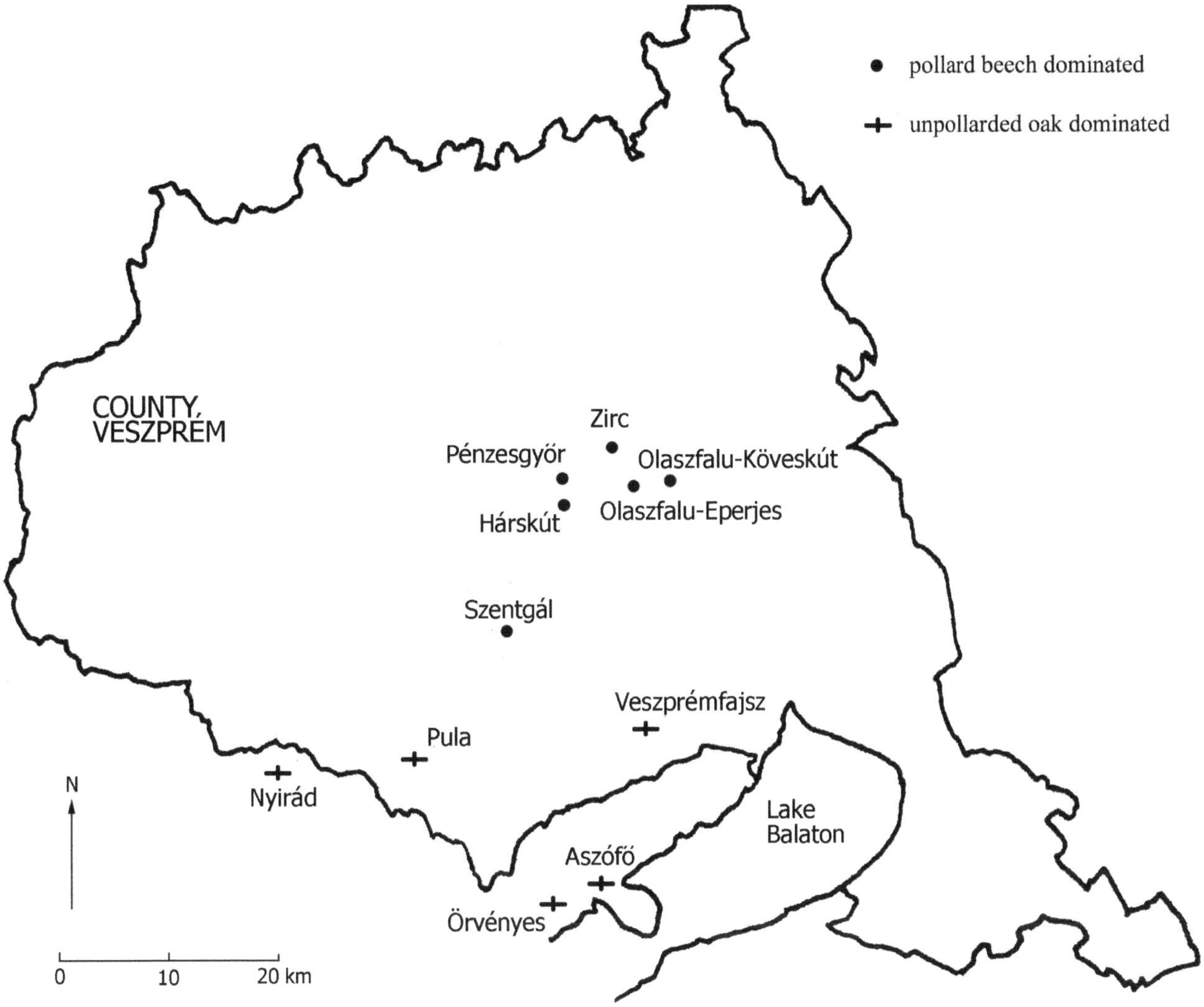

Fig. 59. Wood-pasture in present-day Bakony.

Rather unexpectedly, two very distinct types appeared, differentiated both in their appearance and geographically. In the north, around the present-day settlements of Zirc, Pénzesgyőr, Hárskút, Olaszfalu, and Szentgál, one finds mighty pollards, mostly beech, with the occasional oak, hornbeam, and lime (*Figs. 60–62*). In the south (Nyirád, Pula, Veszprémfajsz, Aszófő, Örvényes), however, wood-pasture is dominated by oak, and the trees are *not* pollarded (*Fig. 63*). In general, these oaks appear younger (or at least they are smaller in girth), although in Nyirád I have measured one that was 510 cm around. The size of the pollard beeches in the north is impressive, one often finds trees with a girth of five metres or more (*Fig. 64*).[10]

How old is the tradition of pollarding in Bakony? Living trees in most cases do not seem older than a few hundred years, although I could not do ringcounts to confirm this impression. Nonetheless, the discontinuity of the landscape (that came, just as in Pilis, with the Ottoman occupation) suggests that the tradition is older than the surrounding settlements. These settlements are almost all recent (founded in the eighteenth- and nineteenth-centuries). Even though they often keep the name of the previous medieval village (Pula, Fajsz, Nyirád) they have little in common with them, in many cases even the location has changed. The most interesting place for us is the large wood-pasture between today's Pénzesgyőr and Hárskút. Not only are these settlements newly founded, they had no medieval predecessors, either. Consequently, this area was depicted on the first OS map as being covered in trees, which the modern settlers would have transformed into wood-pasture with pollard beeches. Some trees, however, contradict this. One can debate about the age of a pollard beech that is 520 cm in girth, but it is probably older than two hundred years. What does this mean? It is likely that parts of this wood-pasture are older than the settlements. They were taken over or renewed from an unknown past by the modern settlers. As there was a period of very sparse population in the region in the sixteenth and seventeenth centuries, their origins may well predate that time.

Medieval written sources make it clear that wood-pasture was an integral part of the landscape. There are many references to pannage, starting with a charter that claimed to be the foundation charter of Bakonybél from 1037,[11] in which the monastery obtained the right to pasture its pigs in Bakony Forest. A charter of 1456 informed us that two nobles from Ajka had the right to keep their animals in

[10] A project has started at Pangea to catalogue all pollards in Pénzesgyőr. So far (April 2002) they have listed some 250 trees.

[11] It is in fact a thirteenth-century forgery. *DHA*, 113–119.

[12] Dl 66 273. "pecudes et pecora" I wonder if these referred to cows and sheep, or whether they were just fancy expressions. For what cows and sheep were usually called, see the next footnote.

Fig. 60. Wood-pasture with pollard beeches near Pénzesgyőr. The fact that the trees are little visible shows that the territory is nowadays only grazed by wild animals. April 2002.

Fig. 61. Pollard beech near Pénzesgyőr with some fresh cuts. This tree is probably older than the settlement with which it is now associated. April 2002.

Fig. 62. A grotesque pollard small-leaved lime near Olaszfalu. April 2002.

Fig. 63. A wood-pasture of non-pollarded oaks (Quercus cerris) at Nyirád. Most of the area is being infilled, which is good for the next generation of trees but threatens the meaning of the place. April 2002.

Bakony.[12] When founding the Pauline monastery in Porva, Ladislaus of Gara provided an even more detailed list: the monks could pasture their pigs, sheep, oxen, horses, and other animals.[13] Many *acta potentiaria* were concerned with pasturing animals in the fifteenth century. In 1432, forty pigs of the monastery of Bakonybél were driven away by the

[13] Dl 14 424. "porcos eorum, oves, boves, equos at alia animalia" What else could there be? Perhaps goats? These animals are usually considered too destructive for woodland, but I have not seen that mentioned in medieval Hungarian documents.

Fig. 64. An improbable tree: a wild-cherry at Pénzesgyőr. It is 440 in girth and is in perfect health. April 2002.

people of neighbouring noblemen. Three years before this, nobles of Rum pannaged their pigs in the woods of Ganna and Polány (northern Bakony).[14] In 1499, the people of the bishop of Veszprém killed a pig in Litér, although this may well have been during a hunt, because the pig was termed "wild," and was deposited – for some unknown reason, but obviously demonstrating something – in front of the door of Emeric Himfi, owner of the village.[15] Eleven years later in the same village of Litér, the people of the bishop were pannaging their pigs in the local acorn-bearing wood, when the villagers attacked them and seized the pigs.[16] The numerous acorn-bearing woods of the estimations also make reference to pannage. Another reference to this type of land-use is the high number of *virgulta* and *rubeta* in the region, which, as I have tried to demonstrate in Chapter 6, were technical terms for wood-pasture.

Wood-pasture is inseparable from pollards. My interpretation (Chapter 7) is that in the Middle Ages pollards were termed *troncus*. Six *tronci* were mentioned in perambulations of the Bakony region, with the addition of a walnut that was a *stips*, that is, described with a word that appears to be the synonym to *troncus*.[17] Three of these pollards were oak, two undefined, one walnut, and one "fruit-tree." The dominance of oak, as ever, does not mean that this was the tree most often pollarded, but rather that it was most often mentioned.

How do present wood-pasture and medieval sources relate to each other? Although no particular wood-pasture can be shown to have existed since the Middle Ages, many different source types confirm that the combination of pollards and grassland was at least as visible in the landscape of 1400 as it is today. Nothing in written sources, however, informs us whether the two distinct wood-pasture types of our days were already there in the Middle Ages. A fairly good case could be argued for oaks not being pollarded because of acorn-production, but perambulations do mention pollard oaks, albeit in numbers that neither prove nor disprove the above assumption.

15. 5. Conclusions

In Bakony, sources on trees and woodland had a different structure from those in Pilis. While in the latter there were few written sources and more archaeological evidence of high quality, in the former, archaeology only comprised my own fieldwork, but there was more written material of various kinds. Thus, the two analyses are particularly useful side by side, because they complement each other.

Perambulations from Bakony again showed the dominance of oak, walnut, and willow, although it was interesting to notice that when compared to the Vkpmlt corpus, the Bakony charters featured walnut to a lesser extent. The data from estimations suggested that the Bakony region was approximately sixty percent woodland in the fifteenth century. This is less than what was imagined before, which becomes especially interesting in comparison to the area west of Bakony, which was more wooded. Mentions of individual woods were numerous in charters, however, their distribution map required caution in its analysis. Such mentions appeared somewhat similar to the mentions of trees in perambulations in the sense that they could not be interpreted straightforwardly. Just like many oaks in perambulations did not signal the dominance of oak in woods, many woods in charters did not mean the dominance of woods in the landscape. As a special feature, Bakony seems to have been a stronghold of wood-pasture, also discovered in fieldwork, although written sources and field evidence are hard to connect as yet.

[14] Sörös 1903, 363, 365.

[15] Df 102 677.

[16] Df 102 696.

[17] 1234: *PRT*, vol. 8, 256–260; 1352: Df 200 915; 1372: *PRT*, vol. 8, 387–389; 1380: Df 201 029; 1381: Df 201 030 "stips nucis"; 1399: Df 201 125; 1454: Df 201 343. For the equation between *troncus* and *stips*, see for example *PRT*, vol. 8, 387–389: "stips sive truncum arboris pomifere." The classical form of the word is *stipes*.

Chapter 16

Settlements in Bakony

16. 1. Written Sources and the Archaeological Topography Revisited

In Chapter 12, I have outlined the major problems that one has to be aware of when interpreting the results of field surveys. Now, we have to focus more on the differences between the individual volumes of *MRT*, and also on the quantity of written sources and their influence on the results presented in the *Archaeological Topography*. Bakony, as we have seen in the previous two chapters, is far better equipped with written evidence than the Forest in the *medium regni*. As a consequence, we should expect the majority of the settlements that existed in the fifteenth century to be known from written records. Probably the most important source is the tax conscription of Co. Veszprém from 1488, one of the few such documents to cover a whole county.[1] Many more documents were processed in the *Local Historical Lexicon of Co. Veszprém*,[2] which described each settlement one by one. Therefore, when the field surveys started, the archaeologists knew almost too well what they were looking for,[3] and archival research remained rather limited. This was partly justified, however, there was plenty of material left to study, some of which would have revealed many finer details in the history of the individual settlements.

The Bakony volumes were the first four in the *MRT* series.[4] As expected, they present a number of problems. Most importantly, the earliest volumes tried to draw a dividing line between "settlements" and "sites," the latter being finds not found in sufficient numbers to warrant being called the former. This turned out to be a cul-de-sac and was given up later. Thus, I myself interpreted every site in each volume as a "settlement," understanding this term in the widest possible sense. Another problem, especially when compared to the volumes covering Pilis, is that thirty years ago dating of medieval pottery was very imprecise. In the Bakony volumes, in most cases all the information we have is that the potsherds were either "Árpádian period" or "late medieval." One considerable advantage of working on Bakony was that István Éri, one of the authors of the pertinent volumes of the *Archaeological Topography*, has already tried to summarise the topographical information provided by *MRT*.[5]

16. 2. Settlement System

My original intention was to repeat the kind of analysis I carried out in Pilis.[6] However, given the circumstances described above, this turned out to be impossible, especially in respect to those settlements that never appeared in written documents. To somewhat compensate for the loss in data quality, I decided to present the results generated from the individual volumes of *MRT* separately. This allowed for some, even if rudimentary, observations about the spatial patterns of different processes.

In *Figure 65*, I have refrained myself from presenting all settlements individually (with an overall number of 208 that would have been pointless) but rather counted existing settlements for every half-century. The pattern reveals that the medieval settlement system of the Bakony region was established at an early date, and was very stable throughout the centuries. Low numbers for the eleventh and twelfth centuries are rather hard to interpret because of the lack of sources. There are more and more known settlements in the thirteenth century, and after 1300 the total number remains stable. The real questions to emerge from this dataset are whether the thirteenth century was, in fact, the time of rapid growth in the number of settlements or whether that is just an illusion created by the written sources, and how all these factors relate to the eleventh and twelfth centuries. Archaeological predating of settlements (as carried out in Pilis) might bring us closer to the answer, but, as outlined above, it was not available in this case. Nonetheless, although the dating of pottery proved unsatisfactory for revealing finer details of chronology, there are still twenty-four occasions where a settlement that first appeared in the sources in the fourteenth century was described as having Árpádian period potsherds in it.[7] The precise meaning of "Árpádian period" is

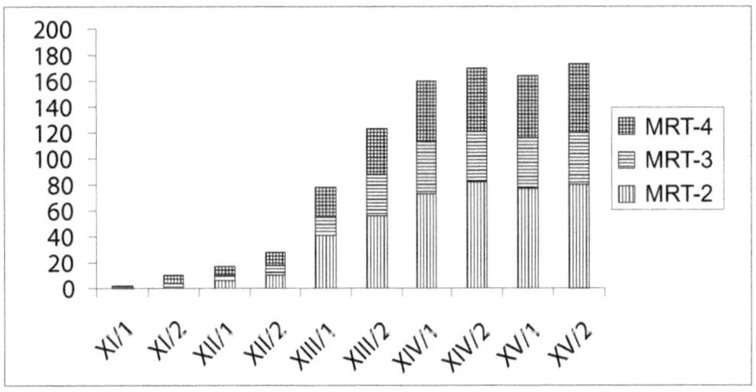

Fig. 65. Settlements with written sources in Bakony. Roman numerals stand for centuries, Arabic ones for the first or the second half of the given century. (XII/2 is the second half of the twelfth century).

[1] Solymosi 1984. The source was published on pages 198–211.

[2] Ila and Kovacsics 1964. This work drew heavily on an important source publication: *VeszpReg*.

[3] Although the 1488 tax conscription was not published until 1984.

[4] *MRT-2*; *MRT-3*; *MRT-4*. *MRT-1* also covers a part of Co. Veszprém, but this region falls outside of what I defined as Bakony.

[5] Éri 1969. He considered this a preliminary work, which was to be followed by a county monograph, but that has yet to happen. Earlier settlements were summarised in a similar manner by Torma 1993–1994.

[6] The point in having a series of topographical books is that, in theory, they provide *compatible* data from different places.

[7] *MRT-2*: 14/29 Túzoktelek, 23/1 Hímháza, 23/4 Horhi, 23/6 Szentistván, 23/8 Herend, 24/4 Alsóhidegkút, 31/2 Mesteri, 33/1 Nagyvázsony, 33/18 Leányfalu, 37/11 Lovas, 41/1 Pula, 49/5 Tikolülése, 49/25 Réti, 51/70 Tekeres; *MRT-3*: 14/2 Patony, 35/3 Etej, 42/6 Tegye; *MRT-4*: 9/1 Ság, 9/3 Kisdém, 10/1 Szentiván, 13/3 Szakácsi, 29/6 Zimán, 52/4 Veim, 70/1 Szápár.

not clear, but such finds must be from the thirteenth century at the latest. This means that there must have been almost as many settlements in this century as there were later.

Figure 66 is even simpler than the previous table, yet it offers some conclusions. Most importantly, although not much is revealed here about the continuity of individual sites, one has to consider István Éri's remark that "Árpádian period and late medieval pottery finds almost always occurred together."[8] In other words, the many deserted Árpádian period sites, so characteristic of most of the kingdom (including Pilis) are missing from Bakony. This, however, is not true for the entire region. Early desertion appears to have been very much present in north-east Bakony, the parts covered in *MRT-4*.

Regardless of the distribution of these settlements in space and time, it must be stressed that their overall number was very low. Interpreting all sites as individual settlements (certainly an exaggeration) there were only 169 such settlements as opposed to the 208 villages with written records. This is in strong contrast with Pilis, where there were about three times as many settlements known from field surveys compared to those mentioned in written sources. No one has done calculations for the whole area covered in all volumes of *MRT*, but my impression is that Pilis is much closer to the average for the country than Bakony. Why should this be so? An obvious solution may be that it is the result of insufficient or inefficient fieldwalking. While there is no reason to doubt that the methods and techniques of field surveys underwent considerable development in the years that passed between the making of the Bakony and the Pilis volumes of *MRT*, differences appear too large to be blamed on the archaeologists. A more probable explanation is that István Szabó's thirteenth- and fourteenth-century "settlement disintegration and spread"[9] never happened in the Bakony region. In opposition to this idea, one could claim that our impression is created by the abundance of written sources in Bakony, in which all those little settlements that usually remain anonymous were mentioned. Let us remember, however, that the usual state of affairs was that a fairly fluid settlement system was replaced by a lower number of nucleated settlements from the fourteenth century onwards. These disappearing small settlements are the ones missing from Bakony, in other words, a settlement type that was unique. These sites, irrespective of written sources, had their archaeologically recognisable characteristics.[10] If they had been present, we should be able to identify them purely by using archaeological typology. The only difference between these settlements and others elsewhere would be that in Bakony we would also know their names.

The spatial distribution of settlements demonstrates that people lived in a densely settled environment in most places (*Fig. 67*). Settlements with and without written records apparently do not form distinct patterns. As the reader may remember, in Pilis there was a tendency for the latter

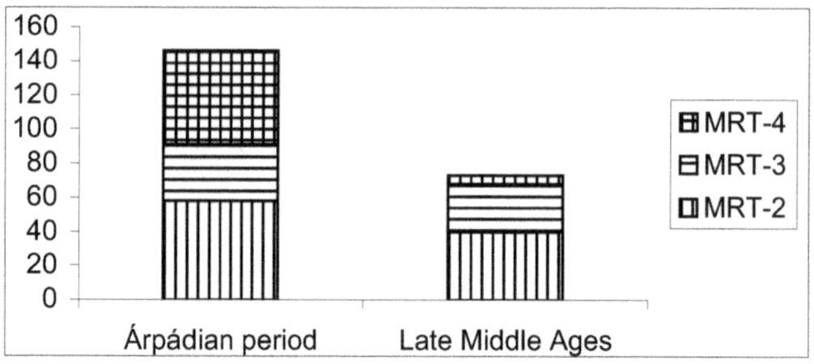

Fig. 66. Settlements without written documents in Bakony.

to cluster in marginal regions as though suggesting faint attempts at colonisation. Here, nothing of this sort is visible. Both types were frequent at almost every location. The dividing line between the settlements and the uninhabited regions was fairly sharp. Earlier I have suggested that this was mainly the result of the existence of the Royal Forest, but the answer cannot be this simple. Would medieval people have settled everywhere in Bakony had it not been for the Forest? Geography undoubtedly played its part in the story, hand in hand with administration. Bakony is rarely so hostile as to make living impossible, however, the territory of the Forest (as much as we can follow it) intentionally coincided with higher altitude, less habitation-friendly regions.[11] This is best demonstrated by the fate of the valley that separates the northern and southern parts of Bakony. This region controls both sides of the Forest and the important road that crosses it. Indeed, it was here that Charles Robert formed the three privileged villages of hunters, the new focal place of Bakony. Neither before, nor after, however, was there any royal intention to exclude other forms of settlement from the valley: there would not have been any point. Settlements were allowed as long as they did not interfere with what was expected from the Forest.

16. 3. Conclusions

All evidence points to the conclusion that the settlement system of the Bakony region was complete by the thirteenth century (or possibly earlier) and changed very little until the sixteenth century.[12] There was no internal colonisation, because places that were considered habitable already had people in them, and those remaining were in a Royal Forest and were probably not intended to be populated. The Mongol invasion of 1241 is invisible in this picture, we can take it for granted that it did not have a lasting effect on the region. The spatial distribution of the settlements could not demonstrate any pattern of changes, either. What the sources (both written and archaeological) show is a system that developed at a time before these sources were made.[13] Then, the different settlement stories of Pilis and Bakony were dramatically brought together by the Ottoman invasion. Bakony was also deserted: some continuity up to the present

[8] Éri 1969, 203.

[9] Szabó 1966, 186.

[10] Such as being distributed along streams. Laszlovszky 1986b, 137–138. I identified such settlements in Pilis.

[11] These were hardly inhabited in prehistory, either. The only notable exception is the Late Bronze Age, when there were 480 settlements in Co. Veszprém, and many of these were inside the woods up to heights of 600 metres. Torma 1993–1994, 67–68.

[12] Also remarked by Éri 1969, 203. He further pointed out that by the Late Middle Ages, especially in the west of the county, doubling or tripling of individual settlements were characteristic as a sign of overpopulation.

[13] However, a re-examination and more precise dating of the pottery finds may alter this picture.

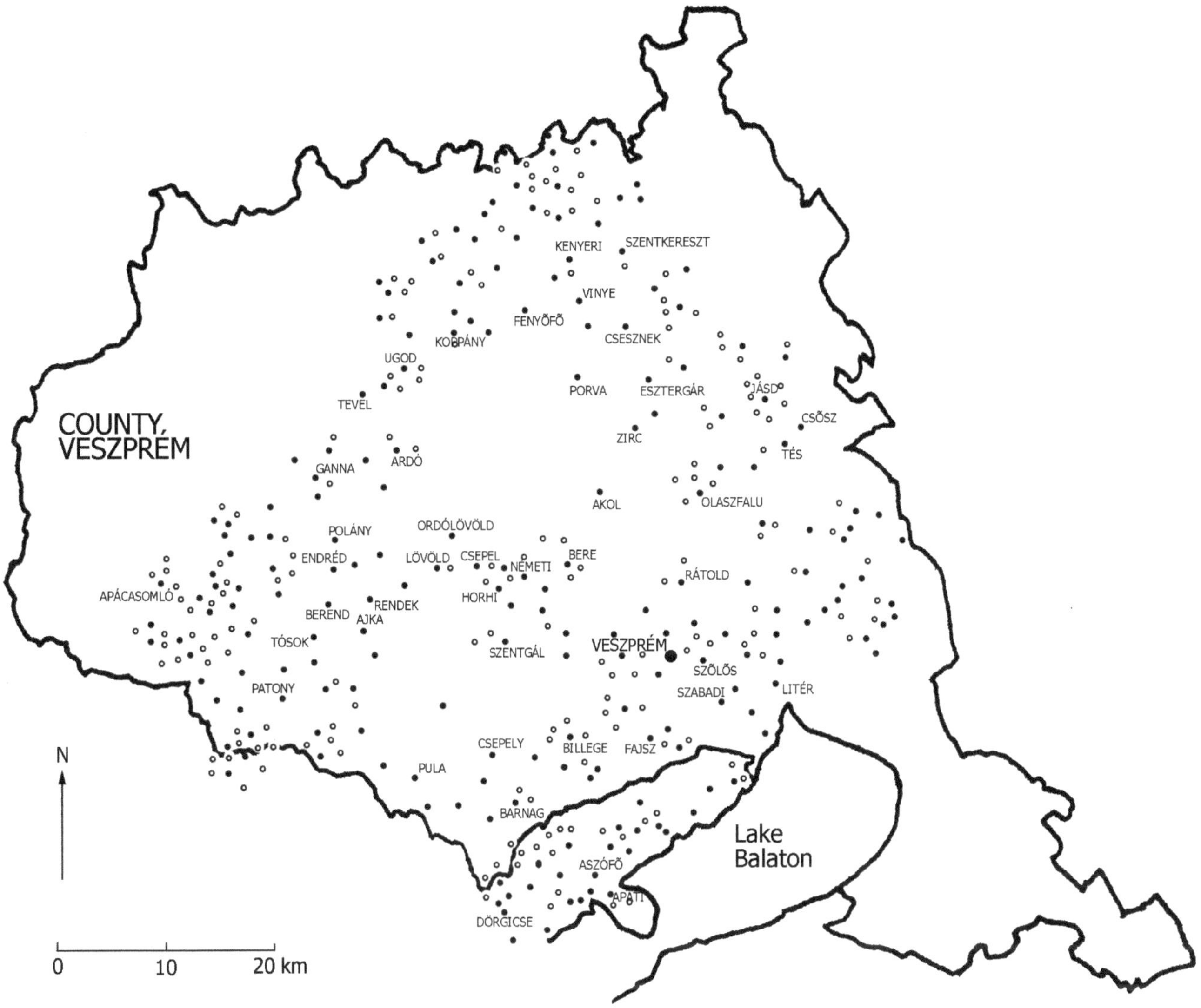

Fig. 67. Medieval settlements in Bakony. Those that appear in the text are named, but this has nothing to do with importance. Dots: settlements mentioned in written documents. Circles: settlements located through fieldwalking, not mentioned in written documents.

can be claimed for a mere ten percent of all settlements. It is surprising that the vast majority of these settlements were on the northern shore of Lake Balaton, that is to say, in an area worse affected by the Ottoman wars than the north of the county.

In sum, Bakony was very different from what is understood as the general line of development of the medieval Hungarian settlement system. Why this was so is a difficult question, which becomes even more challenging when compared to the regular development of the Pilis region. I have already suggested that this phenomenon may have been in connection with the early history of the Royal Forest. It will only be possible to approach this period more closely through an examination of the history of the monastic orders in Bakony, especially the Benedictine abbey of Bél, which will be dealt with in the next chapter.

Chapter 17

Monasteries in Bakony

There were sixteen monasteries in the Middle Ages in the Bakony region, however, not all of them concern us directly. Franciscans and Dominicans (especially the former) undoubtedly had an influence on how people related to their environment, but they were urban orders, therefore their houses in Veszprém and Várpalota have little to offer in terms of the history of Bakony Forest. An early and famous[1] nunnery in Veszprémvölgy (literally "the valley of Veszprém,") later a Cistercian nunnery,[2] will not be discussed, either, together with three Benedictine monasteries (Tihany,[3] Jásd, and Apácasomló) and a Premonstratensian house in Rátold.

As in Pilis, I chose only those monasteries that had *real connections with the Forest*. In Bakony, these included the Benedictine monastery in Bél, the Cistercian monastery in Zirc, the Pauline monastery in Porva, and the Carthusian monastery in Lövöld (*Fig. 68*). Similarly to settlements, however, what to include was not quite as obvious as in Pilis. The interactions between the Forest and the individual orders comprise a complicated story, which will sometimes (especially in the case of the Paulines) lead us to places outside the region discussed so far.

17. 1. The Benedictines

17. 1. 1. Bél

While in Pilis the first monastic foundation was that of the Cistercians,[4] in Bakony, a Benedictine monastery was established in the middle of the Forest in the early eleventh century.[5] I have described the foundation of this monastery in some detail in Chapter 14. It started out as the hermitage of St. Günther and later St. Gerald.[6] By the time the latter left for Csanád in 1030, the monastery had most probably been founded by St. Stephen and the construction works were well on their way. The foundation charter, although it once undoubtedly existed, does not survive. What we have today is an early thirteenth-century forgery that has nothing to do with the original.[7] There are, however, some early documents, notably the 1086 conscription of the holdings of the monastery.[8] Written sources are unfortunately all we have, for there have not yet been any excavations at the present Baroque church where the medieval monastery once stood.[9]

What was the landscape like around the site in the year 1000, and how did the monastery change it (if at all)? The legends of St. Stephen and St. Gerald stressed that Bakony was an appropriate place for hermits, but did not go too far with the image. Gerald, says his *legenda minor*, wanted to "get away from the noise of the crowd,"[10] but we do not read about *vasta solitudo* or some similar topos. Quite the contrary, the saint's life in Bakony was almost pleasant. He contemplated and wrote. Just like his earlier and later counterparts (for example St. Blaise, St. Giles, or St. Godric of Finchale),[11] he befriended and nursed animals such as a deer and a wolf. All these, so some argue, may be connected with the image of a new type of hermit, which Gerald brought with himself from Italy.[12] In this tradition, woodland was peaceful, not frightful. Without even trying to unravel the physical environment from hagiography, we do not learn whether the location Günther and Gerald came to was unsettled in the first place. The question is, as in all monasteries in Royal Forests, whether a royal (hunting) residence had existed there previously. Without explicit written sources and excavations, we are left with learned guesses. György Györffy proposed first that there had been a royal residence in Bél, although his arguments were rather weak.[13] Recently, Hervay observed that the 1086 conscription recorded a many-sided and well-functioning system of estates, which must have been inherited from a previous royal organisation.[14] Whether this is true or not, the examination of this document appears to be the only way to get closer to the landscape of Bél in the eleventh century.

---

[1] For having what is possibly the only original charter issued by St. Stephen. Original here means that the words survive unchanged in a later copy. Szentpétery 1938.

[2] Hervay 1984b, 192–199.

[3] Including an early Basilite monastery nearby that functioned until 1170.

[4] Providing that we accept Hervay's view and that the early settlement in Szentkereszt was not a Benedictine monastery. See Chapter 13.

[5] An exhaustive book on the medieval history of the monastery is *PRT*, vol. 8. Complementing the high quality charter collection at the end of this book, the author (Pongrác Sörös) published some more sources in 1903. All research on Bakonybél has been based on these two works.

[6] Their hermitages may have been in different, although nearby, places. Sólymos 2001. Archaeologists appear to see this in a different way. See *MRT-4*, 2/15.

[7] *DHA*, 113–119.

[8] *DHA*, 247–255. This also used to be considered a forgery, however, recent research (see the *DHA* commentaries) has demonstrated that its first part is original, while the first and second additions are from the early twelfth century.

[9] The only archaeological study I know of (also based on Sörös' material) is Végh 2001. See also *MRT-4*, 2/1. Written sources, on the other hand, are numerous, because the archives of the monastery were transported to Pannonhalma before the arrival of the Ottomans, and survive more or less intact.

[10] *SRH*, vol. 2, 472. "tumultum populi devitans"

[11] For St. Blaise and St. Giles, see Iacopa da Varazze 1998, vol. 1, 180–187, vol. 2, 887–890. For St. Godric, see Clay 1914, 28.

[12] Imre Takács 2001, 58–60. These attitudes do change over time. In the early sixteenth century, a Carthusian monk dwelling in Bakony (in Lövöld) preferred to imagine the Bakony of the eleventh century as a "wild woodland" (vadon erdő). Volf 1876, 325. I would like to thank Marina Miladinov for sharing her thoughts with me on hermits.

[13] Györffy 1969. He listed three reasons on page 214: 1) Veszprém was very close to Bél 2) St. Günther was a relative of Queen Gizella 3) The queen herself donated some land to the monastery. However, I do not see why the existence of a royal residence should follow from these facts.

[14] Imre Takács 2001, 479.

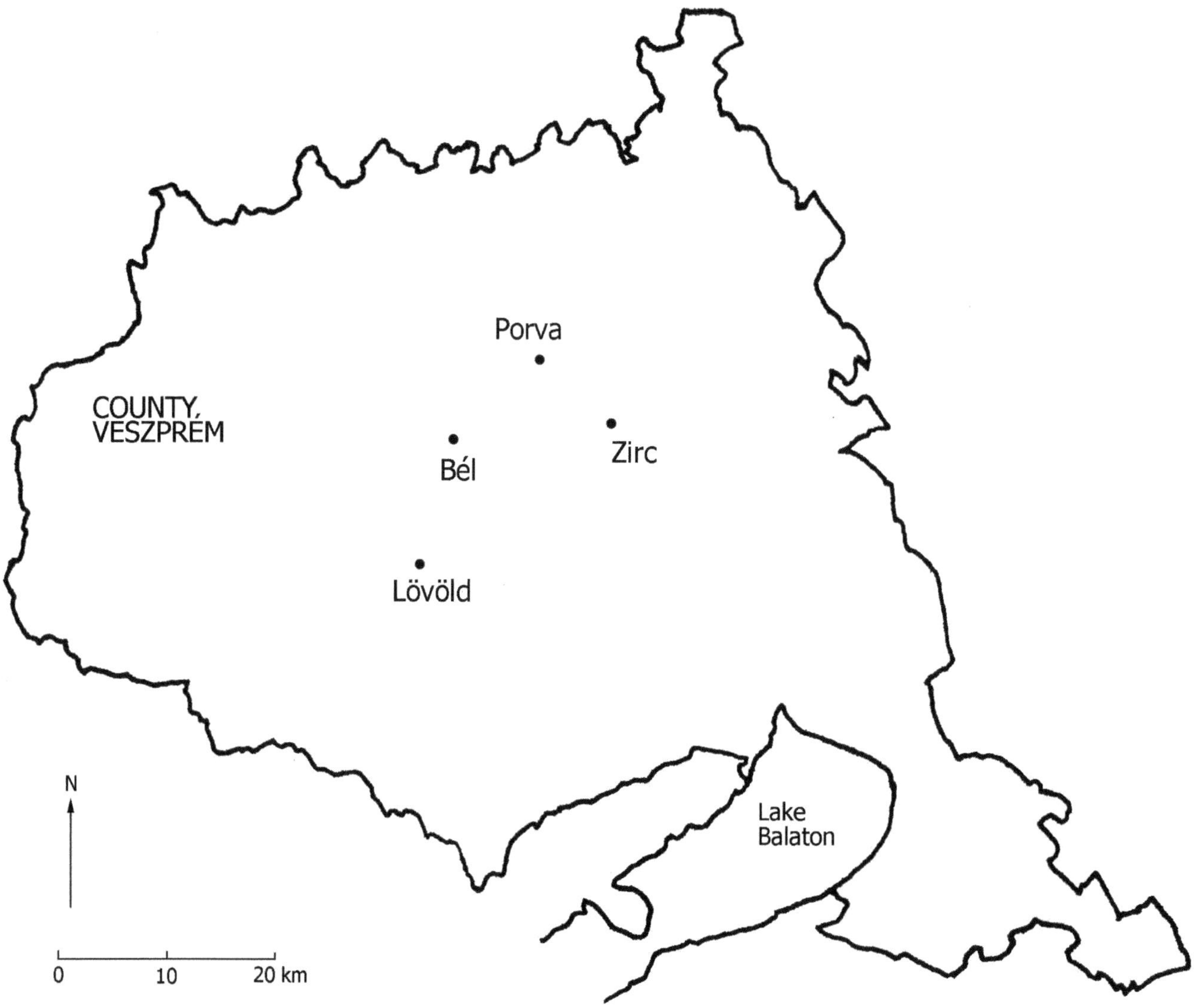

Fig. 68. Monasteries in Bakony.

I have already noted that lands in the immediate surroundings of Bél were missing from its holdings, which may have been connected to the fact that it was founded inside a Forest. Its most important villages, however, did not lie far from it. Koppány, Tevel, and Endréd formed a wide circle around Bél. We know from the perambulations of these villages (included in the 1086 conscription) that their area encompassed an entirely cultural landscape. Trees, in fact, were used as boundary signs without further marks (such as crosses) on them. What lay within the ring between these villages and Bél is revealed by the laconic description of Akol, a settlement in this region, which was "bordered by woodland all around."[15] This was already exceptional in the eleventh century, and a glance at *Figure 67* reveals that Akol remained an isolated settlement throughout the Middle Ages. All this, then, confirms the above idea of a pre-existing royal residence and estate complex, which was taken over by the Benedictines. If the *raison d'être* of the original system was the Royal Forest, it becomes understandable that the settlements were distributed in a way that would preserve the Forest (both by not clearing and by surrounding, thus protecting it). The Benedictine monastery of Pannonhalma (which I have previously mentioned in contrast to Bakonybél as possessing all the land surrounding it) also had some villages in the vicinity of Bakony as early as the eleventh century.[16] These, especially Vinye and Szentkereszt, appear to be in such a similar position as the villages of Bél that I am tempted to think that at one point they were part of the same system, whence some went to Pannonhalma and others to Bél. These two monasteries have always had close connections, also in a physical sense: the ancient road that once connected them is partly visible in the landscape today (*Fig. 69*).

For my last observation, we have to return to the hermitage of St. Gerald. This, as confirmed by archaeology,[17] was at a comfortable walking distance (some fifteen minutes) from the building of the monastery. Why did they not make the monastery on that holy spot? There may be several answers, and at least one of them would reinforce the possible existence of a royal residence. The justification for any answer depends on the starting date of the actual building process, and, to a lesser extent, on the location of the hermitage of St. Günther. As a starting point, let us suppose that from a technological point of view there was not much of a difference between the location of the medieval monastery and that of the hermitage;

[15] *DHA*, 252. "circum terminatur silvis" The place-name "Akol" is most probably connected to the keeping of animals. *FNESz*, s. v. "Aklipuszta."

[16] Solymosi 1996.

[17] *MRT-4*, 2/15.

in other words, that the monastery could have been built at either place. In one version of the events, St. Gerald chose the location of his hermitage because he wanted to avoid the royal residence, but when the construction works of the monastery began, they naturally made use of the existing buildings. If, however, we suppose that the site of the monastery had been selected before St. Gerald's arrival, it seems logical that he settled down a bit further away and we get no information about the royal residence. St. Günther's hermitage comes into the play because it would be hard to suppose that the building of the monastery preceded it. It is not supported archaeologically, but tradition has it that his hermitage was also at some distance from the monastic site. This, again, could only be explained by the existence of some earlier building where the medieval monastery stood. We should also note, however, that the co-existence of eremitism and coenobitism was not at all unusual in this period. For example St. John of Rila, one of the most famous Bulgarian monks, was also first a hermit then a monastic founder in the eighth and ninth centuries. Such stories abound in contemporary Byzantine hagiographic sources, as well.[18]

The Benedictine monastery of Bél was, in conclusion, by no means a colonising agency in Bakony Forest, but possibly the heir to an existing system. I have already quoted Beatrix Romhányi's observation about the active role Benedictines played in eleventh-century colonisation in Hungary.[19] Bél, however, did not partake in this process. Why was it founded, then, at this particular location? The answer lies partly in the Benedictine eremitic tradition.[20] That Bakony was a good place for a hermit, is, after all, not a mere topos. We have no reason to doubt that St. Stephen was in fact moved by Günther and Gerald's advice to establish a monastery. If he really had a residence in Bél, his religious sentiments may have been strong enough to make him offer it to the church. All the more so, because this was not the only residence he had in Bakony, as we shall see later on.

17. 1. 2. Economy

The Benedictine monastery of Bél was a rich institution. Early royal land donations were followed shortly by private donations, some of which concerned settlements (such as Ganna or Polány) that further strengthened the circle around the wooded ring that surrounded the monastery.[21] The kings did not cease to care for the monks, although Coloman is said to have taken away some of their fishing places.[22] His father, however, enriched the monastery by donating large amounts of salt to it.[23] Andrew II also provided the monks with thirty marks from his own salt income.[24] All these and more made Bél a well-off place that lived entirely on its incomes in money and kind. There was never any settlement around the monastery.[25] Sörös mentioned

Fig. 69. The ancient holloway connecting Bél and Pannonhalma. It is even today called Barátok útja ("Monks' road.") September 2001.

the remains of a building that probably had some economic function,[26] and a few sources talked about a certain "village" (*possessio*) in Bél, therefore we can infer the existence of an economic unit within the monastic complex. Whether this was a simple storage facility (for produce supported by the villages) or had a more complex function in organising the agricultural – or even industrial – production of the region, still needs to be clarified through archaeological work.

The central importance of estates and the increasingly difficult situation of the monastery over the next centuries can be easily followed by the many charters that concerned disputes over land. From the fourteenth century onwards, Bél was quarrelling with almost all of its neighbours, including the Carthusian and the Cistercian monasteries. In this period, the Benedictine order as a whole was in decline in Hungary,[27]

[18] Kostova 2002, 58–78.

[19] Romhányi 1993–1994, 195.

[20] There is vast literature on this subject, which I do not intend to cover here. A good starting point is Lawrence 1989, 19–148. For the archaeological side, see for example Gilchrist 1995, especially chapter 5 on "A Desert Place: The Archaeology of Hermits" on pages 157–208.

[21] *PRT*, vol. 8, 274–277.

[22] *PRT*, vol. 8, 33.

[23] *PRT*, vol. 8, 32.

[24] *ÁÚO*, vol. 1, 292–293.

[25] Field surveys covered the accessible territory of the village and found nothing. *MRT-4*, 2/1. Neither do later written sources mention any settlement. Ila and Kovacsics 1964, vol. 1, 107.

[26] *PRT*, vol. 8, 16–17.

[27] Hervay 2001, esp. 467.

although in 1488, Bél still had eighty households in seven villages in Co. Veszprém alone.[28] By 1504, however, the abbot remained alone in the monastery, and the building itself was near collapsing. The monks fled before the Ottoman invasion, and the building lay in ruins by the seventeenth century.[29]

17. 2. The Cistercians

17. 2. 1. Zirc

In 1182, a Cistercian monastery was founded in Zirc, some ten kilometres east of Bél.[30] The reader may remember from the description of the Pilis Cistercian monastery that Zirc was also part of Béla III's grand enterprise to promote the Cistercians in Hungary. The deliberate connection between Forests and the order is reflected in the names of monasteries: just as the Pilis monastery had this straightforward name, the monastery in Zirc was called (although not exclusively) *monasterium Bochoniensis*. The modern researcher is in a somewhat more difficult position in the case of Zirc than with the Pilis monastery. The archives of Zirc also perished during the Ottoman wars,[31] and, although there were archaeological excavations at the site,[32] these concerned only the church and building of the monastery. The precinct (where in Pilis such important remains of industrial activity were discovered) has not yet been studied.[33] Remig Békefi (who analysed the history of the Pilis monastery) planned to write a book on Zirc as well, but he did not live to accomplish this. His notes, however, were used by Konstantin Horváth in his *History of Zirc*.[34]

Perhaps the most important aspect of the site is the definite presence of an early royal residence and settlement. While in Bél I could only conjecture at the existence of such a centre, here there is abundant evidence, both written and archaeological. The fourteenth-century chronicle composition reported that King Andrew I, when wounded in a battle with his brother Béla in 1060, was taken to his *curtis* in Zirc, where he died.[35] That the story may be true is reinforced by the fact that the king was buried in the nearby monastery of Tihany that he had founded. This story is mysteriously similar to that of Dömös, where King Béla died, and here, just as in Dömös, the possible remains of the actual residence were excavated.[36] The interpretation of these remains, however, is rather ambiguous. There certainly had been a church in Zirc before the construction of the monastery began. This church was built in the eleventh century, and postholes excavated in it suggest that there might have been an even earlier, wooden-frame church on the site. Archaeologists discovered a roughly contemporaneous house on the northern side of this church. This is what Koppány identified as the *curtis*, regretting that the excavations only partly uncovered the site, establishing neither the construction phases nor the full extent of this building.

A special interest in Zirc is that the Cistercian monastery was *not* built, as was usual, upon the earlier royal residence. We cannot be sure why this happened, but it is perhaps important that, in contrast to all monasteries discussed so far in this book, there existed a settlement separate from the monastery all throughout the Middle Ages in Zirc.[37] The early church served as the parish church of this settlement. The building identified as the residence was abandoned, while on the other side of the church a new house was built in the Late Middle Ages. The most important question is the relationship between the early royal residence and the later village, whether or not there was any continuity. Written sources first mentioned the village only in the thirteenth century,[38] which would allow the village to have been founded by the Cistercians. The location of the monastery, however, suggests the opposite. The royal residence itself can be imagined as a small settlement with a group of buildings including a church, houses, and outbuildings. It seems that by the twelfth century a settlement grew out of these beginnings and it was big enough to make the Cistercians think that their monastery should be located somewhere else. This, by the way, contradicts the traditional image of the order. The village was theirs from the moment of their foundation. They could have chased the locals away to create a proper *solitudo*. They decided, however, not to settle inside, but approximately two kilometres away from the existing settlement, which remained prosperous until the Ottoman invasion.[39]

17. 2. 2. Economy

Archaeology does not help in this question, and written evidence, as argued above, is sparse. Horváth tried to define the economy of the monastery,[40] and, although influenced by Cistercian ideology, his picture is still relevant. As observed in the case of the Pilis monastery, Zirc was also "primarily active in the secondary economy." It owned a number of settlements, and the tolls on the Danube at Győr – sharing this with the Pilis monastery! From the thirteenth century on, it was also involved in salt trade.[41] All through the thirteenth century, the monastery was wealthy and important. From 1204, the abbot often represented the Hungarian king or the Pope in political missions.[42] The symbol of this prosperity (both economic and intellectual) was John of Limoges, who came from France to be abbot of Zirc from 1208 to 1218, and then returned home

[28] Solymosi 1984, 184.

[29] *MRT-4*, 2/1.

[30] The best survey on the history of the monastery is Hervay 1984b, 208–216. Also see Lackner 1982.

[31] Hervay 1984b, 212.

[32] Hümpfner 1964; *MRT-4*, 81/2.

[33] Now that it is inside a botanical garden, it is likely to remain so.

[34] Horváth 1930.

[35] *SRH*, vol. 1, 357.

[36] Koppány 1972.

[37] *MRT-4*, 81/1.

[38] *HOkl*, 7; *RA*, vol. 2/3, 176.

[39] In 1488, Zirc (the village) paid 18 forints, slightly more than the average. Solymosi 1984, 183.

[40] Horváth 1930, 7–14.

[41] *MES*, vol. 1, 294; Hervay 1984b, 213.

[42] Horváth 1930, 265–291.

to become prior of Clairvaux, one of the most important Cistercian monasteries.[43] By the fourteenth century, however, things changed for the worse. In 1356, the *visitator* sent by the order to Hungary could not get to Zirc, which had a bad reputation. Seventy year later, the abbot obtained permission not to go the yearly general Cistercian assembly to France because the monastery was poor and could not afford it.[44] In the 1540s, the monks left the monastery because of the repeated Ottoman attacks. The building itself did not simply decay, but was blown up in 1738, in order to provide building material for the present Baroque church.[45]

17. 3. The Paulines

At the beginning of this chapter, I listed only one Pauline monastery in Bakony: Porva. Close by, however, were located eight more (Szentjakab, Henye, Vallus, Tomaj, Kékkút, Tálod, Uzsa, and Vázsony).[46] Some of these were among the earliest Pauline monasteries, roughly contemporaneous with the ones in Pilis.[47] (The monasteries in Pilis and Bakony, in fact, formed the first two groups of Pauline monasteries in the whole kingdom.)[48] In 1263 – reported Gyöngyösi, the early-sixteenth-century chronicler of the order – Paul, bishop of Veszprém, conscribed the Pauline monasteries in his diocese.[49] Szentjakab, Tomaj, and Kékkút definitely appeared in this conscription. A certain "Idegsyt" was also mentioned in the list, which is usually considered unidentifiable.[50] A recent study, however, has drawn our attention to some aspects that permit this name to be identified with Tálod.[51] In any case, Tálod was established at some point before 1324 (when it was first mentioned) by the Rátolt kindred, who founded Szentjakab.[52] Henye was also a thirteenth-century foundation,[53] and although Vallus first appeared in written sources in 1429, its archaeological remains suggest an earlier (possibly thirteenth-century) origin.[54] These monasteries clustered in a small area just above Lake Balaton, and had nothing to do with the Royal Forest of that time. This is also reflected in their foundations: they were all established by noble families.[55] Compared to the Pilis region, where the king was founding Pauline monasteries at about the same time, there was little sense of "imitation" involved. That is, the nobles of Co. Zala did not follow the example set by the king, but rather there were two separate and contemporary waves of Pauline foundations. While the one in Pilis could be connected to the Royal Forest, the motivation behind the acts of the nobles of Co. Zala are more difficult to explain. It is perhaps important in this regard that the nobility, as a social group, was unusually strong in Co. Zala.[56]

17. 3. 1. Porva

Porva, on the other hand, was, in a sense very similar to the Pauline monasteries in Pilis.[57] The settlement, on which it was later founded, was known to be located within the Forest, as explicitly stated by Prince Stephen in the thirteenth century, and repeated by King Louis in the fourteenth.[58] If we are right to identify the "Purua" that appeared in the 1086 perambulation of Koppány with our Porva, we should also consider that an early royal hunting residence existed in it. The next piece of information the sources provide us with is the charter of Prince Stephen (1260), in which he donated a chapel in Porva dedicated to St. Emeric to the Benedictine monastery of Pannonhalma with as much land as can be reached by three bowshots from the cemetery in four directions.[59] All elements involved carry some extra meaning. To start with, we should remember the foundation of Dömös in Pilis. As argued there, there is no way to find out why Prince Álmos had authority over that particular piece of land, other than the fact that he was a member of the royal family. In Porva, we again cannot explain why the chapel belonged to Prince Stephen. The dedication of the chapel (St. Emeric, son of St. Stephen) may easily refer to royal origins. That the cemetery was also mentioned makes it certain that there was more than a simple chapel on the site.[60] The obvious solution, then, is that there was an early royal residence in Porva, which, in the thirteenth century, consisted of a chapel, a cemetery, and some buildings. This was something definitely comparable to nearby Zirc. The date also fits the general pattern, especially that of Pilis, where it was in the mid-thirteenth century that changes became obvious in the system of the Forest. We do not know whether the prince donated only the chapel and kept some sort of residence for himself, but it is a fact that King Béla IV visited Porva in 1264.[61] It is likely that there remained

[43] Horváth 1930, 29–36.

[44] Horváth 1930, 28.

[45] *MRT-4*, 81/2.

[46] I do not include Palota, originally a Franciscan friary, which was held shortly by the Paulines in the fifteenth century.

[47] Vázsony, a late fifteenth-century foundation, was rather similar to the monasteries that will be described below. Because of the time gap, however, it will not be discussed either with these, or separately. For more on Vázsony, see *MRT-2*, 33/6; *DAP*, vol. 3, 201–211.

[48] Compare Hervay 1984a, 170.

[49] Gyöngyösi 1988, 45.

[50] It appears as such in Romhányi 2000, 148–149.

[51] Guzsik 2000. Firstly, following the itinerary of the conscription, Tálod would be a logical location. Secondly, we know that the Tálod monastery was dedicated to St. Elisabeth (a rare title in Co. Zala), as was the mysterious Idegsyt. Thirdly, the ending *-syt* might refer to the stream Séd, which flows near Tálod.

[52] For Tálod, see *MRT-2*, 41/7; *DAP*, vol. 3, 1.

[53] *DAP*, vol. 1, 308.

[54] *DAP*, vol. 3, 200; Guzsik 2000.

[55] See Romhányi 2000, 62, 30, 70, 68, 38.

[56] Holub 1929.

[57] *MRT-4*, 67/1; *DAP*, vol. 2, 305–308.

[58] *PRT*, vol. 2, 311, 483–484: "possessio Porwa vocata intra silvam Bokoniensem habita"

[59] *PRT*, vol. 2, 311.

[60] It does not strictly belong here, but one also wonders why Pannonhalma, and not Bél, got the chapel.

[61] He issued a charter here. *ÁÚO*, vol. 8, 94.

some inhabitants other than monks, because one hundred years later, there was a settlement in Porva,[62] although in 1392 (when Porva was given to the Gara family) it was reported to be deserted.[63]

The second phase in the life of Porva started with the rule of the Gara. It seems that both the chapel and the settlement were empty, and the Benedictines of Pannonhalma lost control over Porva for good. I have already discussed the foundation of the Pauline monastery in Chapter 14. The most striking feature of this was that the Gara, *mutatis mutandis*, repeated what the royal family did in Pilis Forest in the thirteenth century. By 1450, the "new" Porva comprised a small chapel, some stone houses and a *curia* for the hunters, very much in the way we imagine early royal residences.[64] We cannot tell whether they kept the building of the original chapel, but the dedication changed from St. Emeric to the Holy Virgin. The foundation charter informs us that the houses and the *curia* were built by the Gara, and, indeed, in 1431, the abbot of Pannonhalma was complaining about such activities in Porva.[65] The Gara were also hunting in "their" Forest, and exactly as the king gave his hunting lodges to the Paulines in Pilis in the thirteenth century, the Gara gave them their lodge in Porva. A further similarity is that there was no settlement around the Pauline monastery.

17. 3. 2. Economy

The economy of the Porva monastery was based on its settlements, vineyards, and mills. Before the actual construction began, Ladislaus of Gara (the father of the issuer of the 1450 foundation charter) donated five settlements to the monastery.[66] In 1450, only one of these (Újfalu, Co. Veszprém) was mentioned, but there were many more things. The monks received two mills (and a promise for a third one to be built by Ladislaus), a house in the town of Pápa, a vineyard in Lovászpatonya, rights to use woodland in Szerecsen (Co. Győr)[67] and in Bakony, and two ploughs of arable around Porva.[68] From the second half of the fifteenth century, vineyards and mills became more and more important. Strangely, the monastery was continuously getting vineyards and buying mills. Between 1454 and 1520, the monastery was donated seven vineyards, and bought[69] two mills (and possibly constructed a third one).[70] We should remember that mills also appeared important for the monastery of Szentlászló, in Pilis.

Woodland was a significant, but restricted element in this system. Already before 1450, the monks obtained rights to cut wood in a certain "Aszóerdő."[71] Then, as if Bakony had been somehow considered insufficient, Ladislaus donated to them the above-mentioned rights in the woods of Szerecsen, some forty kilometres from Porva. This would suggest commercial activity, especially because the monks were strictly forbidden to sell wood from Bakony. The same, however, seems to have applied to the *silva usualis* in Szerecsen as well. It was to provide wood for the inhabitants of Újfalu and no one else.[72] It is noteworthy that there were still twenty kilometres between Újfalu and Szerecsen. It seems that that the whole of Újfalu belonged to the Gara (and then to the Paulines), thus, the villagers would have had access to woodland if there had been any around. How this could happen in the vicinity of Bakony is a far-reaching question.

17. 4. The Carthusians

The Carthusians were the last to arrive in the Bakony region. They were latecomers to Hungary, in general.[73] Before the Mongol invasion, they had one monastery (Ercsi, Co. Fejér) originally a Benedictine, then a Cistercian house, which existed only for a few years. Their first lasting foundations were Letanovce (Sk; Co. Szepes) and Červeny Klaštor (Sk; Co. Szepes) in 1299 and 1319, respectively.[74] These monasteries were established in a region of high mountains that largely resembled the environment of the Grande Chartreuse in the Alps. St. Bruno would probably have been somewhat less content with the location of the next monastery, Tárkány (Co. Borsod, 1332) which was surrounded by mountains, but the Bükk Mountains were far from providing the kind of seclusion the Carpathians or the Alps could offer.

Lövöld was the last Carthusian foundation in Hungary,[75] but the first royal one. Its foundation charter was issued in 1378, but most probably it was founded decades earlier.[76] Construction workers and the prior himself were involved in a case of *acta potentiaria* in 1369, and the monastery's estates were mentioned in 1357. Péter Németh proposed that the foundation of the monastery could be connected to the death

---

[62] *PRT*, vol. 2, 483–484.

[63] *HO*, vol. 7, 428.

[64] Dl 14 424. Note that the charter is barely legible where it talks about the stone houses. It is hard to decide whether one should read "certa domus lapidea" or "certe domus lapidee," that is, one or more houses. The regesta prepared in OL decided on the latter.

[65] *PRT*, vol. 3, 267.

[66] We know this from a charter of King Ladislaus (1441) in which he confirmed the donation. *DAP*, vol. 2, 305.

[67] Or occasionally Co. Veszprém. *Csánki*, vol. 3, 560.

[68] These two ploughs were reported to be wooded land (*terra silvosa*) that the monks had to clear. This reinforces the idea that the earlier settlement had disappeared.

[69] Or the mill was impignorated to them.

[70] *DAP*, vol. 2, 305–306, 308. The existence of the third mill is uncertain, because it may be the second mill in Pápa, mentioned in the foundation charter. In 1450 there is a mention of two mills in Pápa on the Tapolca stream, the second of which was not for grain, although the monks could have transformed it if they had wanted to. In 1456, Ladislaus of Gara provided an opportunity for the monks to build a second mill on the Tapolca right next to the one they already had. Were these two one and the same?

[71] Note the name. For a treatment of *aszó*, see Chapter 4.

[72] Dl 14 424. "ad usus eorum proprios dumtaxat et non aliorum"

[73] For a history of Carthusians in Hungary, see Török and Legeza 2001; Dedek 1889; Mályusz 1971, 224–247.

[74] Romhányi 2000, 23, 40, 43.

[75] Although a faint attempt was made in 1494 to give the Premonstratensian monastery of Oradea (Ro; Co. Bihar) to the Carthusians. Romhányi 2000, 70.

[76] The best study on the history of the monastery is Németh 1993–1994. See also Dedek 1889, 137–143; Török and Legeza 2001, 38–42. For the archaeological part, see *MRT-2*, 48/1; Csengel and Gere 1996. For the foundation charter, see *CD*, vol. 9/5, 239–244.

of Prince Andrew, brother of Louis I.[77] 1378 signified the end of the long foundation and building process. In the charter, the king went on for lines and lines, praising the chastity and purity of the order. His feelings were, apparently, shared by others. In 1372 – as Gyöngyösi reported – two Pauline priors complained to the bishop of Esztergom that the monks were leaving their monasteries to become Carthusians in Lövöld.[78] The Paulines, however, were not breaking the rules. In 1329, Pope John XXII ordered that the Paulines, if they wanted to leave the order, could only go to Carthusian houses.[79] We should note that King Louis was the greatest benefactor of the Pauline order. It is perhaps justifiable to infer that there was a connection between him and the monks both turning away from the Paulines and towards the Carthusians in the second half of the fourteenth century.

Geographically speaking, the monastery did not quite have the isolated position that is, even today, associated with it.[80] On the contrary, it was next to a busy market-town and just off the main east-west road crossing Bakony.[81] King Louis related in the foundation charter that he had had the castle of Hölgykő demolished and given to the monks. The monastery was, in fact, built close to the former castle. The location would become more understandable if we assumed that the castle was destroyed to provide building material for the monastery. However, it was not unusual in the *Alemaniae Superioris* province of the Carthusians (where all Hungarian monasteries belonged) for the monasteries to be next to larger towns or main commercial routes. (Prague and Brno speak for themselves, but for example Maurbach or Aggsbach were located along the Danube.)[82] We should not forget that the king could have settled the Carthusians in more solitary locations – although this would probably have met with some resistance from the recently established royal hunters of Szentgál, Németi, and Csepel.[83]

Speaking of royal hunters, I have already mentioned in Chapter 14 that Hölgykő belonged to the *honor* of the Royal Forest, and the name itself refers to the existence of hunters. Similarly, Ordólövöld (one of the three settlements that belonged to Hölgykő in 1378) was once populated by Forest-guards. If somewhat distantly, this still resembles thirteenth-century Pauline foundations in Pilis, or twelfth-century Cistercian foundations in Bakony and Pilis, when the monasteries took over the system established to manage the Forest. In Lövöld, however, there was little continuity: Ordólövöld was deserted before it was given to the monks.[84]

The Lövöld Carthusian monastery was very rich. So rich, in fact, that modern researchers were sometimes reluctant to believe the data demonstrating this wealth. In 1498, the prior was compelled to provide two hundred horsemen for the army of the kingdom, the same number as, for example, the bishops of Győr and Veszprém.[85] Sörös thought that this must have been a mistake for one hundred, although the number two hundred was repeated in 1528.[86] The monastery's wealth was based on its settlements. During its two-hundred-year existence, the monastery possessed (fully or partially) thirty-two villages, and two houses in Óbuda and Pápa. As a latecomer in the landscape, however, it was involved in continuous confrontations to preserve its estates. From *acta potentiaria* we also know that in the fifteenth century the monks were also involved in cattle and wine trade.[87] All this money permitted cultural investments to be made, as well. The most famous book produced here was the work of an unknown monk, usually referred to as the Érdy codex.[88]

17. 5. The Hermit of Szentkereszt

"No Forest was really complete without a hermit," wrote Rackham about English Forests.[89] There was an official hermit in Bakony Forest as well, and it is his dwelling place that I shall comment upon before finishing the description of monastic presence in the Forest. In Chapter 14, I have described how King Andrew II donated some land to the existing chapel of the Holy Cross (Szentkereszt) near Ganna, and how this decision was questioned by the Forest-guards in 1240. The chapel belonged to the monastery of Bél, but by the early fourteenth century it had fallen out of use. In 1338, Abbot Nicholas issued a charter concerning the chapel.[90] A certain poor brother (*pauper frater*) John, he wrote, came to him and asked him to give him the chapel so that he could live there. Nicholas, considering that there were no hermits living at the place at that time (which probably means that earlier there were) granted John what he wanted upon certain conditions. John had to build a cell for himself, and a house for travellers, because the chapel was "along the road that leads to the monastery of St. Martin."[91] It was, of course, a normal duty of hermits to maintain roads and house travellers all over Europe.[92] We do not hear about John any more, but a perambulation attested that the monastery still owned the land around the chapel (and, if John succeeded, the houses) and maintained its boundaries in 1381.[93]

---

[77] Németh 1993–1994, 370, with references to the sources.

[78] Gyöngyösi 1988, 74.

[79] Török and Legeza 2001, 40.

[80] Török and Legeza 2001, 39.

[81] *CD*, vol. 9/5, 240: "iuxta oppidum nostrum Leweld vocatum" In 1488, Ardóleveld and Nagyleveld together had more inhabitants than any other town in Co. Veszprém, except for Pápa. Solymosi 1984, 191.

[82] For the Carthusians in Europe, see for example Aston 1993a; Lockhart 1985.

[83] For these, see Chapter 14.

[84] *MEO*, vol. 1, no. 72.

[85] Török and Legeza 2001, 41.

[86] *PRT*, vol. 3, 294. Quoted in Németh 1993–1994, 373.

[87] Németh 1993–1994, 373–374.

[88] Volf 1876.

[89] Rackham 1989, 61.

[90] *AnjOkm*, vol. 3, 468–469.

[91] That is, Pannonhalma. "secus viam que ducit ad monasterium beati Martini"

[92] Gilchrist 1995, 208.

[93] *PRT*, vol. 8, 411–412.

17. 6. Conclusions

Monastic presence was more diverse in Bakony than in Pilis. In addition to the Cistercians and the Paulines, the Benedictines and the Carthusians also had monastic houses in the region. The Benedictine monastery in Bél was a foundation that made Bakony Forest similar to early medieval European Royal Forests. Here, there was no direct evidence for the presence of an early royal residence, however, an analysis of the monastery's estate structure in the late eleventh century strongly suggested that the monks had taken over an existing system. The strongest similarity between Bakony and Pilis was the presence of the Cistercians. The Pilis and Bakony monasteries were established roughly at the same time and as parts of the same royal enterprise. In Zirc, the early royal centre was identified both in the written and the archaeological sources. The monasteries were also similar in their dependence on the secondary economy and in the apparent lack of land-colonisation and other such activities, traditionally associated with the order. The Paulines, although present in both Forests, had different histories. They were missing from Bakony Forest until the fifteenth century, but they had many early monasteries nearby. While the Paulines were established in Pilis through royal power, in Bakony (including the above-mentioned early monasteries) foundations were established by the nobility. An interesting aspect of the history of the Porva monastery is that its noble founders imitated, even if not intentionally, thirteenth century royal deeds in the fifteenth century. The appearance of the Carthusians in Bakony in the fourteenth century was, in many respects, a repetition of what had happened to the Paulines in Pilis one hundred years before. Lövöld was a royal foundation, and was intended to revive the ascetic discipline that was reported to have loosened among the Paulines. All monasteries, except for Bél, were similar at least in one respect. They were isolated and in a strategically important position at the same time. In practice, this meant a location comfortably close, but not quite on, a main commercial route. In Bakony, the location of the Carthusian monastery in particular displayed features in which monastic ideology and commercial reality were intricately mixed, and the same held true for the Cistercian house in Zirc.

Far less is known about the economy and possible industrial activity of the orders in Bakony than in Pilis. This is entirely due to the absence of archaeological excavations at the sites. Even in Zirc, where there were excavations, nothing is known about the monastic precinct. In Bél, we are left with written sources alone. Given the very limited nature of the written evidence, if we want to know more about the interactions between *woodland* and the monasteries – for most of what I have argued in this chapter had to do with the *Forest* and the monasteries – we shall have to hope for considerable advances in the monastic archaeology of the region.

Chapter 18

What Does It All Mean?

I have borrowed the light-hearted title of this concluding chapter from Michael Aston's *Interpreting the Landscape*, because I agree with him that the findings of any scholarly work should come together to have a meaning. The title, in its simplicity, also expresses my conviction that this meaning should not be overcomplicated. In some branches of science this cannot be achieved easily, however, historical ecology has the advantage that "most of the results are not too remote from ordinary experience."[1]

In the first part of this book, I have investigated woodland and tree management in medieval Hungary. No one before has approached woodland history using the methodology of historical ecology in Hungary; earlier works interpreted woodland management in comparison to modern forestry. My results have shown that Hungarian woodland management was essentially similar to what has been demonstrated for England, and more recently for other European countries, as well. We can now fairly safely argue that medieval Hungary shared in the common heritage of *traditional European woodland management*, although there were some features specific to the Carpathian Basin. The study of this heritage has a significant position in current historical ecological research in Europe, and my work provides the Hungarian piece in this grand puzzle. Here, however, I do not intend to repeat my earlier conclusions; for these, the reader is referred to Chapter 8. In the second part of my work, I added short conclusions to each chapter, while those in Chapters 14–17 contained comparisons between Pilis and Bakony Forests, as well. These will not be repeated, either. Instead, I shall discuss some general questions about Forest history in Hungary, and about how it relates to the history of medieval woodland. I shall also consider the methodological points that my research has raised, in other words, how this book has added to woodland and Forest history besides dealing with a region unstudied so far.

In connection to woodland, I have pointed out some specifically "Hungarian" phenomena and have also attempted to indicate possible ways in which knowledge of woodland management may have been imported into Hungary. The same questions – which, in this case, mostly overlap – can be asked about Royal Forests. Beyond any doubt, the kind of clear distinction between woodland and Forest that was characteristic of Britain from the eleventh century on did not exist in contemporary Hungary. I have argued in several chapters that the Hungarian system resembled more early medieval Continental Forests, both in its association with marginality (but not only woodland) and in its ambiguous terminology. It seems that in the eleventh and early twelfth centuries, there existed a mixture of ideas connected to woodland, hunting, royal power and representation, which reflected Western European conditions of a somewhat earlier period, in the same way as the whole structure of the Kingdom of Hungary did. The idea of Royal Forests was an essentially western phenomenon. The Hungarian tribes that arrived in the Carpathian Basin in the ninth century brought many things with them. They had, for example, sophisticated animal husbandry and a fair knowledge of agriculture. This eastern heritage can be traced in many fields of life.[2] They probably also knew how to cut trees in different ways to house, feed, and warm themselves. However, managing economically (but not necessarily geographically) marginal lands as Forests, or, in fact, managing economically marginal lands at all, was not part of their eastern heritage. Forests represented a western way of approaching this question. That is why Forests appeared when the (western type) kingdom was organised, and perhaps that is why, on a terminological level, they were connected to counties, which were also, to a large extent, an imported concept.

Changes in this structure came together with social changes on a larger scale in the late twelfth and thirteenth centuries. The new system that crystallised through these changes – the "classical" Hungarian Royal Forest with its *comes*, Forest-guards, territorial integrity, etc. – was very much a product and indicator of local conditions. Most importantly, it demonstrated the strength (or weakness) of royal power. In the early Árpádian period, this power was conspicuous and noticed by foreigners.[3] It is not accidental that no perambulation of a Royal Forest survives. Very few people would have gone so far as to question the king's authority to this extent. After this, when royal power was at a nadir – at the end of the thirteenth century – Forests basically disappeared in an administrative sense. Then in the 1320s, by which time Charles Robert commanded an authority comparable to his twelfth century predecessors, Royal Forests were also revitalised. The different histories of Pilis and Bakony also reflected the influence of royal power in their respective regions. In Pilis, this power was unquestionable because of the closeness of Buda and Visegrád. That is why there were no castles in the region, and monasteries were all royal foundations. In contrast, Bakony was far less dominated by the royal house. There were also castles here, some non-royal. The Pauline monks appeared in the region completely independently from the king both in the thirteenth and the fifteenth centuries. Further studies on other Forests would certainly provide more details of a similar nature. In sum, the *idea* of Royal Forests seems to have arrived in Hungary from the west, however, from the thirteenth century onwards, Hungarian Forests developed independently.

Historical woodland studies can be carried out along several lines, focusing for example on spatial dimensions or types of land ownership. It is possible to examine one

[1] Rackham 1980, viii.

[2] See for example Paládi Kovács 1981.

[3] I refer to the oft-quoted sentence of Otto of Freising about how all the Hungarians obey their king. Gombos 1937, vol. 3, 1767.

particular wood, a region, or a whole country.[4] A usually far less considered aspect is that different types of ownership (royal, ecclesiastical, noble, communal, etc.) can also produce different types of woodland management. When I chose Pilis and Bakony Royal Forests, I opted for case-studies that were *regional* and concentrated on *royal lands*. An advantage of this was the existence of a *longue durée* policy that apparently changed little through the Middle Ages and was kept up later, preserving some of the ancient woodland in the Forests up to the present day. It is a fact that Pilis, where royal presence was stronger, preserved more of its botanical richness than Bakony, although the former lay closer to major towns. I dare say that the successful "conservation efforts" of the royal house – very far from our modern ideas, of course, mostly meaning not allowing *melioratio terre* to occur – played an active role in this. Also, could it be a pure accident that in Bakony, the only environmentally protected areas today are those (around Bakonybél) that were so conspicuously left empty in the eleventh century?[5] Royal ownership was also convenient, because it left documents. We will never be able to say anything about the administrative framework of the management of woods owned by a village community, because in Hungary in the Middle Ages no one wrote down these rules.[6] In contrast, there exists some material about how this was done in the woods of towns.[7] The Church was an eminent record-keeper throughout the Middle Ages, therefore their woods are also a promising target for future research. An important question here is whether certain types of ownership can be connected to certain types of woodland. Did, for example, the Church own more coppice woods and the nobles more acorn-bearing woods? Did land ownership matter in this respect at all?

Finally, let us turn to some wider-ranging implications of this work, both in results and in methodology. In 2000, John McNeill published his renowned book *Something New under the Sun*, in which he used a metaphor to describe the two basic types of ecological behaviour. Species and societies, he wrote, can be either shark-like or rat-like. The strategy of an ecological shark "is supreme adaptation to existing circumstances, which can work well for a while, a long while if circumstances are stable." In contrast, an ecological rat tries to "be adaptable, to pursue diverse sources of subsistence – and to maximize resilience."[8] I find that these attitudes can also be traced in woodland management. Modern forestry is shark-like in the sense that it is dependant upon cheap fuel to move all the machinery it uses in the production and processing of its material and upon a market that values timber highly. In medieval England, the only country where such data are available, we know that most woods were coppiced by the thirteenth century.[9] Coppice woodland management is also partly shark-like because it represents a supreme adaptation to the self-regenerating power that trees have. This is fairly safe: just as sharks can count on the continuous abundance of fish in the sea, there is no reason why trees should lose their ability to coppice. Having standard trees in coppice woods is more rat-like, because these trees form a depository that can be used in times of need. The decline of coppicing in the fossil fuel age showed the drawbacks of the system. With little demand for firewood, the very intensive management that coppice woods require was not maintained, and the woods fell into such neglect that the revival of coppicing is not always possible.[10] My research has shown that medieval Hungarian woodland management was ecologically rather rat-like. There were several types of woodland, and management ranged from extensive to intensive. *Permissoria* provided firewood and timber, acorn-bearing woods could be used for timber and pasture, there was *rubetum* for wood and pasture, and *nemus* to combine water, trees, and animals. If there were environmental "surprises, shocks, and catastrophes"[11] in connection to woodland, medieval Hungarians had a number of ways to be rats and adapt to the new set of circumstances.

As far as methodology is concerned, I would like to emphasise one aspect. When I approached the history of Royal Forests, I came to realise that some of the sources traditionally used for the purpose in Western Europe were not available to me – it is hard to study royal lands without the Royal Archives, for example. Erik Fügedi wrote once in an essay about medieval Hungarian historical demography that "it may sound paradoxical, yet it is true that these deficiencies force Hungarian researchers to apply methodologies that would never occur to scholars in a country well-equipped with written sources."[12] Originally, the analyses of the settlement system and the monastic orders in Pilis and Bakony were to compensate for such "deficiencies." However, the results have turned out to be more than mere gap-filling. They have opened new perspectives in the study of Forests. Involving these source materials allowed for conclusions that would not have been available otherwise, with or without the Royal Archives. These methodological considerations do not apply to Forest history alone. In the study of woodland, coppicing is a typical example. In Western Europe, written documents are mostly connected to the commercialising of coppicing. Coppice cycles are known because it seemed useful in the fourteenth century to write down management plans to see how much income a certain wood could provide, and then to record the actual sales of firewood. Coppicing, nonetheless, does not *have to* go hand in hand with commerce. As a logical and easy way to produce firewood, it already occurred in prehistory. In medieval Hungary, trade in firewood was never quite so important as to leave easily detectable traces in the sadly few economic sources. Supposing that coppicing may still have been present, I had to approach it from a different angle, which – partly – was linguistics. However, this line of investigation did not stop at *permissionalis* woods. With this

[4] Best in this order.

[5] Apart from the basalt region just north of Lake Balaton.

[6] There exists some material from the early modern period, notably from the Sekler areas of Transylvania. Imreh 1973. In Western Europe, such sources were more common in the Middle Ages. For example, an analysis of some Austrian *Weistümer* is Winiwarter 1999.

[7] See, for example, the woods of Sopron in Házi 1933, 117–145, 217–229.

[8] McNeill 2000, xx. By the way, McNeill considers feudal Europe an ecological shark.

[9] This does not include wood-pasture.

[10] Peterken 1996, 410–413.

[11] McNeill 2000, xx.

[12] Fügedi 1981, 388.

method, I have gained valuable information about other types of woodland, which, again, would not have been decipherable from economic documents.

In Chapter 8, I have briefly referred to what I consider tasks for future research in woodland history. Most importantly, more and more individual woods should be studied. Hungarian woods, however, are not always well-defined smaller units. In both Pilis and Bakony, woodland has been dominant for millennia and it does not seem to have had long-lasting boundaries. These woodlands are easier to interpret and understand as components of a region, as I did in the second part of this book. The fact these particular woods lay inside a Royal Forest is a related issue, which has its relevance to their history by providing a larger landscape historical context. There are, however, other woodlands (mostly those in the Hungarian Middle Mountain Range) that appear to be better understandable on a regional basis, and where, consequently, the methods used in my Forest studies will hopefully prove useful.

There remains much to do in the research of Royal Forests themselves. Here again, as many individual Forests should be studied as possible. At present, only Pilis and Bakony are covered in *MRT*,[13] therefore it would be essential to discover adequate ways to approach other Forests archaeologically. More elements of the landscape could also be taken into consideration. For example, the analysis of the road network or water system of Forests may provide further insights into Forest history, complementing the four aspects I have chosen to deal with. Both Forests and woodland offer many more aspects to consider and discoveries to make than could be handled in this book. My work, I hope, has established the foundations and set some possible directions.

[13] This is unlikely to change, as most major medieval Forests (other than my two case studies) now lie outside Hungary and will not be treated in a *Hungarian* topography in the foreseeable future.

APPENDIX

Appendix 1. Data on land-use in estimations.

(Country codes: Au-Austria, Cr-Croatia, H-Hungary, Ro-Romania, Sk-Slovakia, Slo-Slovenia, Sr-Serbia and Montenegro, Ua-Ukraine.)

source	date	settlement	country; county	arable: iugerum	meadow: falcastrum	woodland: iugerum	pasture: iugerum	vineyard: iugerum	remarks
Pesty 1882, 135-137	1378	Potoc (?)	Ro; Krassó	1725	150	600 s.			
ZsOkl, vol. 1, 245-246	1391	Bény	H; Esztergom	1500	58	675 t. saligniose, glandinose, rubeta			
ZsOkl, vol. 1, 245-246	1391	Martin nad Žitavou	Sk; Bars	1350	16	0			
ZsOkl, vol. 1, 677-679	1399	Zsadány	H; Zemplén	375	200	225 s. glandinosa			
ZsOkl, vol. 1, 677-679	1399	Szalonna, Meszes, Martonyi	H; Borsod	1200	150	825 s. glandinosa 600 rubeta			
ZsOkl, vol. 1, 677-679	1399	Perkupa, Dobódél, Varbóc	H; Torna	1200	150	1200 s. 300 rubeta			
ZsOkl, vol. 1, 677-679	1399	Szin	H; Torna	900	100	150 s. glandinosa 150 rubeta			
ZsOkl, vol. 1, 677-679	1399	Jósvafő	H; Torna	900	0	450 rubeta			
ZsOkl, vol. 1, 677-679	1399	Petri	H; Torna	600	6	300 rubeta			
ZsOkl, vol. 1, 677-679	1399	Domolaháza	H; Torna	600	0	0			
ZsOkl, vol. 1, 677-679	1399	Ardó, Kutya, Feketefűz, Silica	Sk; Torna	3750	8	1200 s. glandinosa 1200 rubeta			
ZsOkl, vol. 1, 677-679	1399	Silická Brezová	Sk; Torna	1050	10	300 s. glandinosa 300 rubeta			
ZsOkl, vol. 1, 677-679	1399	Králik	Sk; Gömör	300	26	0			
ZsOkl, vol. 1, 677-679	1399	Závadka	Sk; Sáros	300	20	150 rubeta			
ZsOkl, vol. 2/1, 54-55	1400	Velika Dobrony	Ua; Bereg	450	18	900 s. glandinosa			
Zichy, vol. 12, 81-83	1404	Battyán, Malafalva, Jánosi, Kompoltfalva (?)	H; Tolna	750	27	100 s. glandinosa			arable together with vineyards
ZsOkl, vol. 2/1, 642	1406	Berkesz	H; Szabolcs	1200	20	60 permissoria			
ZsOkl, vol. 6, 511-514	1418	Almaşu Mare, Veresnégyfalva (?), Almaşu de Mijloc, Glod, Nadasd superior, Nadasd inferior	Ro; Hunyad	3000	0	5000 s.			arable together with rubeta, woodland calculated from value
ZsOkl, vol. 6, 511-514	1418	Ilia	Ro; Hunyad	1200	5	1800 bardalia s. glandinosa communis			
ZsOkl, vol. 6, 511-514	1418	Gedenfalva (?)	Ro; Hunyad	1950	0	150 s. communis			woodland calculated from value
Ila and Borsa 1993, 140	1423	Hangony	H; Gömör	150	20	450 s. glandinosa			
Ila and Borsa 1993, 140	1423	Susa	H; Gömör	300	8	37.5 permiss.			arable with nemora

Károlyi, vol. 2, 104-106	1427	Poroszló	H; Heves	2550	600	75 s. salicea			
Károlyi, vol. 2, 104-106	1427	Tilaj	H; Borsod	1800	4	1 permiss.			
Károlyi, vol. 2, 104-106	1427	Ivány	H; Heves	1800	50	10 permiss.			
Károlyi, vol. 2, 104-106	1427	Szőke	H; Heves	1200	20	0			
Dl 95 818	1433	Širákov	Sk; Hont	40	12	64 s.			
Dl 38 210	1436	Bilki, Gyibrivka	Ua; Bereg	450	4	3150 s. glandinosa			unit of measurement for meadow not given
Dl 38 210	1436	Imszticsovo, Lukovo, Velikij Rakovec	Ua; Bereg	450	2	750 s. glandinosa			unit of measurement for meadow not given
Dl 38 210	1436	Lipcsa, Horincsovo	Ua; Mára-maros	300	3	900 s. glandinosa			unit of measurement for meadow not given
Dl 55 062	1436	Bököny (?)	H; Szabolcs	225	0	150 s. cum virgultis et nemoribus			arable together with meadow
Sopron, vol. 2, 241-243	1436	Csér	H; Sopron	19	0	11 permiss. et rubetum utensile			woodland given only in width
Dl 57 678	1438	Solivar, Boltafalva, Dulova Ves, Felsőújfalu, Alsóújfalu, Csalános, Hanušovce nad Topľou, Bystré, Hermanovce nad Topľou, Pavlovce	Sk; Sáros	4950	0	9000 s.			arable together with pasture
Dl 57 678	1438	Sačurov, Davidov, Szent-györgy	Sk; Zemplén	750	17	750 s.			arable together with pasture
Dl 57 678	1438	Nový Kamenec	Sk; Zemplén	375	40	150 permiss. commune			
Dl 57 678	1438	Veľký Horeš	Sk; Zemplén	150	100	300 s. glandinosa communis cum nobilibus de Ymbregh (Brehov)			arable together with pasture
Dl 57 678	1438	Sirník, Brehov, Kisfalud, Cejkov, Zemplínske Jastrabie	Sk; Zemplén	375	101	150 s.			arable together with pasture
Dl 57 678	1438	Mocharosmal (?)	Sk; Zemplén	750	60	450 s. sub venatione			arable together with pasture
Dl 92 876	1438	Nádasd	H; Vas	35	4	300 s. usuales mixtim cum rubetis et vineis (!)			
Dl 92 876	1438	Kukeč	Slo; Vas	10	2	0.5 permiss 10 rubetum cum s. usuali et vineis (!)			
Dl 92 876	1438	Alsópetánc, Felsőpetánc, Petánc predium	Slo; Vas	64	0	12 nemora, rubeta et virgulta			
Békefi 1898, 316-322	1444	Nekézseny	H; Borsod	96	16	450 s. et rubeta		50	

Békefi 1898, 316-322	1444	Jákfalva	H; Borsod	72	4	0			
Békefi 1898, 316-322	1444	Dövény	H; Borsod	54	12	150 permiss. 450 rubeta			
Békefi 1898, 316-322	1444	Hartyán	H; Pest (?)	120	16	300 s. dolabrosa 450 s. usuales et rubeta			
Békefi 1898, 316-322	1444	Megyer	H; Nógrád	126	12	600 s.		200	
Békefi 1898, 316-322	1444	Bátony	H; Heves	45	6	150 s.			
Békefi 1898, 316-322	1444	Dorog	H; Heves	33	18	300 s. et rubeta			
Békefi 1898, 316-322	1444	Pásztó	H; Heves	450	2010	600 s. et rubeta cum s. Matra		200	
Dl 14 261	1449	Telekes	H; Vas	100	4	40 rubetum 130 s. glandinosa			
Dl 94 203	1459	Homokkomár	H; Zala	750	82	750 s. et rubeta			
Dl 94 203	1459	Obronak	H; Zala	80	12	80 s. et rubeta			16 vineyards
Dl 94 203	1459	Budics	H; Zala	80	12	80 s. et rubeta			32 vineyards
Dl 94 203	1459	Kenthyan (?)	H; Zala	80	12	80 s. et rubeta			60 vineyards
Dl 94 203	1459	Homokfalu	H; Zala	80	8	80 s. et rubeta			8 vineyards
Dl 94 203	1459	Szentmiklós	H; Zala	96	20	100 s. et rubeta			16 vineyards
Dl 15 831	1463	Mindszent, Tordafalva	H; Zala	76	26	100 s. et rubeta			
Dl 45 113	1464	Derzs	H; Heves	450	0	450 nemus pirorum et s. salicum			
Dl 86 405	1469	Adásztevel	H; Veszprém	116	8	0			
Dl 67 050	1469	Drahňov	Sk; Zemplén	270	0	300 s. glandinosa	300		
Df 283 678	1476	Bârsa	Ro; Közép-Szolnok	150	200	150 s. glandinosa			
Zichy, vol. 11, 337-342	1483	Csente	Sr; Bodrog	0	15200	450 perm. 1500 s. silicea sub cust. 4500 s. silicea et communis			
Zichy, vol. 11, 337-342	1483	Tótfalu	Sr; Bodrog	3750	60	3000 s. silicea perm. 4500 s. communis	4500		
Zichy, vol. 11, 337-342	1483	Negyven	Sr; Bodrog	2700	100	3300 s. silicea permiss. sub custodia	4200		
Zichy, vol. 11, 337-342	1483	Béküllőd	Sr; Bodrog	0	4600	300 s. silicea et comm. 1500 perm. 2250 nemus vlgo Berek	3000		
Zichy, vol. 11, 337-342	1483	Avató	Sr; Bodrog	2550	2290	1200 perm. 3000 s. communis 1500 nemus	4800		
Zichy, vol. 11, 337-342	1483	Keresztes	Sr; Bodrog	6000	50	0	9000		
Zichy, vol. 11, 337-342	1483	Bezdan	Sr; Bodrog	4500	50	300 rubetum	6000		
Zichy, vol. 11, 337-342	1483	Vanna	Sr; Bodrog	15000	200	0	22500		
Dl 19 607	1489	Tolma	H; Esztergom	129	31	200 s. et rubeta			

Dl 19 607	1489	Kovácsi	H;Esztergom	60	10	10 permiss 50 s. communis et rubeta			
Dl 19 607	1489	Agostyán	H; Komárom	443	21	5 s. prohib. 50 s. et rubeta		28	
Kubinyi 2001, 12-17	1489	19 settlements in the estate of Császárvár	Cr; Varasd	915	737	1780 s., nemora, rubeta et vinee			woodland together with (possibly high amount of) vineyards
Dl 3 022	1492	Dömsöd	H; Solt	1275	500	6 s.	600	16	
Dl 3 022	1492	Veresegyház	H; Pest	4800	120	241 rubeta	630	60	
Dl 3 022	1492	Szada	H; Pest	150	75	225 s. cum alia s. dolabrosa	75	40	
Dl 3 022	1492	Gödöllő	H; Pest	150	25	300 s.	225	20	
Dl 3 022	1492	Lőrinci	H; Pest	900	212	150 s. 150 permiss.	300	150	
Dl 3 022	1492	Kurov	Sk; Sáros	225	4	75 s.			
Dl 3 022	1492	Gaboltov	Sk; Sáros	150	6	75 s. communis			
Dl 3 022	1492	Cigel'ka	Sk; Sáros	3	3	2 s. communis	2		
Dl 3 022	1492	Petrová	Sk; Sáros	10	3	2 s.	10		
Dl 3 022	1492	Vyšný Tvarožec	Sk; Sáros	4	2	1 s. communis	2		
Dl 3 022	1492	Nižný Tvarožec	Sk; Sáros	10	3	40 s.			
Dl 3 022	1492	Varadka	Sk; Sáros	7	2	0			
Dl 3 022	1492	Zborov	Sk; Sáros	75	20	450 s. communis			
Dl 3 022	1492	Andrejová	Sk; Sáros	40	0	30 s. communis			
Dl 3 022	1492	Komárov	Sk; Sáros	150	4	50 s. communis			
Dl 3 022	1492	Bardejovská Nová Ves	Sk; Sáros	150	5	25 s.			
Dl 3 022	1492	Kurima	Sk; Sáros	50	10	150 s.			
Dl 3 022	1492	Cernina	Sk; Sáros	25	4	50 s. communis			
Dl 3 022	1492	Radoma	Sk; Sáros	25	3	40 s.			
Dl 3 022	1492	Dubinné	Sk; Sáros	150	8	25 s.			
Dl 3 022	1492	Mlynárovce	Sk; Sáros	100	2	150 s. communis			
Dl 3 022	1492	Rovné	Sk; Sáros	225	0	75 s. communis			
Dl 3 022	1492	Poliakovce	Sk; Sáros	150	4	100 s.			
Dl 3 022	1492	Štefurov	Sk; Sáros	75	5	100 s.			
Dl 3 022	1492	Tharafalwa	Sk; Sáros	25	0	6 s.			
Dl 3 022	1492	Šašová	Sk; Sáros	225	3	75 s.			
Dl 3 022	1492	Ortuťová	Sk; Sáros	150	2	75 s.			
Dl 3 022	1492	Okrúhle	Sk; Sáros	300	3	150 s.			
Dl 3 022	1492	Soboš	Sk; Sáros	225	3	150 s. communis			
total		157 settlements	26 counties	90238	28869	71509	56144	764	

Appendix 2. Number of settlements and percentages of different types of land-use in estimations by county.

county	no. of settlements	arable (incl. vineyards) %	meadow %	woodland %	pasture %
Bács	5	97.5	0.4	2.0	0.1
Baranya	42	46.4	14.4	35.3	3.9
Bars	1	99.0	1.0	0	0
Bereg	6	22.0	0.4	77.6	0
Bodrog	8	25.0	16.2	19.8	39.0
Borsod	7	55.1	3.1	41.8	0
Csongrád	3	100.0	0	0	0
Esztergom	3	62.0	3.6	34.4	0
Gömör	3	58.1	4.2	37.7	0
Heves	7	61.0	24.5	14.5	0
Hont	1	34.5	10.3	55.2	0
Hunyad	8	47.0	0	53.0	0
Komárom	1	86.1	3.8	10.1	0
Közép-Szolnok	1	30.0	40.0	30.0	0
Krassó	1	70.0	6.0	24.0	0
Máramaros	2	25.0	0	75.0	0
Nógrád	1	34.7	1.3	64.0	0
Pest	5	64.7	4.5	18.3	12.5
Pozsega	3	44.6	0	53.7	1.7
Sáros	35	41.0	0.7	58.2	0.1
Solt	1	53.9	20.8	0.2	25.1
Sopron	1	63.5	0	36.5	0
Szabolcs	2	86.0	1.3	12.7	0
Szerém	9	62.0	0	37.0	1
Temes	12	55.7	1.5	42.8	0
Tolna	4	85.5	3	11.5	0
Torna	12	60.8	1.8	37.4	0
Valkó	102	46.7	10.3	43.0	0
Varasd	19	26.7	21.4	51.9	0
Vas	18	18.5	3.0	78.0	0.5
Veszprém	16	27.7	5.6	66.7	0
Zala	8	46.3	6.4	47.3	0
Zemplén	13	49.2	8.4	37.5	4.9

Appendix 3. Data on land-use in Co. Somogy from 1229.

village	arable: iugerum	meadow: piece or iugerum	vineyard: piece or iugerum	woodland: iugerum	woodland common with villagers	remarks
Kuisfolwd	600	0	32 ps	40		
Nyn	36	0	4 ps cult, 4 ps inc	4		
Vchma	30	0	3 ps	4		
Kaas	360	0	20 ps	100		
Sumberem	3600	0	40 ps	1200		
Meger	200	7 ps	5 ps cul, 12 ps inc	20	yes	
Seuleus	240	5 ps	9 ps	12		
Told	1800	1 ps	20 ps cul, 20 ps inc	1000		
pr. iuxta Told	720	4 ps	0	60		
Azalar	2400	20 ps	15 ps cul, 20 ps inc	200	yes	
Vlwes	480	23	8 ps cul, 5 ps inc	200		
Wodoy	2040	20	15 ps	1800		habent in consuetudine extirpare silvas et implantare vineas
Baba	360	10	0	240		
Muchula	480	20	12 ps cul, 13 ps inc	120		
Merena	1800	0	30 ps	140		
Merena inferior	4800	40	0	1200		
Tumurken	360	satis	0	25		
Egis	720	30	0	40		
Mehus	480	0	0	20		
Kara	1680	satis	0	100		
Bev	360	4 ps	0	50		
Oztupan	600	10	0	35		
Quer et Bard	600	30	0	20		
Palyan	360	30	0	100		
Kulked	1440	40	0	50		
Tord	360	20	40 ps cul, 20 ps inc	20		
Tur	390	44	42 ps + 30	32		
Wysch	10	1 ps	4 ps cul, 2 ps inc	2		
Fer	600	12	14 ps cul, 10 ps inc	23		
Chyopok	50	0	4 ps	0		
Rad	600	10 ps + 10	45 ps cul, 34 ps inc	32		

Eusced	483	3	21 ps cul, 4 ps inc	3		
Cup	360	3	22 ps	0		
Vrs	920	52	49 ps cul, 58 ps inc	73		
Sti Nicolai de Zoulat	1220	28	123 ps	1		
Nezda	160	8	23 ps cul,18 ps inc	71		
Keccha superior	540	8 ps	32 ps	70		
Keccha inferior	360	8 ps	25 ps	8		
Feeyreghaz	360	20	35 ps	12		
Chepel	390	20	47 ps	60		
Karad	480	20	6 ps	20		
Teluki	480	20	60 ps cul, 40 ps inc	65		
Zarozozou	492	0	96 ps cul, 40 ps inc	29		
Keurushyg	1440	4 ps + 8	90 ps cul, 34 ps inc	50		
Kerequi	295	3 ps + 4	78 ps cul, 24 ps inc	45		
Kereky	273	7 ps	36 ps cul, 9 ps inc	42		same as previous?
Lyuba	122	3 ps	25 ps cul, 2 ps inc	2		
Chega	340	3 ps + 4	30 ps cul, 10 ps inc	8		
Baluanus	240	3 ps	14 ps cul, 7 ps inc	15		
Egis	60	0	1 ps	0		same as line 19?
Berky	3600	10 ps	82 ps cul, 70 ps inc	220		
Gyogu	540	9 ps	9 ps cul, 13 ps inc	10		
Endred	477	3 ps + 20	48 ps cul, 2 ps inc	78		
Scemes	120	0	4 ps cul, 3 ps inc	10		
Copul	120	0	2 ps cul, 2 ps inc	0		
Endred inferior	862	10 ps	61 ps cul, 24 ps inc	62		
Endred superior	727	13 ps	91 ps cul, 32 ps inc	37		terra cum villa
Zamard	1035	1 ps	109 ps cul, 33 ps inc	63		
alia Zamard	600	0	53 ps cul, 26 ps inc	16		
Kiralturky	480	5 ps	20 ps cul, 7 ps inc	15		
Joud	3600	60	200 ps	1200		fenetum ibi et in Gan
Clety	720	0	0	0		
Berey	720	8 ps	81 ps cul, 3 ps inc	28		
Gan	30	3 ps + 8	10 ps cul, 5 ps inc	20		
Chaba	260	4 ps	7 ps cul, 4 ps inc	11		
total	51462	151 ps + 617	686 ps, 1266 ps cul, 610 ps inc, 30 iug	9233		

Appendix 4. Different woodland types in estimations.

woodland type	**percentage**	**total (hectares)**	**Gara-Szécs (hectares)**	**other (hectares)**
silva	47.4%	48573	19831	28742
silva glandinosa	16.5%	16931	7113	9818
permissoria	10.3%	10554	2096	8458
rubeta	8.4%	8641	4858	3783
silva et rubeta	5.4%	5551	2706	2845
nemus	3.5%	3576	30	3546
silva et silva glandinosa	3.0%	3033	3033	
nemus, rubeta et virgulta	1.5%	1513		1513
silva glandinosa et rubeta	1.0%	1036	1036	
silva dolabrosa et glandinosa	0.6%	668	668	
virgulta	0.6%	601	601	
silva saligniosa, glandinosa et rubeta	0.5%	570		570
silva dolabrosa	0.4%	404	151	253
silva sub venatione	0.4%	380		380
silva cum rubetis et vineis	0.2%	262		262
virgulta et nemora	0.1%	127		127
silva et virgulta	0.1%	28	28	
silva salicea	0.1%	63		63
permissoria et rubeta	0.0%	9		9
total	100.0%	102520	42151	60369

Appendix 5. Illegal woodcutting 1400–1420.

year	month, day	ZsOkl, vol., no.	remarks
1400	2/20	II/1. 94.	
1400	5/28	II/1. 295.	
1401	1/1	II/1. 792.	200 cartloads
1401	8/23	II/1. 1205	
1402	4/26	II/1. 1596	
1402	10/18	II/1. 1995	
1402	12/8	II/1. 2102	
1403	1/1	II/1. 2901	
1403	2/23	II/1. 3005	
1404	11/1	II/1. 3943	date of charter: 1405.5.5
1404	12/6	II/1. 4307	date of charter: 1405.12.18
1405	3/14	II/1. 3714	
1405	3/17	II/1. 3727	
1405	6/20	II/1. 3994	
1405	6/29	II/1. 4080	
1406	2/24	II/1. 4476	
1406	9/22	II/1. 5008	
1406	12/21	II/1. 5168	
1406	12/31	II/1. 5182	
1407	1/8	II/2. 5222	
1407	4/19	II/2. 5440	
1407	6/5	II/2. 5547	
1407	6/24	II/2. 5941	date of charter: 1408.2.14
1407	11/29	II/2. 6092	date of charter: 1408.5.8
1408	1/13	II/2. 5905	
1408	6/10	II/2. 6191	date of charter: 1408.7.3
1409	3/25	II/2. 6990	date of charter: 1409.8.19
1409	4/14	II/2. 6735	
1409	8/2	II/2. 6953	actual date: "a few days ago"
1409	10/31	II/2. 7157	
1410	4/28	II/2. 7529	
1410	10/1	II/2. 7952	
1410	11/8	II/2. 8025	
1410	12/2	II/2. 8067	
1411	10/5	III. 1024	
1411	11/20	III. 1246	
1411	12/28	III. 1435	
1412	1/12	III. 1581	
1412	4/25	III. 2019	
1413	1/10	IV. 33	actual date: "a few days ago"
1413	2/14	IV. 162	
1413	2/17	IV. 183	
1413	2/20	IV. 195	
1413	4/24	IV. 483	
1413	11/13	IV. 1280	
1414	1/19	IV. 1597	
1414	1/20	IV. 1609	

1414	1/27	IV. 1629	
1414	4/8	IV. 1966	date of charter: 1414.5.11
1414	4/8	IV. 2356	
1414	6/30	IV. 2196	
1414	9/3	IV. 2430	
1414	9/8	IV. 2455	
1414	11/5	IV. 2644	
1414	12/4	IV. 2762	
1415	1/22	V. 116	
1415	2/9	V. 201	
1415	3/6	V. 323	
1415	3/20	V. 393	
1415	3/28	V. 428	
1415	9/7	V. 1034	
1415	11/12	V. 1216	
1415	11/14	V. 1234	
1415	12/19	V. 1337	
1415	12/24	V. 1533	date of charter: 1416.2.13. Actual date: "around Christmas"
1416	2/20	V. 2293	date of charter: 1416.9.15. Actual date: "last Lent"
1416	2/20	V. 2356	date of charter: 1416.10.11. Actual date: "this Lent"
1416	3/18	V. 1673	
1416	6/3	V. 1955	
1416	9/3	V. 2250	
1416	9/30	V. 2329	Actual date: "a few days ago"
1416	11/30	V. 2473	
1417	2/6	VI. 76	
1417	4/12	VI. 330	date of charter: 1417.4.18
1417	6/14	VI. 577	
1417	7/16	VI. 689	
1417	8/31	VI. 862	
1417	11/10	VI. 1270	date of charter: 1417.12.21
1417	12/15	VI. 1270	date of charter: 1417.12.21
1417	12/17	VI. 1270	date of charter: 1417.12.21
1418	3/30	VI. 1700	
1418	4/27	VI. 1817	
1418	6/16	VI. 2495	date of charter: 1418.11.8
1418	6/24	VII. 332	date of charter: 1419.4.21
1418	11/6	VI. 2483	
1419	3/19	VII. 206	
1419	5/8	VII. 408	
1419	5/26	VII. 521	
1419	12/24	VII. 1697	date of charter: 1420.5.6. Actual date: "around Christmas"
1420	2/20	VII. 1881	date of charter: 1420.6.24. Actual date: "last Lent"
1420	11/7	VII. 2303	
1420	12/20	VII. 2408	

Appendix 6. Perambulations in the Vkpmlt corpus.

Df number	year	remarks
283,210	1203	
201,312	1215	
200,630	1231	early modern copy
200,902	1233	Billege and Vázsony
200,902	1233	Seg
200,903	1233	forgery
200,631	1235	
201,086	1256	early modern copy
200,661	1263	
200,660	1266	
200,684	1275	early modern copy
201,106	1284	
200,726	1292	early modern copy
200,757	1299	
200,761	1302	early modern copy
200,815	1323	
201,137	1326	
200,833	1329	
200,849	1335	early modern copy
201,202	1338	
200,859	1338	
200,865	1339	
200,869	1340	
200,870	1340	
200,891	1343	
200,896	1343	
200,915	1352	
200,926	1354	
200,928	1354	
200,979	1355	
200,955	1361	
200,956	1361	
200,962	1363	
200,963	1364	
200,968	1364	
200,980	1367	
200,993	1368	
201,009	1368	
201,027	1379	
201,036	1379	
201,029	1380	
201,030	1381	
201,040	1383	
283,220	1383	
201,061	1386	
201,079	1389	
201,091	1392	
201,109	1394	
201,118	1395	
201,123	1397	
201,125	1399	
201,229	1423	
201,255	1430	
201,256	1431	
201,268	1435	early modern copy
201,275	1436	in Hungarian
201,343	1454	
201,405	1466	
201,409	1466	
283,239	1489	early modern copy
201,566	1507	201 564 early modern copy of same
201,573	1508	

Appendix 7. Boundary signs in the Vkpmlt perambulations.
(mt-meta terrea)

boundary sign	occurrences
meta	466
meta terrea	348
via	129
vallis	62
ilex et mt	60
via magna	55
mons	53
vinea	49
terra arabilis	49
lapis	39
silva	35
ilex	31
fluvius	29
monticulus	26
molendinus	23
pirus et mt	17
rivulus	16
aqua	16
pirus silvestris et mt	15
nemus	14
virgultum	14
pratum	13
pirus	12
tölgy et mt	12
bérc	11
fenetum	11
meta lapidea	11
via publica	10
via erbosa	10
rubetum	10
nux et mt	10
puteus	9
salix	9
salix et mt	9
sessio	9
quercus et mt	8
semita	8

boundary sign	occurrences
ulmus et mt	7
fons	6
arbor	6
gyertyán et mt	6
insula	5
nux	5
spinetum (1379=mege)	5
fagus	5
fossatum	5
campus	5
quercus	4
pirus silvestris	4
fagus et mt	4
sorbellus	4
via graminosa	3
alveus	3
pomus	3
ortus	3
fovea	3
berkenye et mt	3
tilia	3
tilia et mt	3
magyal	3
pons	3
lacus	2
stagnum	2
dumus spinarum	2
ilicis dumus	2
ilicis troncus et mt	2
pomus et mt	2
prunus	2
cerasus et mt	2
avelle dumus	2
platea	2
collis	2
dumus	2
populus et mt	2

boundary sign	occurrences
cser et mt	2
iuniperis dumus	2
sorbellus et mt	2
castanea	2
castanea et mt	2
palus	2
via vindemialis	1
silva regis	1
molendini clausura	1
nemus pro canapis	1
rupes	1
rubeti dumus	1
gyertyán dumus	1
ilex et mlapidea	1
ilicis troncus	1
ilicis frutex	1
querci dumus et mt	1
nucis stips	1
salicis dumus	1
arbor fructifer	1
pomus silvestris	1
pomi silvestris dumus et mt	1
ornus	1
pirus et lapis	1
piri silvestris dumus et mt	1
dumus tölgy et mt	1
pomerium (vadkert)	1
ravazlyuk	1
castrum	1
viminarum dumus	1

boundary sign	occurrences
terra firma et spinosa	1
parlag	1
nidus accipitris	1
porta seu apertum arbustorum	1
fraxinus	1
fraxini dumus et mt	1
gyümölcsény dumus	1
gyümölcsény dumus et mt	1
juhar	1
curia	1
frutex	1
ripa	1
populus	1
populi frutex	1
alnus et mt	1
cser dumus	1
barkóca et mt	1
iuniper	1
iuniperis dumus et mt	1
ulmi rubus	1
cimiterium	1
magyal et mt	1
arundinetum	1
domus	1
troncus	1
troncus et mt	1
pinus et mt	1
rubusculum	1
cini silvestris rubus spinarum	1
cini dumus et mt	1
total	**1901**

Appendix 8. Woods in Bakony mentioned in written sources.

settlement	date	name of wood	type of wood	size of wood	source	type of source
Ajka	1292		silua		HO, vol. 1, 88-90	perambulation
Arács	1231	Kövecsét	silua		Df 200 630	perambulation
Arács	1231	Zalia	silua		Df 200 630	perambulation
Ardó	1512	Arday eresthwen	permissoria		Dl 102 698	acta potenciaria
Árkibánszörcsök	1299		virgultum		Df 200 757	perambulation
Árkibánszörcsök	1299	Kywsberek	nemus		Df 200 757	perambulation
Árkibánszörcsök	1299	Fywzkerth	virgultum		Df 200 757	perambulation
Árkibánszörcsök	1299	Tumpatova	virgultum		Df 200 757	perambulation
Barnag	1284		virgultum		Zala, vol. 1, 93-95	perambulation
Barnag	1284		silua		Zala, vol. 1, 93-95	perambulation
Barnag	1284		virgultum		Zala, vol. 1, 93-95	perambulation
Barnag	1284		virgultum		Zala, vol. 1, 93-95	perambulation
Barnag	1284		nemus		Zala, vol. 1, 93-95	perambulation
Barnag	1284		silua		Zala, vol. 1, 93-95	perambulation
Barnag	1284		virgultum		Zala, vol. 1, 93-95	perambulation
Bercsény	1395	Guzud	silua		Df 201 118	perambulation
Bere	1454		silua		Df 201 343	perambulation
Betereg	1082 (ca. end of thirteenth c.)		silua		HO, vol. 4, 1-8	conscription
Billege	1233	Csepel	nemus		CD, vol. 7/1, 235-236; HO, vol. 4, 14-19	perambulation
Billege	1233		nemus		CD, vol. 7/1, 235-236	perambulation
Billege	1233	longum nemus	nemus		HO, vol. 4, 14-19	perambulation
Billege	1233		silua		HO, vol. 4, 14-19	perambulation
Billege	1233	Thadeuca	silua		HO, vol. 4, 14-19	perambulation
Bogdány	1299		virgultum		Df 200 757	perambulation
Csékút	1470	Láz	silua		Dl 102 567	legal case
Csetény	1478		silua glandinosa		Df 18 145	estimation
Csetény	1478		rubetum	778 iugera	Df 18 145	estimation
Csopak	1380		virgultum		Df 201 029	perambulation
Csopak	1471	Badarfou	silua		Df 283 235	terrier
Endred	1256		nemus		PRT, vol. 8, 293-295	perambulation
Endred	1256		nemus		PRT, vol. 8, 293-295	perambulation

Endred	1256		silua		PRT, vol. 8, 293-295	perambulation
Fajsz	1082 (ca. end of thirteenth c.)	Soulomos	silua		HO, vol. 4, 1-8	conscription
Fenyőfő	1258		pinetum		ÁÚO, vol. 2, 315-315	legal case
Ganna	1366		silua germinabilis	10 aratra	Dl 69 984	estimation
Gic	1353		silua		PRT, vol. 2, 408-410	perambulation
Gyimót	1355	Nyri	silua		Df 200 979	perambulation
Haraszti	1372		virgultum		PRT, vol. 8, 387-389	perambulation
Haraszti	1372		silua		PRT, vol. 8, 387-389	perambulation
Jutas	1354		rubetum		Df 200 928	perambulation
Jutas	1426, 1486, 1518, 1521		(permissoria)		Df 201 237, Df 201 362, Df 201 604, Df 201 609	terrier
Káloz	1383		virgultum		Df 283 220	perambulation
Kemecse	1292		nemus		Df 200 726	perambulation
Kér	1363		nemus		Df 200 962	perambulation
Kér	1363		nemus		Df 200 962	perambulation
Kisberény	1232/1281	Hygmuguy	silua	1 aratrum	Df 200 705	property sale
Koppány	1437, 1438		silua		Sörös 1903, 355-372	acta potentiaria
Kövesd	1446		silua		Df 201 306	terrier
Litér	1510		silua glandinosa		Dl 102 696	acta potentiaria
Markó	1399		silua		Df 201 125	perambulation
Mencsel	1460	Kysberek	nemus		Df 201 360	terrier
Mencsel	1460		silua		Df 201 360	terrier
Noszlop	1332	Dyonisii	silua		CD, vol. 8/3, 579-590; PRT, vol. 8, 315-316	perambulation
Noszlop	1332		nemus		CD, vol. 8/3, 579-590	perambulation
Noszlop	1332		nemus		CD, vol. 8/3, 579-590	perambulation
Noszlop	1342		silua		PRT, vol. 8, 315-316	perambulation
Noszlop	1342	Merete	silua		PRT, vol. 8, 315-316	perambulation
Noszlop	1342		nemus		PRT, vol. 8, 315-316	perambulation
Noszlop	1342		nemus		PRT, vol. 8, 315-316	perambulation
Oszlop	1478		silua glandinosa		Df 18 145	estimation
Oszlop	1478		rubetum	125 iugera	Df 18 145	estimation
Örs	1082 (ca. end of thirteenth c.)		silua		HO, vol. 4, 1-8	conscription

Pécsely	1082 (ca. end of thirteenth c.)		silua		HO, vol. 4, 1-8	conscription
Pécsely	1381		rubetum		Df 201 030	perambulation
Pécsely	1431	Nagyfaluerdeje / Monyoros-hegye	permissoria		CD, vol. 10/7, 373-380	perambulation
Pécsely	1431	Kulcsar	virgultum		CD, vol. 10/7, 373-380	perambulation
Pécsely	1431	Chabahegye	virgultum		CD, vol. 10/7, 373-380	perambulation
Pécsely	1431	Chordauthegye	virgultum		CD, vol. 10/7, 373-380	perambulation
Pécsely	1431	Diastheleke	silua		CD, vol. 10/7, 373-380	perambulation
Pécsely	1431		silua		CD, vol. 10/7, 373-380	perambulation
Polyán	1381		silua		PRT, vol. 8, 411-412	perambulation
Ség	1233		virgultum		HO, vol. 4, 14-19	perambulation
Ség	1233	"quod dicitur fagus"	nemus		HO, vol. 4, 14-19	perambulation
Ség	1233		nemus		HO, vol. 4, 14-19	perambulation
Ség	1233		silua		HO, vol. 4, 14-19	perambulation
Ség	1233		virgultum		HO, vol. 4, 14-19	perambulation
Ség	1233		silua		HO, vol. 4, 14-19	perambulation
Sikátor	1478	Lesthye	silua glandinosa		Df 18 145	estimation
Sikátor	1478		rubetum	325 iugera	Df 18 145	estimation
Szabadi	1508		rubetum		Df 201 573	perambulation
Szabadi	1508		silua usualis		Df 201 573	perambulation
Szabadi	1508	Monyoros	silua		Df 201 573	perambulation
Szabadi	1508	Parra.hyertche (?)	silua		Df 201 573	perambulation
Szabadi	1314		nemus		Df 200 789	terrier
Szabadi	1314		silua		Df 200 789	terrier
Szabadi	1314		silua		Df 200 789	terrier
Szabadi	1383		silua	60 iugera	Df 201 059	terrier
Szabadi	1383		silua seu rubeta		Df 201 059	terrier
Szabadi	1383	Berek	nemus	2 iugera	Df 201 059	terrier
Szárberény	1386		silua		Df 201 061	perambulation
Szárberény	1386, 1389	Konyhaerdő	silua	2 aratra	Df 201 061, Df 201 079	perambulation
Szentkirály	1478		silua glandinosa	33 iugera	Df 18 145	estimation
Szőlős	1309	Barduta	silua communis		Df 200 779	terrier
Teszér	1478		silua glandinosa	357 iugera	Df 18 145	estimation
Teszér	1478		rubetum	870 iugera	Df 18 145	estimation

Vámhegy	1478		silua glandinosa	240 iugera	Df 18 145	estimation
Vámhegy	1478	Gerendawagas	silua glandinosa		Df 18 145	estimation
Vámos	1304	Naraad	nemus		Df 200 767	terrier
Vászoly	1433	Eresvin	(permissoria)		PRT, vol. 8, 479-481	terrier
Vászoly	1433		silua		PRT, vol. 8, 479-481	terrier
Vászoly	1433		silua		PRT, vol. 8, 479-481	terrier
Vászoly	1433		rubetum		PRT, vol. 8, 479-481	terrier
Vászoly	1433	Keleartusfolde	silua		PRT, vol. 8, 479-481	terrier
Vászoly	1433		silua		PRT, vol. 8, 479-481	terrier
Veim	1478		silua glandinosa	604 iugera	Df 18 145	estimation
Veim	1478		rubetum	294 iugera	Df 18 145	estimation
Veszprém	1313/1392, 1446, 1314, 1383,1332	Ispansagerdew	silua		Df 201 098, Df 201 306, Df 200 789, Df 201 059, Df 200 933	legal case, terrier
Veszprém	1492	Barát erdő	silua		Df 201 513	legal case
Veszprém	1332	Monorous	rubetum		Df 200 933	terrier
Veszprém	1483	Monyoros seu Kwththethew	silua		Dl 102 626	acta potentiaria

Bibliography

Andersen, Svend Th. 1988. "Changes in Agricultural Practices in the Holocene Indicated in a Pollen Diagram from a Small Hollow in Denmark." In *The Cultural Landscape: Past, Present and Future*, ed. Hilary H. Birks, H. J. B. Birks, Peter Emil Kaland, and Dagfinn Moe, 395–407. Cambridge: Cambridge University Press.

Andrásfalvy, Bertalan. 1973. *A Sárköz ősi ártéri gazdálkodása* (Ancient flood-plain management in Sárköz). Budapest: VIZDOK.

Andrásfalvy, Bertalan. 1975. *A Duna mente népeinek ártéri gazdálkodása Tolna és Baranya megyében az ármentesítés befejezéséig* (The flood-plain management of the people living around the Danube in Co. Tolna and Co. Baranya, until the completion of the river regulations). Szekszárd: Tolna megyei Tanács Levéltára.

Andrásfalvy, Bertalan. 1989. "A magyarság gyümölcsészete" (The fruit production of the Hungarians). Doctoral diss., Budapest.

Andrén, Anders. 1997. "Paradise Lost: Looking for Deer Parks in Medieval Denmark and Sweden." In *Visions of the Past: Trends and Traditions in Swedish Medieval Archaeology*, ed. Hans Andersson, Peter Carelli, and Lars Ersgård, 469–490. Stockholm: Almqvist & Wiksell.

Archéologie de la France rurale de la préhistoire aux temps modernes. 1986. Paris: Belin.

Arnoux, Mathieu. 1990. "Forges et forêts au Moyen Age: l'exemple normand." In *Forges et forêts: recherches sur la consommation proto-industrielle de bois*, ed. Denis Woronoff, 213–218. Paris: ESEH.

Astill, G. G. 1993. *A Medieval Industrial Complex and its Landscape: The Metalworking Watermills and the Workshops of Bordesley Abbey*. York: Council for British Archaeology.

Astill, G. G. and Wendy Davies. 1997. *A Breton Landscape*. London: UCL Press.

Aston, Michael. 1985. *Interpreting the Landscape: Landscape Archaeology in Local Studies*. London: Batsford.

Aston, Michael. 1993a. "The Development of the Carthusian Order in Europe and Britain: A Preliminary Survey." In *In Search of Cult*, ed. Martin Carver, 139–151. Woodbridge: Boydell and Brewer.

Aston, Michael. 1993b. *Monasteries*. London: B. T. Batsford.

Aubrun, Michel. 1988. "Droits d'usages forestiers et libertés paysannes (XIe – XIIIe siècle)." *Revue Historique* 280: 377–386.

Austad, Ingvild. 1988. "Tree Pollarding in Western Norway." In *The Cultural Landscape: Past, Present and Future*, ed. Hilary H. Birks, H. J. B. Birks, Peter Emil Kaland, and Dagfinn Moe, 11–29. Cambridge: Cambridge University Press.

Babos, Károly. 1994. *Faanyagismeret és fafaj-meghatározás restaurátoroknak* (Information on wood and tree species identification for restorers). Budapest: MNM.

Bacsó, Nándor. 1960. *Magyarország éghajlata* (The climate of Hungary). Budapest: Akadémiai Kiadó.

Baillie, M. G. L. 1982. *Tree-ring Dating and Archaeology*. London: Croon, Helm.

Bak, János M., György Bónis, Leslie S. Domonkos, and James Ross Sweeney, ed. 1989–1996. *Decreta Regni Mediaevalis Hungariae*. 3 vols. Idyllwild, California: Charles Schlacks, Jr. Publisher.

Bakay, Kornél. 1989. *Feltárul a múlt* (The past revealed). Budapest: Múzsák.

Balás, Vilmos. 1961. *Az alföldi hosszanti sáncok* (Longitudinal ditches in the Alföld). Régészeti füzetek II/9. Budapest: n.p.

Balás, Vilmos. 1963. "Die Erdwälle der Ungarischen Tiefebene." *Acta Archaeologica Academiae Scientiarum Hungaricae* 15: 309–336.

Balassa, Iván. 1973. *Az eke és a szántás története Magyarországon* (A history of the plough and ploughing in Hungary). Budapest: Akadémiai Kiadó.

Balassa, M. Iván. 1985. *A parasztház évszázadai: A magyar lakóház középkori fejlődésének vázlata* (Centuries of peasant houses: An outline of the medieval development of Hungarian vernacular buildings). Békéscsaba: n. p.

Balázs, Péter, ed. 1976. *Guide to the Archives of Hungary*. Budapest: Archival Board of the Ministry of Culture.

Bálint, Alajos. 1960–1962. "A középkori Nyársapát lakóházai" (Houses in medieval Nyársapát). *Móra Ferenc Múzeum Évkönyve*: 39–115.

Bálint, Csanád. 1991. *Südungarn im 10. Jahrhundert*. Budapest: Akadémiai Kiadó.

Banyó, Péter. 2000. "Birtoköröklés és leánynegyed: Kísérlet egy középkori jogintézmény értelmezésére" (Inheritance of landed property and the filial quarter: An attempt to interpret a medieval Hungarian legal institution). *Aetas* no. 3 of 2000: 76–92.

Baráz, Csaba. 2000. *Kaptárkövek: Szakrális kőemlékek a Bükkalján* (*Kaptárkövek*: Sacral stone monuments in Bükkalja). Eger: Kaptárkő Közművelődési és Tájvédelmi Egyesület.

Bárczi, Géza. 1951. *A tihanyi apátság alapítólevele mint nyelvi emlék* (The foundation charter of the abbey of Tihany). Budapest: Akadémiai Kiadó.

Bárczi, Géza, Loránd Benkő, and Jolán Berrár. 1978. *A magyar nyelv története* (A history of the Hungarian language). Budapest: Tankönyvkiadó.

Bártfai Szabó, László, ed. 1938. *Pest megye történetének okleveles emlékei 1002–1599-ig* (Charters of the history of Co. Pest from 1002 to 1599). Budapest: published by the author.

Bartha, Dénes. 1994. *Magyarország faóriásai és famatuzsálemei* (Large and ancient trees in Hungary). Erdészettörténeti közlemények 15. Budapest and Sopron: Országos Erdészeti Egyesület, Erdészettörténeti Szakosztály.

Bartha, Dénes. 1997. *Fa- és cserjehatározó* (Tree and shrub identification). Budapest: Mezőgazda Kiadó.

Bartha, Dénes. 2000. "Erdőterület-csökkenések, fafajváltozások a Kárpát-medencében" (The decline of wooded areas and the changes in tree species in the Carpathian Basin). In *Táj és történelem* (Landscape and history), ed. Ágnes R. Várkonyi, 11–24. Budapest: Osiris.

Bartha, Dénes, ed. 2001. *A természetszerű erdők kezelése, a kultúr- és származékerdők megújítása* (Management of semi-natural woodland, renewal of secondary woodland and plantations). Budapest: TermészetBÚVÁR Alapítvány Kiadó.

Bartha, Dénes and Piroska Esztó. 2001. "Az országos Erdőrezervátum-hálózat bemutatása az Országos Erdőállomány-adattár alapján" (Presentation of the Woodland Reserve Network in Hungary, based on the Hungarian Forest Resource Database). *ER* 1, no. 1: 21–44.

Bartosiewicz, László. 1999. "Animal Husbandry and Medieval Settlement in Hungary: A Review." *Beiträge zur Mittelalterarchäologie in Österreich* 15: 139–155.

Beck, Corinne and Robert Delort, ed. 1993. *Pour une histoire de l'environnement*. Paris: CNRS Edition.

Beck, Patrice, Philippe Braunstein, and Michel Philippe. 1998. "Wood, Iron, and Water in the Othe Forest in the Late Middle Ages." In *Technology and Resource Use in Medieval Europe: Cathedrals, Mills and Mines*, ed. Elizabeth B. Smith and Michael Wolfe, 173–184. London: Ashgate.

Behre, Karl-Ernst, ed. 1986. *Anthropogenic Indicators in Pollen Diagrams*. Rotterdam: Balkema.

Beke, Ödön. 1913. "A -va, -ve és -ván, -vén képzőről" (About the suffixes *-va*, *-ve* and *-ván*, *-vén*). *Magyar Nyelvőr* 42: 193–199.

Békefi, Remig. 1891–1892. *A pilisi apátság története 1184–1814* (A history of the Pilis monastery 1184–1814). 2 vols. Pécs: n. p.

Békefi, Remig. 1898. *A pásztói apátság története 1190–1702* (A history of the Pásztó monastery 1190–1702). Budapest: Horánszky Viktor.

Belényesi, Károly. 2004. *Pauline Monasteries in the Zemplén Hegyalja Region*. Forthcoming.

Belényesy, Márta. 1954. "A földművelés fejlődésének alapvető kérdései a XIV. században. I" (Basic features in the development of agriculture in the fourteenth century, part 1). *Ethnogaphia* 65: 387–413.

Belényesy, Márta. 1955a. "A földművelés fejlődésének alapvető kérdései a XIV. században. II" (Basic features in the development of agriculture in the fourteenth century, part 2). *Ethnogaphia* 66: 57–93.

Belényesy, Márta. 1955b. "Le serment sur la terre au moyen âge et ses traditions postérieures en Hongrie." *Acta Ethnographica Academiae Scientiarum Hungaricae* 4: 361–363.

Belényesy, Márta. 1958a. "Der Ackerbau und seine Produkte in Ungarn im XIV. Jahrhundert." *Acta Ethnographica Academiae Scientiarum Hungaricae* 6: 265–321.

Belényesy, Márta. 1958b. "Kerített település és gazdálkodás kapcsolata néhány zalai irtásos falunál egy 1460-as határjárás alapján" (Enclosed settlement and agriculture in some villages in Co. Zala on the basis of a 1460 perambulation). *Ethnographia* 69: 117–138.

Belényesy, Márta. 1960. "La culture permanente et l'evolution du système biennal et triennal en Hongrie medièvale." *Ergon* 2: 311–326.

Belényesy, Márta. 1964. "A parlagrendszer XV. századi kiterjesztése Magyarországon" (The expansion of the fallow system in fifteenth-century Hungary). *Ethnographia* 75: 321–346.

Bellon, Tibor. 1996. "Ártéri gazdálkodás az Alföldön az ármentesítések előtt" (Flood-plain management in the Alföld before the river regulations). In *A Kárpát-medence történeti földrajza* (A historical geography of the Carpathian Basin), ed. Sándor Frisnyák, 311–321. Nyíregyháza: MTA and BGYTKF Földrajz Tanszéke.

Bencze, Zoltán and György Szekér. 1993. *A budaszentlőrinci pálos kolostor* (The Pauline monastery of Budaszentlőrinc). Budapest: Budapesti Történeti Múzeum.

Bencze, Zoltán, Ferenc Gyulai, Tibor Sabján, and Miklós Takács. 1999. *Egy Árpád-kori veremház feltárása és rekonstrukciója* (The excavation and reconstruction of an Árpádian period semi-subterranean house). Budapest: Budapesti Történeti Múzeum.

Benkő, Elek. 1980. "A középkori Nyársapát" (Medieval Nyársapát). *Studia Comitatensia* 9: 315–424.

Benoît, Paul. 1994. "L'espace industriel cistercien à la lumière des exemples bourguignons et champenois." In *L'espace cistercien*, ed. Léon Pressouyre, 378–390. Paris: Comité des travaux historiques et scientifiques.

Berrár, Jolán and Sándor Károly, ed. 1984. *Régi magyar glosszárium* (Old Hungarian glossary). Budapest: Akadémiai Kiadó.

Bertsch, Karl. 1951. *Geschichte des deutschen Waldes*. Jena: G. Fischer.

Birks, Hilary H., H. J. B. Birks, Peter Emil Kaland, and Dagfinn Moe, ed. 1988. *The Cultural Landscape: Past, Present and Future*. Cambridge: Cambridge University Press.

Birrel, Jean. 1987. "Common Rights in the Medieval Forest: Disputes and Conflicts in the Thirteenth Century." *Past and Present* 117: 22–49.

Bodó, Sándor. 1992. *A Bodrogköz állattartása* (Animal husbandry in Bodrogköz). Miskolc: HOM.

Bogdán, István. 1978. *Magyarországi hossz- és földmértékek a XVI. század végéig* (Measures of length and territory in Hungary until the end of the sixteenth century). Budapest: Akadémiai Kiadó.

Bogdán, István. 1991. *Magyarországi űr-, térfogat-, súly- és darabmértékek 1874-ig* (Hungarian cubic, solid, weight, and piece measures until 1874). Budapest: Akadémiai Kiadó.

Boissière, Jean. 1986. "La longue vie de la forêt." In *Archéologie de la France rurale de la préhistoire aux temps modernes*, 138–142. Paris: Belin.

Bóna, István, ed. 1993. *Hunok, gepidák, langobárdok: történeti régészeti tézisek és címszavak* (Huns, gepids, and lombards: Historical archaeological theses and catchwords). Szeged: JATE M. Östört. Kutcsop.

Bónis, György. 1972. *Középkori jogunk elemei* (Elements of Hungarian medieval law). Budapest: Közgazdasági és Jogi Könyvkiadó.

Borbás, Vince. 1881. "Békésvármegye flórája" (The flora of Co. Békés). *Értekezések a természettudományok köréből* 18: 1–105.

Borsa, Iván. 1984. "A szenyéri uradalom Mohács előtti iratanyagának vizsgálata" (An analysis of the pre-Mohács documents of the estate of Szenyér). In *Mályusz Elemér emlékkönyv* (Studies in honour of Elemér Mályusz), ed. Éva H. Balázs, Erik Fügedi, and Ferenc Maksay, 59–76. Budapest: Akadémiai Kiadó.

Bosl, Karl. 1963. "Pfalzen und Forsten." In *Deutsche Königspfalzen: Beitrage zu ihrer historischen und archäologischen Erforschung*, 1–29. Göttingen: Vanderhoek and Ruprecht.

Bowden, Mark, ed. 1999. *Unravelling the Landscape: An Inquisitive Approach to Archaeology*. Stroud: Tempus.

Bökönyi, Sándor. 1974. *History of Domestic Mammals in Central and Eastern Europe*. Budapest: Akadémiai Kiadó.

Bölöni, János. 2001. "Főbb erdőtársulás csoportok részaránya az Országos Erdőrezervátum-hálózatban" (Proportion of the main native woodland types in the Hungarian Woodland Reserves Network). *ER* 1, no. 1: 45–52.

Brandl, Helmut and Uwe Eduard Schmidt. 1998. "Forstgeschichte in Deutschland: Forschung und Lehre." *News of Forest History* 27: 1–39.

Broekmeyer, Mirjam A. E., ed. 1993. *European Forest Reserves: Proceedings of the European Forest Reserves Workshop*. Wageningen: PUDOC.

Broekmeyer, Mirjam A. E.. 1994. "European Forest Reserves: a Review." In *Conservation of Forests in Central Europe*, ed. Jozef Paulenka and Ladislav Paule, 15–24. Zvolen: Arbora Publishers.

Burton, Janet. 1994. *Monastic and Religious Orders in Britain 1000 – 1300*. Cambridge: Cambridge University Press.

Buzás, Gergely. 1995. "The Remains of the Royal Palace of Visegrád from the Angevin period." In *Medieval Visegrád*, ed. József Laszlovszky, 9–19. Budapest: ELTE.

Clay, Rotha Mary. 1914. *The Hermits and Anchorites of England*. London: Methuen.

Coles, John M. 1973. *Archaeology by Experiment*. London: Hutchinson.

Constable, Giles. 1996. *The Reformation of the Twelfth Century*. Cambridge: Cambridge University Press.

Crone, Anne and Fiona Watson. 2003. "Sufficiency to Scarcity: Medieval Scotland, 500–1600." In *People and Woods in Scotland: A History*, ed. Chris Smout, 60–81. Edinburgh: Edinburgh University Press.

Csánki, Dezső and Antal Fekete Nagy, ed. 1890–1941. *Magyarország történelmi földrajza a Hunyadiak korában* (A historical geography of Hungary in the age of the Hunyadi). 5 vols. Budapest: Magyar Tudományos Akadémia.

Csapodi, Csaba. 1978. *Az Anonymus-kérdés története* (A history of the Anonymus-debate). Budapest: Magvető.

Csendes, László. 1980. *Térképhistória* (A history of maps). Budapest: Magvető.

Csengel, Péter and László Gere. 1996. "Előzetes jelentés a városlődi karthauzi kolostor kutatásáról" (Preliminary report on the research on the Carthusian monastery in Városlőd). *Műemlékvédelmi Szemle* 6: 53–84.

Csontos, P., J. Tamás, and T. Kalapos. 2001. "Correlation between Age and Basal Diameter of Fraxinus ornus in Three Ecologically Different Habitats." *Acta Botanica Hungarica* 43: 127–136.

Csontos, Péter. 1996. "Az aljnövényzet változásai cseres-tölgyes erdők regenerációs szukcessziójában" (Regeneration succession of sessile oak - turkey oak woods: processes in the shrub layer). *Synbiologia Hungarica* 2, no. 2.

Csőre, Pál. 1980. *A magyar erdőgazdálkodás története: Középkor* (A history of Hungarian forestry: The Middle Ages). Budapest: Akadémiai Kiadó.

Czájlik, Péter. 1994. "Az erdőrezervátumok hasznosításáról" (The utilisation of woodland reserves). *Környezet és fejlődés* 5, no. 2: 29–36.

Czuczor, Gergely and János Fogarasi, ed. 1862–1874. *A magyar nyelv szótára* (A dictionary of the Hungarian language). 6 vols. Pest: Emich Gusztáv.

Danszky, István, ed. 1973. *Erdőművelés* (Forest cultivation). 2 vols. Budapest: Mezőgazdasági Kiadó.

Dányi, Dezső and Vera Zimányi. 1989. *Soproni árak és bérek a középkortól 1750-ig* (Prices and wages in Sopron from the Middle Ages to 1750). Budapest: Akadémiai Kiadó.

Dedek, Lajos Crescens. 1889. *A karthausiak Magyarországon* (The Carthusians in Hungary). Budapest: published by the author.

Devèze, Michel. 1961. *La vie de la forêt française au XVIe siècle*. 2 vols. Paris: Université de Paris.

Devèze, Michel. 1966a. "Forêts françaises et forêts allemandes: Étude historique comparée (1re partie)." *Revue historique* 235: 347–380.

Devèze, Michel. 1966b. "Forêts françaises et forêts allemandes: Étude historique comparée (suite et fin)." *Revue historique* 236: 47–68.

Dickinson, William R. 2000. "Changing Times: The Holocene Legacy." *Environmental History* 5: 483–502.

Dirkx, G. H. P. 1998. "Wood-pasture in Dutch Common Woodlands and the Deforestation of the Dutch Landscape." In *The Ecological History of European Forests*, ed. Keith J. Kirby and Charles Watkins, 53–62. Wallingford and New York: CABI Publishing.

Dóka, Klára. 1981. *Szentendre története* (A history of Szentendre). Szentendre: Pest Megyei Múzeumok Igazgatósága.

Domanovszky, Sándor. 1979. "Mázsaszekér" (Centner cart). In *Gazdaság és társadalom a középkorban* (Economy and society in the Middle Ages), 101–136. Budapest: Gondolat.

Dömötör, Sándor. 1954. "Bakhátas szántás és a nyugat-magyarországi eketípusok" (Ridge-and-furrow ploughing and plough types in Western Hungary). *Néprajzi Múzeum Értesítője* 36: 149–164.

Duby, Georges. 1962. *Rural Economy and Country Life in the Medieval West*, trans. Cynthia Postan. London: Edward Arnold.

Duby, Georges. 1978. *Emberek és struktúrák a középkorban* (People and structures in the Middle Ages), trans. László Vekerdi. Budapest: Magvető Kiadó.

Duceppe Lamarre, François. 1998. "L'archéologie du paysage à la conquête des milieux forestiers, ou l'objet paysage vu par l'archéologue de l'environnement." *Travaux de l'École doctorale d'Histoire, Publications de la Sorbonne* 2: 85–94.

E. Kovács, Péter. 1998. "A grebeni uradalom 1522-es összeírása" (The 1522 survey of the estate of Greben). In *Tanulmányok Borsa Iván tiszteletére* (Essays in honour of Iván Borsa), ed. Enikő Csukovits, 131–172. Budapest: Magyar Országos Levéltár.

Égető, Melinda. 1980. "Középkori szőlőművelésünk kérdéséhez" (Regarding medieval Hungarian vine-growing). *Ethnographia* 91: 53–78.

Égető, Melinda. 1983. "A lugasos szőlőművelés vizsgálata" (An analysis of vine-growing in bowers). *Népi kultúra – népi társadalom* 13: 119–49.

Engel, Pál. 1977. *Királyi hatalom és arisztokrácia viszonya a Zsigmond-korban* (The relationship between royal power and the aristocracy in the Sigismund period). Budapest: Akadémiai Kiadó.

Engel, Pál. 1981. "A honor: A magyarországi feudális birtokformák kérdéséhez" (The *honor*: Regarding the question of feudal property types in Hungary). *Történelmi Szemle* 24: 1–19.

Engel, Pál. 1982. "Honor, vár, ispánság" (*Honor*, castle, county). *Századok* 116: 880–921.

Engel, Pál. 1985. "Ung megye településviszonyai és népessége a Zsigmond-korban" (Settlements and population of Co. Ung in the Sigismund period). *Századok* 119: 941–1005.

Engel, Pál. 1988. "Az ország újraegyesítése: I. Károly küzdelmei az oligarchák ellen, 1310 – 1323" (Reunification of the country: Charles I's fights against the oligarchs, 1310 – 1323). *Századok* 122: 89–146.

Engel, Pál. 1989. *Kamarahaszna-összeírások 1427-ből* (Conscriptions of *lucrum camere* from 1427). Új történelmi tár 2. Budapest: Akadémiai Kiadó.

Engel, Pál. 1996. *Magyarország világi archontológiája 1301 – 1457* (Lay archontology of Hungary 1301 – 1457). Budapest: História and MTA Történettudományi Intézete.

Engel, Pál. 1998. *A nemesi társadalom a középkori Ung megyében* (Noble society in medieval Co. Ung). Budapest: MTA Történettudományi Intézete.

Engel, Pál. 2001a. *The Realm of St. Stephen: A History of Medieval Hungary*. London: I.B. Tauris.

Engel, Pál. 2001b. *Középkori magyar genealógia* (Medieval Hungarian genealogy). Arcanum. [CD-ROM].

Engel, Pál. 2001c. *Magyarország a középkor végén. Digitális térkép és adatbázis a középkori Magyar Királyság településeiről* (Hungary at the end of the Middle Ages: Digital map and database about the settlements of the medieval Kingdom of Hungary). Budapest: Térinfó BT and MTA Történettudományi Intézete. [CD-ROM].

Eötvös, Károly. 1909. *A Bakony*. 2 vols. Budapest: Révai.

Erdélyi, László and Pongrác Sörös, ed. 1902–1916. *A pannonhalmi Szent-Benedek-rend története* (A history of the Benedictine order of Pannonhalma). 12 vols. Budapest: Pannonhalmi Szent-Benedek-rend.

Éri, István. 1969. "Veszprém megye középkori településtörténeti vázlata" (A medieval settlement history of Co. Veszprém). *Veszprém Megyei Múzeumok Közleményei* 8: 199–212.

Érszegi, Géza, ed. 2000. *Középkori oklevelek a Szabolcs-Szatmár-Bereg megyei Levéltárban (1300–1525)* (Medieval charters in the archives of Co. Szabolcs-Szatmár-Bereg (1300–1525)). Nyíregyháza: Szabolcs-Szatmár-Bereg megyei Levéltár.

The Exeter Book Riddles, tr. Kevin Crossley-Holland. 1993. London: Penguin Books.

Faegri, Knut, Peter Emil Kaland, and Knut Krzywinski. 1989. *Textbook of Pollen Analysis*. Chicester: Wiley.

Fejér, György, ed. 1829–1844. *Codex Diplomaticus Hungariae Ecclesiasticus ac Civilis*. 11 vols. Buda: n.p.

Fejérpataky, László, ed. 1885. *Magyarországi városok régi számadáskönyvei* (Old account books of Hungarian towns). Budapest: MTA.

Fekete, Gábor. 1964. *A Bakony növénytakarója* (The vegetation of Bakony). Veszprém: Bakonyi Múzeum.

Fekete, Zoltán. 1906. "Növényföldrajzi megfigyelések a Magas-Tátrában" (Plant-geographical studies in the High Tatras). *Erdészeti Lapok* 44: 161–172.

Félegyházi, Enikő. 1998. "Kutatási jelentés a visegrádi régészeti anyag palynologiai vizsgálatáról" (Research report on the palynological analysis of the archaeological material in Visegrád). Manuscript, Debrecen.

Fodor, István. 1994. "Árpád-kori boronaház nyomai Tiszaszigeten" (Remains of an Árpádian period log-cabin at Tiszasziget). In *A kőkortól a középkorig* (From the Stone Age to the Middle Ages), ed. Ottó Trogmayer, 421–432. Szeged: Csongrád Megyei Múzeumok Igazgatósága.

Forster, Gyula, ed. 1900. *III. Béla magyar király emlékezete* (In memoriam Béla III, King of Hungary). Budapest: n. p.

Földes, László. 1978. "Szilárd telekrendszerű irtásfalu a Szepességben" (Village with a stable plot system founded on cleared land in the Szepes area). *Agrártörténeti Szemle* 20: 357–378.

Frank, Norbert. 2001. "Adatok a soproni Dudlesz-erdő történetének ismeretéhez" (Regarding the history of Dudlesz wood in Sopron). *Soproni Szemle* 55: 149–162.

Fügedi, Erik. 1970. *A 15. Századi magyar arisztokrácia mobilitása* (Mobility of the fifteenth-century Hungarian aristocracy). Budapest: Művelődési Minisztérium Levéltári Osztálya.

Fügedi, Erik. 1977. *Vár és társadalom a 13–14. századi Magyarországon* (Castle and society in Hungary in the thirteenth-fourteenth centuries). Budapest: Akadémiai Kiadó.

Fügedi, Erik. 1981. "A középkori Magyarország történeti demográfiájának mai állása" (Recent developments in the historical demography of medieval Hungary). In *Kolduló barátok, polgárok, nemesek* (Mendicants, burghers, nobles), 387–397. Budapest: Magvető.

Fügedi, Erik. 1982. "Királyi tisztség vagy hűbér?" (Royal office or fief?) *Történelmi Szemle* 25: 483–509.

Fügedi, Erik. 1986. "Verba volant…Oral Culture and Literacy among the Medieval Hungarian Nobility." In *Kings, Bishops, Nobles and Burghers in Medieval Hungary*, VI, 1–25. London: Variorum Reprint.

Fügedi, Erik. 1998. *The Elefánthy: The Hungarian Nobleman and His Kindred*. Budapest: CEU Press.

Garády, Sándor. 1931. "Mátyás király budai vadaskertje" (King Matthias' park in Buda). *Historia*: 40–60.

Gencsi, László and Rudolf Vancsura. 1997. *Dendrológia* (Dendrology). Budapest: Mezőgazda Kiadó.

Gerald of Wales. 1978. *The Journey Through Wales*, trans. Lewis Thorpe. London: Penguin Books.

Géresi, Kálmán, ed. 1882–1897. *A nagy-károlyi gróf Károlyi család oklevéltára* (Charters of the Károlyi family of Nagykároly). 5 vols. Budapest: n. p.

Gerevich, László. 1983. "The Royal Court (villa), the Provost's Residence and the Village at Dömös." *Acta Archaeologica Academiae Scientiarum Hungaricae* 35: 385–409.

Gerevich, László. 1984. *A pilisi ciszteri apátság* (The Cistercian monastery of Pilis). Szentendre: Pest megyei múzeumok igazgatósága.

Gerevich, László. 1992. "Dömös." *Műemlékvédelem* 36: 73–80.

Gerlai, Arnold Gusztáv, ed. 1936. *A magyar erdészeti irodalom könyvészete 1934-ig* (Bibliography of Hungarian writings on forestry until 1934). Sopron: M. Kir. Erdészeti Kutató Intézet.

Gilchrist, Roberta. 1995. *Contemplation and Action: The Other Monasticism*. London and New York: Leicester University Press.

Gombocz, Zoltán. 1900. "Adalékok a magyar nyelv török elemeihez" (Data on the Turkish elements of Hungarian). *Magyar Nyelvőr* 29: 53–56.

Gombos, Albin, ed. 1937. *Catalogus Fontium Historiae Hungaricae*. 3 vols. Budapest: Szent István Akadémia.

Green, E. E. 1993. "Observations on the Importance of Old Trees and Their Management." In *Ancient Woodlands: Their Archaeology and Ecology*, ed. Pauline Beswick and Ian D. Rotherham, 91–92. Sheffield: Landscape Conservation Forum.

Grenville, Jane. 1997. *Medieval Housing*. London and Washington: Leicester University Press.

Grove, A. T. and Oliver Rackham. 2001. *The Nature of Mediterranean Europe: An Ecological History*. New Haven and London: Yale University Press.

Grynaeus, András. 1996. "Progress of Dendrochronological Research in Hungary." *Dendrochronologia* 14: 223–226.

Grynaeus, András. 1997. "Dendrokronológiai kutatások Magyarországon" (Dendrochronological research in Hungary). PhD diss., Eötvös Loránd University, Budapest.

Grynaeus, András and Tamás Grynaeus. 2001. "The Geobotany of Medieval Hungary: A Preliminary Report." *Medium Aevum Quotidianum* 44: 78–93.

Grynaeus, András. 2003. "Dendrochronology and Environmental History." In *People and Nature in Historical Perspective*, ed. József Laszlovszky and Péter Szabó, 175–193. Budapest: CEU Medieval Studies and Archaeolingua.

Grynaeus, Tamás and József Papp. 1977. "Régi magyar (gyógy)növénynevek, 15.–17. század" (Old Hungarian names of (medicinal) plants, fifteenth-seventeenth centuries). *Orvostörténeti Közlemények*, Supplementum 9–10: 31–49.

Guilaine, Jean, ed. 1991. *Pour une archéologie agraire*. Paris: Armand Colin.

Gutheil, Jenő. 1979. *Az Árpád-kori Veszprém* (Veszprém in the Árpádian period). Veszprém: Veszprém Megyei Levéltár.

Guzsik, Tamás. 2000. "A pálos rend 'születési anyakönyvi kivonata'" (The 'birth certificate' of the Pauline order). *Architectura Hungariae* 2/1: http://arch.eptort.bme.hu/epitesz5.html.

Gyöngyösi, Gergely. 1983. *Arcok a magyar középkorból* (Portraits from the Hungarian Middle Ages), trans. Vince Árva, Béla Csanád, and Ferenc Csonka. Budapest: Szépirodalmi Könyvkiadó.

Gyöngyösi, Gergely. 1988. *Vitae Fratrum Eremitarum Ordinis Sancti Pauli Primi Eremitae*, ed. Ferenc L. Hervay. Budapest: Akadémiai Kiadó.

Györffy, György. 1956. "Adatok a Pilis megyei monostorok középkori történetéhez" (Regarding the medieval history of monasteries in Co. Pilis). *Művészettörténeti Értesítő* 4: 280–285.

Györffy, György. 1963–1998. *Az Árpád-kori Magyarország történeti földrajza* (A historical geography of Hungary in the Árpádian period). 4 vols. Budapest: Akadémiai Kiadó.

Györffy, György. 1969. "A magyar egyházszervezés kezdeteiről újabb forráskritikai vizsgálatok alapján" (On the beginnings of Hungarian church organisation in light of recent source critical studies). *MTA Filozófiai és Történettudományi Osztályának Közleményei* 18: 199–225.

Györffy, György. 1983. *István király és műve* (King Stephen and his work). Budapest: Gondolat.

Györffy, György, ed. 1992. *Diplomata Hungariae Antiquissima*. Budapest: Akadémiai Kiadó.

Györffy, György and Bálint Zólyomi. 1994. "A Kárpát-medence és Etelköz képe egy évezred előtt" (The landscape of the Carpathian Basin and Etelköz a millenium before). In *Honfoglalás és régészet* (The Conquest and archaeology), ed. László Kovács, 13–37. Budapest: Balassi Kiadó.

Györffy, György. 1997. *Pest-Buda kialakulása* (The formation of Pest-Buda). Budapest: Akadémiai Kiadó.

Győri, Róbert. 2000. "Vadvízországtól a fokgazdálkodásig" (From wildwater-country to *fok* management). *Korall* 1: 20–26.

Gyulai, Ferenc. 2001. *Archaeobotanika: A kultúrnövények története a Kárpát-medencében a régészeti-növénytani vizsgálatok alapján* (Archaeobotany: A history of cultivated plants in the Carpathian Basin based on archaeobotanical research). Budapest: Jószöveg.

Halász, Aladár. 1994. *A magyar erdészet 70 éve számokban. 1920–1990* (Seventy years of Hungarian forestry in numbers: 1920–1990). Budapest: FM Erdőrendezési Szolgálat.

Haraszthy, László, Ferenc Márkus, and László Bank. 1997. *A fás legelők természetvédelme* (Nature protection of wood-pastures). Budapest: WWF Magyarország.

Harvey, Paul D. 1980. *The History of Topographical Maps: Symbols, Pictures and Surveys*. London: Thames and Hudson.

Hausen, Wilhelm. 1984. *Kalendarminiaturen der Stundenbücher: Mittelalterliches Leben im Jahreslang*. Munich: Callwey.

Házi, Jenő. ed. 1933. *Sopron királyi város története* (A history of the free royal town of Sopron). Vol. 2/3. *Különféle számadások és kimutatások 1432-től 1455-ig* (Various accounts from 1432 to 1455). Sopron: Székely és társa Könyvnyomdája.

Hæggström, Carl-Adam. 1983. "Vegetation and Soil of the Wooded Meadows in Nåtö, Åland." *Acta Botanica Fennica* 120: 1–66.

Hæggström, Carl-Adam. 1994. "Pollards in Art." *Botanical Journal of Scotland* 46: 682–687.

Hæggström, Carl-Adam. 1998. "Pollard Meadows: Multiple Use of Human-Made Nature." In *The Ecological History of European Forests*, ed. Keith J. Kirby and Charles Watkins, 33–41. New York: CABI Publishing.

Heckenast, Gusztáv. 1970. *A fejedelmi (királyi) szolgálónépek a korai Árpádkorban* (Princely (royal) servitor populations in the early Árpádian period). Budapest: Akadémiai Kiadó.

Hefty, Gyula Andor. 1911. "A térszini formák nevei a magyar népnyelvben" (Names of topographical units in the Hungarian vernacular). *Magyar Nyelvőr* 40: 155–169.

Hegyi, Imre. 1978. *A népi erdőkiélés történeti formái (Az északkeleti Bakony erdőgazdálkodása az utolsó 200 évben)* (Historical forms of traditional woodland exploitation (Woodland management in north-east Bakony in the past 200 years)). Budapest: Akadémiai Kiadó.

Hennebicque, Régine. 1980. "Espaces sauvages et chasses royales dans le Nord de la Francie, VIIe – IXe siècles." *Revue du Nord* 62: 35–57.

Herkely, Károly. 1941. "Népi erdőgazdálkodás Veszprém megyében" (Traditional woodland management in Co. Veszprém). *Ethnographia* 52: 54–58.

Hervay, Ferenc. 1984a. "A pálos rend elterjedése Magyarországon" (The spread of the Pauline order in Hungary). In *Mályusz Elemér emlékkönyv* (Essays in honour of Elemér Mályusz), ed. Éva H. Balázs, Erik Fügedi, and Ferenc Maksay, 159–172. Budapest: Akadémiai Kiadó.

Hervay, Ferenc. 1984b. *Repertorium Historicum Ordinis Cisterciensis in Hungaria*. Roma: Editiones Cisteriensis.

Hervay, Ferenc. 1985. "Részletek a pilisi apátság történetéből" (Details from the history of the Pilis monastery). *Studia Comitatensia* 17: 595–601.

Hervay, Ferenc. 1991. "A ciszteri rend története Magyarországon" (A history of the Cistercian order in Hungary). In Lajos Lékai, *A ciszteriek: Eszmény és valóság*, 470–492. Budapest: Szent István Társulat.

Hervay, Ferenc. 2001. "Bencés élet a középkori Magyarországon" (Benedictine life in medieval Hungary). In *Paradisum plantavit: Bencés monostorok a középkori Magyarországon* (Paradisum plantavit: Benedictine monasteries in medieval Hungary), ed. Imre Takács, 462–475. Pannonhalma: Bencés Főapátság.

Higounet, Charles. 1965. "Les forêts de l'Europe occidentale du Ve au XIe siècle." *Settimane di Studio* 13: 343–398.

Hilf, Richard B. 1938. *Der Wald in Geschichte und Gegenwart*. Potsdam: Akad. Verl. – Anst. Athenaion.

Hiller, István. 1993. "Ökológiai ismeretek és ökológiai szemlélet a Selmecbányai Erdészeti Akadémián" (Ecological knowledge and ecological approach at the Selmecbánya Academy of Forestry). In *Európa híres kertje: Történeti ökológiai tanulmányok Magyarországról* (The famous garden of Europe: Historical-ecological essays on Hungary), ed. Ágnes R. Várkonyi and László Kósa, 184–200. Budapest: Orpheusz Könyvkiadó.

Hoffmann, Richard C. 2001. "Frontier Foods for Late Medieval Consumers: Culture, Economy, Ecology." *Environment and History* 7: 131–167.

Holdsworth, William. 1938. *A History of English Law*. London: Methuen and Co. Ltd.

Holl, Imre. 1979. "Sarvaly középkori lakóházai" (Medieval houses at Sarvaly). *Archaeologiai Értesítő* 106: 33–51.

Holl, Imre and Nándor Parádi. 1982a. *Das mittelalterliche Dorf Sarvaly*. Budapest: Akadémiai Kiadó.

Holl, Imre and Nándor Parádi. 1982b. "Nagykeszi középkori falu ásatása" (Excavations at the medieval village of Nagykeszi). *Veszprém Megyei Múzeumok Közleményei* 16: 181–202.

Holl, Imre. 1990. "A középkori Szentmihály falu ásatása II. A házak, a falukép, az írásos adatok tanúsága" (The excavation of the medieval village of Szentmihály, part 2: The houses, the village, the written sources). *Zalai Múzeum* 2: 189–207.

Holl, Imre. 2000. *Funde aus dem Zisterzienserkloster von Pilis*. Budapest: Paulus-Publ. Verlag.

Holt, James Clarke. 1982. *Robin Hood*. London: Thames and Hudson.

Holub, József. 1929. *Zala megye története a középkorban* (A history of Co. Zala in the Middle Ages). Vol. 1. Pécs: n. p.

Hóman, Bálint. 1916. *Magyar pénztörténet 1000 – 1325* (Hungarian monetary history 1000 – 1325). Budapest: MTA.

Hooke, Della. 1992. "The Use of Early Medieval Charters as Sources for the Study of Settlement and Landscape Evolution." In *The Transformation of the European Rural Landscape: Methodological Issues and Agrarian Change 1770–1914*, ed. Antoon Verhoeve and Jelier A.J. Vervloet, 39–47. Wageningen: DLO Winand Staring Centre.

Hooke, Della. 1999. *Warwickshire Anglo-Saxon Charter Bounds*. Woodbridge: Boydell Press.

Horánszky, András. 1964. *Die Wälder des Szentendre-Visegráder Gebirges*. Budapest: Akadémiai Kiadó.

Horváth, Konstantin. 1930. *Zirc története* (A history of Zirc). Veszprém: Egyházmegyei Nyomda.

Horváth, Lajos. 1995. *Pest-Pilis-Solt vármegye kialakulása és működése 1659-ig* (The formation of Co. Pest-Pilis-Solt and its management until 1659). Budapest: Pest megyei Levéltár.

Hoskins, William. 1955. *The Making of the English Landscape*. London: Hodder and Stoughton.

Hunyadi, Zsolt. 1997–1998. "...scripta manent. Archival and Manuscript Resources in Hungary." *Annual of Medieval Studies at CEU* 5: 231–240.

Hümpfner, Tibor. 1964. "A zirci apátsági templom ásatása (1912–1913)" (The excavations of the monastic church in Zirc (1912–1913)). *Veszprém Megyei Múzeumok Közleményei* 2: 119–140.

Iacopa da Varazze. 1998. *Legenda Aurea*, ed. Giovanni Paolo Maggioni. 2 vols. Florence: Edizioni del Galluzo.

Ila, Bálint and József Kovacsics, ed. 1964. *Veszprém megye helytörténeti lexikona* (The local historical lexicon of Co. Veszprém). Budapest: Akadémiai Kiadó.

Ila, Bálint and Iván Borsa, ed. 1993. *Az Abaffy család levéltára, A Dancs család levéltára, a Hanvay család levéltára* (Archives of the Abaffy, Dancs, and Hanvay families). Budapest: Akadémiai Kiadó.

Imreh, István, ed. 1973. *A rendtartó székely falu* (By-laws of the Sekler village). Bucharest: Kriterion.

Iversen, Johannes. 1956. "Forest Clearance in the Stone Age." *Scientific American* 194: 36–41.

Jakó, Zsigmond, ed. 1990. *A kolozsmonostori konvent jegyzőkönyvei* (Protocols of the convent of Kolozsmonostor). 2 vols. Budapest: Akadémiai Kiadó.

Jankó, Annamária. 1995. "Végvárak az I. és II. katonai felmérés térképein" (Castles in the first and second Ordinance Survey maps). In *Végvár és környezet* (Castle and environment), ed. Tivadar Petercsák and Ernő Pető, 33–54. Eger: Heves Megyei Múzeumi Szervezet.

Jankovich, B. Dénes. 1985. "Archaeological Topography: Theoretical and Practical Lessons." *Mitteilungen des Archäologischen Instituts der Ungarischen Akademie der Wissenschaften* 14: 283–92.

Jankovich, B. Dénes. 1986. "Középkori okleveleink régészeti hasznosításának kérdései" (On the question of the archaeological utilization of medieval Hungarian charters). In *Középkori régészetünk újabb eredményei és időszerű feladatai* (Recent results and current issues in Hungarian medieval archaeology), ed. István Fodor and László Selmeczi, 443–452. Budapest: Művelődési Minisztérium.

Jankovich, B. Dénes. 1993. *A felszíni leletgyűjtés módszerei és szerepe a régészeti kutatásban* (The methods and role of fieldwalking in archaeological research). Budapest: Magyar Nemzeti Múzeum and Magyar Tudományos Akadémia Régészeti Intézet.

Járai Komlódi, Mária. 1987. "Postglacial Climate and Vegetation History in Hungary." In *Holocene Environment in Hungary*, ed. Márton Pécsi, 37–47. Budapest:

Geographical Research Institute, Hungarian Academy of Sciences.

Jarnut, Jörg. 1985. "Die frühmittelalterliche Jagd unter rechts- und sozialgeschichtlichen Aspekten." *Settimane di Studio* 31: 765–808.

Jones, Eustace W. 1945. "The Structure and Reproduction of the Virgin Forest of the North Temperate Zone." *New Phytologist* 44: 130–148.

Jones, M. 1986. "Coppice Wood Management in the Eighteenth Century: An Ecological Account." *Irish Forestry* 43: 15–31.

Kalapos, Tibor. 1998. "A magyarföldi husáng (*Ferula sadleriana*) Pilis-tetői populációjának dinamikája" (Population dynamics of *Ferula sadleriana* on Pilis peak). In *Sziklagyepek szünbotanikai kutatása* (Ecological studies of rocky grassland), ed. Péter Csontos, 41–56. Budapest: Scientia Kiadó.

Kalász, Elek. 1932. *A szentgotthárdi apátság birtokviszonyai és a ciszteri gazdálkodás a középkorban* (The estates of the monastery of Szentgotthárd and the Cistercian economy in the Middle Ages). Budapest: n. p.

Karácsonyi, János. 1900–1901. *A magyar nemzetségek a XIV. század közepéig* (Hungarian kindreds until the middle of the fourteenth century). 3 vols. Budapest: Magyar Tudományos Akadémia.

Karácsonyi, János. 1902. "Mit jelent e helynév 'Ardó'?" (What does the place-name 'Ardó' mean?) *Ethnographia* 13: 347–350.

Karácsonyi, János and Samu Borovszky, ed. 1903. *Az időrendbe szedett váradi tüzesvaspróba-lajstrom az 1550-iki kiadás hű hasonmásával együtt* (The trials by hot iron of Várad in chronological order along with the facsimile of the 1550 edition). Budapest: n.p.

Kaspers, Heinrich. 1957. *Comitatus nemoris: die Waldgrafschaft zwischen Maas und Rhein.* Düren: Dürener Geschichtsverein.

Kázmér, Miklós. 1970. *A "falu" a magyar helynevekben. XIII–XIX. század* ("Falu" in Hungarian place-names: thirteenth – nineteenth centuries). Budapest: Akadémiai Kiadó.

Kerner, Anton. 1863. *Das Pflanzenleben der Donauländer.* Innsbruck: Wagner Verlag.

Kircsi, Andrea, ed. 2000. *III. Erdő és klíma konferencia: Debrecen 2000. június 7–9* (Third conference on woodland and climate: Debrecen, 7–9, July, 2000). Debrecen: DE TTK Meteorológiai Tanszék.

Kiss, Lajos, ed. 1988. *Földrajzi nevek etimológiai szótára* (An etymological dictionary of geographical names). 2 vols. Budapest: Akadémiai Kiadó.

Knapp, Éva. 1994. "Pálos gazdálkodás a középkori Baranya megyében" (Pauline economy in medieval Co. Baranya). In *Pálos rendtörténeti tanulmányok* (Essays on the history of the Pauline order), ed. Gábor Sarbak, 62–103. Csorna: Árva Vince.

Knauz, Nándor. 1863. "Az esztergomi érsekség okmányai" (Charters of the bishopric of Esztergom). *Magyar Sion* 1: 129–36.

Kolossváry, Szabolcsné, ed. 1966. *A magyar erdészeti irodalom bibliográfiája 1945–1964* (Bibliography of Hungarian writings on forestry 1945–1964). Budapest: Akadémiai Kiadó.

Kolossváry, Szabolcsné, ed. 1975. *Az erdőgazdálkodás története Magyarországon. (Tanulmányok)* (The history of forestry in Hungary (Essays)). Budapest: Akadémiai Kiadó.

Koppány, Tibor. 1972. A XI. századi udvarház maradványai Zircen (The remains of the eleventh-century residence in Zirc). *Veszprém Megyei Múzeumok Közleményei* 11: 139–147.

Kostova, Rossina. 2002. "Monasticism in Bulgaria Ninth to Tenth Centuries: Interpreting the Archaeological Evidence." PhD diss., CEU, Budapest.

Kovalovszki, Júlia. 1960. "Ásatások Szarvas környéki Árpád-kori falvak helyén" (Excavations at the Árpádian period settlements around Szarvas). *Archaeologiai Értesítő* 87: 32–40.

Kovalovszki, Júlia. 1969. "Ásatások Csepelyen" (Excavations at Csepely). *Veszprém Megyei Múzeumok Közleményei* 8: 235–250.

Kovalovszki, Júlia. 1975. "Előzetes jelentés a dobozi Árpád-kori faluásatásról" (Preliminary report on the excavations of the Árpádian period settlement at Doboz). *Archaeologiai Értesítő* 102: 204–223.

Kovalovszki, Júlia. 1992a. "A Pálos remeték Szent Kereszt-kolostora (Méri István ásatása Klastrompusztán)" (The Holy Cross monastery of the Pauline hermits (István Méri's excavations at Klastrompuszta)). *Communicationes Archaeologiae Hungariae*: 173–207.

Kovalovszki, Júlia. 1992b. "A Pálos remeték Szent Kereszt-kolostora Klastrompusztán" (The Holy Cross monastery of the Pauline hermits at Klastrompuszta). *Műemlékvédelem* 36: 11–116.

Köblös, József. 1997. *A Dunántúli Református Egyházkerület Levéltárának magyar vonatkozású középkori oklevelei. 1265–1525* (Hungarian related medieval charters in the Archives of the Transdanubian Diocese of the Reformed Church). Pápa: Dunántúli Református Egyházkerület Tudományos Gyűjteményei.

Körmendy, Adrienn. 1974. "A soltész ("more scultetorum") telepítette falvak a Szepességben" (Settlements founded by *sculteti* ("more scultetorum") in Szepesség). *Agrártörténeti Szemle* 16: 305–348.

Körmendy, Adrienn. 1995. *Melioratio Terrae: Vergleichende Untersuchungen über die Siedlungsbewegung im östlichen Mitteleuropa im 13.–14. Jahrhundert.* Poznan: Publishing House of the Poznan Society for the Advancement of the Arts and Sciences.

Körmendy, József. 1980. "A veszprémi püspöki és káptalani levéltár Mohács előtti oklevelei" (The pre-Mohács charters of the episcopal and chapter archives of Veszprém). *Levéltári Szemle* 30: 463–489.

Kredics, László and László Solymosi, ed. 1993. *Urbarium episcopatus Vesprimiensis anno MDXXIV.* Új történelmi tár 4. Budapest: Akadémiai Kiadó.

Kredics, László, Lajos Madarász, and László Solymosi, ed. 1997. *A veszprémi káptalan számadáskönyve 1495 – 1534: krónika 1526 – 1558: javadalmasok és javadalmak 1550, 1556* (Account book of the chapter of Veszprém 1495 – 1534: chronicle 1526 – 1558: prebendaries and prebends 1550, 1556). Veszprém: Veszprém Megyei Levéltár.

Kristó, Gyula. 1976. "Szempontok korai helyneveink történeti tipológiájához" (Aspects of the historical typology of early Hungarian place-names). *Acta Universitatis Szegediensis de Attila József Nominatae. Acta Historica* 55.

Kristó, Gyula. 1986. *Csák Máté* (Matthew Csák). Budapest: Gondolat.

Kristó, Gyula. 1988. *A vármegyék kialakulása Magyarországon* (The formation of counties in Hungary). Budapest: Magvető.

Kristó, Gyula, ed. 1990–. *Anjou-kori Oklevéltár* (Charters from the Angevin period). Several volumes not numbered continuously. Budapest and Szeged: n.p.

Kristó, Gyula, ed. 1994a. *Kun László emlékezete* (In memoriam Ladislaus the Cuman). Szeged: Szegedi Középkorász Műhely.

Kristó, Gyula, ed. 1994b. *Korai magyar történeti lexikon (9–14. század)* (Early Hungarian historical lexicon (ninth-fourteenth centuries)). Budapest: Akadémiai Kiadó.

Kristó, Gyula. 2002a. "Néhány vármegye kialakulásának kérdéséhez" (Regarding the formation of some counties). *Századok* 136: 469–484.

Kristó, Gyula. 2002b. "Rendszerjelleg a korai középkor kutatásában" (Systematism in the research of the early Middle Ages). *Századok* 136: 686–690.

Kriza, János. 1872. "Helynevek Marosszéken" (Place-names in Marosszék), *Magyar Nyelvőr* 1: 383–384.

Kubinyi, András. 1965. "Csepel története a Vasgyár alapításáig," (A history of Csepel until the foundation of the Ironworks). In *Csepel története* (A history of Csepel), 6–31. Budapest: Csepel Vas- és Fémművek Pártbizottsága.

Kubinyi, András. 1975. "A királyi kancellária és udvari kápolna a XII. század közepén" (The royal chancellery and the royal chapel in the middle of the twelfth century). *Levéltári Közelmények* 46: 59–121.

Kubinyi, András. 1986. "A nagybirtok és jobbágyai a középkor végén az 1478-as Garai-Szécsi birtokfelosztás alapján" (The estate and its peasants at the end of the Middle Ages, as seen in the 1478 Gara-Szécs estate division). *Veszprém Megyei Múzeumok Közleményei* 18: 197–225.

Kubinyi, András. 1997. "A magyar királyság népessége a 15. század végén" (Population of the kingdom of Hungary at the end of the fifteenth century). In *Magyarország történeti demográfiája 896–1995* (Historical demography of Hungary 896–1995), ed. József Kovacsics, 93–110. Budapest: KSH.

Kubinyi, András. 2001. "A császárvári uradalom közbecsü összeírása 1489-ből" (The estimation of the estate of Császárvár from 1489). *Történelmi Szemle* 43: 3–17.

Kumorovitz, Lajos Bernát, ed. 1953. *Veszprémi regeszták* (Regesta from Veszprém). Budapest: Akadémiai Kiadó.

Kumorovitz, Lajos Bernát, ed. 1987. *Budapest történetének okleveles emlékei* (Charters of the history of Budapest). Vol. 3. Budapest: Budapesti Történeti Múzeum.

Kunoss, Endre. 1853. *A selymészet kézikönyve vagyis népszerű útmutatás a szederfák ültetése, ápolása és selyemhernyók tenyésztése ügyében* (A handbook of silk production, or popular guide to the planting and management of mulberries, and the breeding of silkworms). Pest: n.p.

Kvassay, Judit. 1996. "Régészeti adatok a Zala-megyei 15–16. századi háztípusokhoz" (Archaeological data on house types in the fifteenth-sixteenth centuries in Co. Zala). In *Nyugat-Dunántúl népi építészete* (Vernacular architecture in the western parts of Transdanubia), 183–196. Szentendre and Szombathely: Szentendrei Szabadtéri Néprajzi Múzeum.

Lackner, Beda. 1982. "The Abbey of Zirc, 1182–1982, Historical Survey." *Cistercian Studies* 17: 350–365.

Lamb, Hubert H. 1982. *Climate, History and the Modern World*. London and New York: Methuen.

Lambert, George. 1996. "Recherche de signaux anthropiques dans les séries dendrochronologiques du Moyen Âge: exemple des séquences de Charavines-Colletière." In *L'homme et la nature au Moyen Âge: Paléoenvironnement des sociétés occidentales*, 143–152. Paris: Editions Errance.

László, Gyula. 1993. *A Szent László-legenda középkori faliképei* (Wall-paintings of the medieval St. Ladislaus legend). Budapest: TKM Egyesület.

Laszlovszky, József. 1982. "Karámok Árpád-kori falvainkban" (Bawns in Hungarian Árpádian period villages). *Acta Archaeologica Academiae Scientiarum Hungaricae* 109: 281–285.

Laszlovszky, József. 1986a. "Dedi etiam terram, que adiacet circa aquam, que vocatur Tiza: Adatok az 1075-ös garamszentbenedeki oklevél helyneveinek lokalizálásához" (*Dedi etiam terram, que adiacet circa aquam, que vocatur Tiza*: Regarding the localization of the place-names in the 1075 charter of Garamszentbenedek). *Zounuk* 1: 9–24.

Laszlovszky, József. 1986b. "Tanyaszerű települések az Árpád korban" (Farm-steads in the Árpádian period). In *Falvak, mezővárosok az Alföldön* (Villages and towns in the Great Hungarian Plain), ed. László Novák and László Selmeczi, 131–151. Nagykőrös: Arany János Múzeum.

Laszlovszky, József. 1994. "Angol-magyar kapcsolatok a 12. század második felében" (Anglo-Hungarian relations in the second half of the twelfth century). *Századok* 128: 223–253.

Laszlovszky, József. 1997. "Hungarian University Peregrinatio to Western Europe in the Second Half of the Twelfth Century." In *Universitas Budensis 1395–1995*, ed. László Szögi and Júlia Varga, 51–60. Budapest: ELTE.

Laszlovszky, József. 1999. "Field Systems in Medieval Hungary." In *The Man of Many Devices Who Wandered Full Many Ways: Festschrift in Honor of János M. Bak*, ed. Balázs Nagy and Marcell Sebők, 432-444. Budapest: CEU Press.

Laszlovszky, József and Péter Szabó, ed. 2003. *People and Nature in Historical Perspective*. Budapest: CEU Medieval Studies and Archaeolingua.

Lawrence, Clifford Hugh. 1989. *Medieval Monasticism*. London and New York: Longman, 2d edition.

Lázár, Sarolta. 1992. "A pilisszentléleki volt pálos kolostortemplom kutatása 1985–86" (Excavations in the former Pauline monastery at Pilisszentlélek, 1985–86). *Komárom-Esztergom Megyei Múzeumok Közleményei* 5: 493–518.

Le Goff, Jacques. 1985. "Le désert-forêt dans l'Occident médiéval." In *L'imaginaire médiéval*. Paris: Édition Gallimard.

Lékai, Louis. 1977. *The Cistercians: Ideals and Reality*. N.p.: The Kent State Press.

Lexicon Latinitatis Medii Aevi Hungariae. 1992–1999. 5 vols. Budapest: Argumentum.

Linnard, William. 1982. *Welsh Woods and Forests: History and Utilization*. Cardiff: National Museum of Wales.

Lockhart, Robin Bruce. 1985. *Halfway to Heaven: The Hidden Life of the Sublime Carthusians*. London: Thames Methuen.

Longnon, Jean and Raymond Cazelles. 1989. *The Très Riches Heures of Jean, Duke of Berry*. New York: Braziller.

Lőrinczy, Gábor, ed. 1993. *Az Alföld a 9. században* (The Great Hungarian Plain in the ninth century). Szeged: Móra Ferenc Múzeum.

Lukcsics, József. 1908. *A veszprémi káptalan a XVI. században* (The chapter of Veszprém in the sixteenth century). Veszprém: Egyházmegyei ny.

Lukcsics, Pál. 1930. "A veszprémi székeskáptalan levéltára" (The archives of the chapter of Veszprém). *Levéltári Közlemények* 8, no. 3–4: 151–181.

Lukinich, Imre, ed. 1937–1943. *A podmanini Podmaniczky-család oklevéltára* (Charters of the Podmaniczky family of Podmanin). 5 vols. Budapest: Magyar Tudományos Akadémia.

Magdics, Iván. 1888. *Diplomatarium Ráczkeviense*. Székesfehérvár: n.p.

Magyar, Eszter. 1983. *A feudalizmus kori erdőgazdálkodás az alsó-magyarországi bányavárosokban* (Forestry in the feudal period in the mining-cities of Lower-Hungary). Budapest: Akadémiai Kiadó.

Magyar, Eszter. 1984. "A királyi erdőkezelés kiváltságolt népcsoportjainak továbbélése a bakonyi erdőispánság területén" (The survival of privileged groups of society of royal Forest administration in the territory of Bakony Forest). In *Tanulmányok Veszprém megye múltjából* (Essays from the past of Co. Veszprém), ed. László Kredics, 101–118. Veszprém: Veszprém megyei Levéltár.

Magyar, Eszter. 1993. "Erdőgazdálkodás a 18. századi Magyarországon" (Forestry in eighteenth-century Hungary). In *Európa híres kertje: Történeti ökológiai tanulmányok Magyarországról* (The famous garden of Europe: Historical-ecological essays on Hungary), ed. Ágnes R. Várkonyi and László Kósa, 141–163. Budapest: Orpheusz Könyvkiadó.

Magyar, Eszter. 1998. *Visegrád története 1684–1756* (A history of Visegrád 1684–1756). Budapest: Pest Megyei Levéltár.

A magyar nyelv történeti-etimológiai szótára (A historical-etymological dictionary of the Hungarian language). 1967–1984. 4 vols. Budapest: Akadémiai Kiadó.

Magyarország régészeti topográfiája (The archaeological topography of Hungary). 1966. Vol. 1. *Veszprém megye régészeti topográfiája: A keszthelyi és tapolcai járás* (The archaeological topography of Co. Veszprém: Keszthely and Tapolca districts), ed. Károly Sági. Budapest, Akadémiai Kiadó.

Magyarország régészeti topográfiája (The archaeological topography of Hungary). 1969. Vol. 2. *Veszprém megye régészeti topográfiája: A veszprémi járás* (The archaeological topography of Co. Veszprém: Veszprém district), ed. István Éri. Budapest: Akadémiai Kiadó.

Magyarország régészeti topográfiája (The archaeological topography of Hungary). 1970. Vol. 3. *Veszprém megye régészeti topográfiája: A devecseri és sümegi járás* (The archaeological topography of Co. Veszprém: Devecser and Sümeg districts), ed. Kornél Bakay. Budapest: Akadémiai Kiadó.

Magyarország régészeti topográfiája (The archaeological topography of Hungary). 1972. Vol. 4. *Veszprém megye régészeti topográfiája: A pápai és zirci járás* (The archaeological topography of Co. Veszprém: Pápa and Zirc districts), ed. István Torma. Budapest: Akadémiai Kiadó.

Magyarország régészeti topográfiája (The archaeological topography of Hungary). 1979. Vol. 5. *Komárom megye régészeti topográfiája: Az esztergomi és dorogi járás* (The archaeological topography of Co. Komárom: Esztergom and Dorog districts), ed. István Torma. Budapest, Akadémiai Kiadó.

Magyarország régészeti topográfiája (The archaeological topography of Hungary). 1986. Vol. 7. *Pest megye régészeti topográfiája: A budai és szentendrei járás* (The archaeological topography of Co. Pest: Buda and Szentendre districts), ed. István Torma. Budapest, Akadémiai Kiadó.

Magyarország régészeti topográfiája (The archaeological topography of Hungary). 1989. Vol. 8. *Békés megye régészeti topográfiája* (The archaeological topography of Co. Békés), ed. László Gerevich. Budapest: Akadémiai Kiadó.

Majer, Antal. 1973. "Természetes erdőfelújítási eljárások" (Natural woodland renewal processes). In *Erdőművelés* (Forest cultivation), ed. István Danszky, vol. 1, 642–654. Budapest: Mezőgazdasági Kiadó.

Majer, Antal. 1980. *A Bakony tiszafása* (The yew-wood of Bakony). Budapest: Akadémiai Kiadó.

Majer, Antal. 1988. *Fenyves a Bakonyalján* (A pine-wood in Bakonyalja). Budapest: Akadémiai Kiadó.

Major, Jenő. 1959. "Szempontok a faluépítési hagyományok kutatásának módszeréhez" (Methodological considerations of the research on traditions in village formation). *Településtudományi Közlemények* 11: 3–16.

Maksay, Ferenc. 1957. "Urbáriumok." In *A történeti statisztika forrásai* (Sources of historical statistics), ed. József Kovacsics, 119–144. Budapest: Közgazdasági és Jogi Könyvkiadó.

Mályusz, Elemér. 1922. *Turóc megye kialakulása* (The formation of Co. Turóc). Budapest: Budavári Tudományos Társaság.

Mályusz, Elemér and Iván Borsa, ed. 1951–2001. *Zsigmondkori Oklevéltár* (Charters from the Sigismund period). 7 vols. Budapest: Akadémiai Kiadó and Magyar Országos Levéltár.

Mályusz, Elemér. 1967. "1526 előtti okleveleink forrásértéke" (Source value of Hungarian pre-1526 charters). *Történelmi Szemle* 10: 416–429.

Mályusz, Elemér. 1968. "Les problèmes des sources de l'histoire médiévale hongroise." *Acta Historica Academiae Scientiarum Hungaricae* 14: 179–197.

Mályusz, Elemér. 1971. *Egyházi társadalom a középkori Magyarországon* (Ecclesiastical society in medieval Hungary). Budapest: Akadémiai Kiadó.

Marsina, Richard, ed. 1971–1987. *Codex Diplomaticus et Epistolaris Slovaciae*. 2 vols. Bratislava: Academia Scientiarum Slovaca.

Matolcsi, János. 1975–1977. "A bakonyi sertés XV–XVI. századi csontleletei Sarvalyon" (Fifteenth and sixteenth-century bone finds of the Bakony-type swine at Sarvaly). *Magyar Mezőgazdasági Múzeum Közleményei*: 331–355.

Mátyás, Csaba, ed. 1996. *Erdészeti ökológia* (Forestry ecology). Budapest: Mezőgazda Kiadó.

Mátyás, Flórián. 1865. "Magyar nyelvritkaságok" (Hungarian language curiosities). *Nyelvtudományi Közlemények* 4: 188–208

Mayer, Hannes. 1989. *Urwaldreste, Naturwaldreservate und schützenswerte Naturwälder in Österreich*. Vienna: IUFRO.

McNeill, John. 2000. *Something New under the Sun*. London: Allen Lane The Penguin Press; Penguin Books 2001.

Medium Regni: Medieval Hungarian Royal Seats. 1999. Budapest: Nap.

Merényi, Lajos. 1900. "Nagykárolyi vadászati és erdészeti utasítások" (Hunting and woodland management instructions from Nagykároly). *Magyar Gazdaságtörténeti Szemle* 7: 416–421.

Méri, István. 1952. "Beszámoló a tiszalök-rázompusztai és túrkeve-mórici ásatások eredményeiről. I." (Report on the results of the excavations at Tiszalök-Rázompuszta and Túrkeve-Móric: part 1). *Archaeologiai Értesítő* 79: 49–65.

Méri, István. 1954. "Beszámoló a tiszalök-rázompusztai és túrkeve-mórici ásatások eredményeiről. II." (Report on the results of the excavations at Tiszalök-Rázompuszta and Túrkeve-Móric: part 2). *Archaeologiai Értesítő* 81: 138–154.

Méri, István. 1962. "Az árkok szerepe Árpád-kori falvainkban" (The role of ditches in Hungarian Árpádian period villages). *Archaeologiai Értesítő* 89: 211–218.

Mészöly, Gedeon. 1908. "A -vány, -vény képző eredete" (The origins of the suffix *-vány, -vény*). *Magyar Nyelv* 4: 410–414.

Midriak, Rudolf. 1994. "Carpathian Biosphere Reserves in Slovakia: Trends of Ecological Research." In *Conservation of Forests in Central Europe*, ed. Jozef Paulenka and Ladislav Paule, 51–54. Zvolen: Arbora Publishers.

Miklós, Zsuzsa. 1997. "Falvak, várak, kolostorok a Dél-Börzsönyben" (Villages, castles, monasteries in South-Börzsöny). *Váci könyvek* 8: 46–49.

Mikos, József. 1935. "A fehérvári keresztesek 1193. évi oklevele mint magyar nyelvemlék" (The 1193 charter of the hospitallers of Fehérvár). *Magyar Nyelv* 31: 152–167, 243–258, 288–309.

Molnár, Zsolt. 1996. "Interpreting Present Vegetation Features by Landscape Historical Data: An Example from a Woodland-Grassland Mosaic Landscape (Nagykőrös Wood, Kiskunság, Hungary)." In *The Ecological History of European Forests*, ed. Keith J. Kirby and Charles Watkins, 241–269. Wallingford: CABI Publishing.

Molnár, Zsolt and András Kun. 2000. *Alföldi erdőssztyepp-maradványok Magyarországon* (Remnants of woodland-steppe in Hungary). Budapest: WWF Magyarország.

Montanari, Massimo. 1979. *L'alimentazione contadina nell'alto Medioevo*. Naples: Liguori Editore.

Montanari, Massimo. 1984. *Campagne medievali : strutture produttive, rapporti di lavoro, sistemi alimentari*. Turin: Einaudi.

Monumenta Ecclesiae Strigoniensis. 1874–1999. 4 vols. Esztergom: n. p.

Moreno, Diego, Pietro Piussi, and Oliver Rackham, ed. 1982. *Boschi. Storia e archaeologia I. Quaderni Storici* 49.

Moreno, Diego, ed. 1986. *Boschi. Storia e archaeologia II. Quaderni Storici* 62.

Munkácsi, Bernát. 1883. "Az ugor összehasonlító nyelvészet és Budenz szótára" (Ugric comparative linguistics and the dictionary of Budenz). *Magyar Nyelvőr* 12: 485–499.

Munkácsi, Bernát. 1893. "A magyar népies halászat műnyelve" (The specialised language of Hungarian fishing). *Ethnographia* 4: 161–310.

Müller, Róbert. 1972. "Adatok a Nyugat-Dunántúl középkori népi építészetéhez" (Regarding the medieval vernacular architecture of the western parts of Transdanubia). *Veszprém Megyei Múzeumok Közleményei* 11: 195–212.

Nagy, Gyula, ed. 1887–1889. *A nagymihályi és sztárai gróf Sztáray család oklevéltára* (Charters of the Sztáray family of Nagymihály and Sztára). 2 vols. Budapest: n.p.

Nagy, Imre, ed. 1865–1891. *Hazai Okmánytár / Codex Diplomaticus Patrius*. 8 vols. Győr and Budapest: n. p.

Nagy, Imre, ed. 1889–1891. *Sopron vármegye története. Oklevéltár* (A history of Co. Sopron. Charters). 2 vols. Sopron: Sopron vármegye közönsége.

Nagy, Imre and Gyula Tasnádi Nagy, ed. 1878–1920. *Anjou-kori okmánytár* (Charters from the Angevin period). 7 vols. Budapest: Magyar Tudományos Akadémia.

Nagy, Imre, Farkas Deák, and Gyula Nagy, ed. 1879. *Hazai Oklevéltár 1234–1536* (Hungarian charters 1234–1536). Budapest: Magyar Történelmi Társulat.

Nagy, Imre, Dezső Véghely, and Gyula Nagy, ed. 1886–1890. *Zala vármegye története. Oklevéltár* (A history of Co. Zala. Charters). 2 vols. Budapest: Franklin.

Nagy, Imre, Iván Nagy, Dezső Véghely, Ernő Kammerer, and Pál Lukcsis, ed. 1871–1931. *A zichi és vásonkeöi gróf Zichy-család idősb ágának okmánytára* (Charters of the Zichy family of Zich and Vásonkeö). 12 vols. Budapest: n. p.

Nagyváthy, János. 1821. *Magyar Practicus Termesztő* (Hungarian Practical Cultivator). Pest: Trattner János Tamás; reprint, Szeged: Állami Könyvterjesztő Vállalat, 1984.

Németh, Ferenc. 1998. "Magyarország erdőterületeinek változása 1100 év alatt" (Changes in the wooded areas of Hungary in 1100 years). *Erdészeti Kutatások* 88: 145–161.

Németh, Péter. 1993–1994. "A városlődi karthauzi kolostor története és régészeti emlékanyaga I" (The history and archaeological remains of the Carthusian monastery in Városlőd, part 1). *Veszprém Megyei Múzeumok Közleményei* 19/20: 367–382.

Németh, Péter. 1997. *A középkori Szabolcs megye települései* (Settlements in medieval Co. Szabolcs). Nyíregyháza: Ethnica.

Nováki, Gyula. 1975–1977. "Régi szántóföldek nyomai a Börzsönyben" (Remains of ancient arable fields in Börzsöny). *Magyar Mezőgazdasági Múzeum Közleményei*: 53–77.

Nováki, Gyula. 1984–1985. "Szántóföldek maradványai a XIV–XVI. századból a Sümeg-Sarvalyi erdőben" (Field remains from the fourteenth-sixteenth centuries in the Sümeg-Sarvaly wood). *Magyar Mezőgazdasági Múzeum Közleményei*: 19–31.

Nováki, Gyula. 1990. "A középkori Szentmihály falu földvára és szántóföldjei" (The earthwork castle and fields of the medieval village of Szentmihály). *Zalai Múzeum* 2: 209–219.

Nyári, Albert. 1870. "A modenai Hyppolit codexek" (The Hyppolit codices in Modena). *Századok*: 665.

Nyíri, Antal. 1979. "Az aszó, asszú eredete, hang- és alaktörténete" (The origins of *aszó, asszú*, and their phonetic and morphological history). *Magyar Nyelv* 75: 147–162.

Olahus, Nicolaus. 1938. *Hungaria – Athila*, ed. Kálmán Eperjessy and László Juhász. Budapest: Egyetemi Nyomda.

Ostermann, R. and A. Reif. 2000. "Socioeconomic and Ecological Aspects of Coppice Woods History in the Lower Vosges (France) and the Black Forest (Germany)."

In *Methods and Approaches in Forest History*, ed. Mauro Agnoletti and Steven Anderson, 107–118. Wallingford: CABI Publishing.

Óvári, Lipót. 1890. *A Magyar Tudományos Akadémia Történelmi Bizottságának oklevél-másolatai* (The copies of the charters of the Historical Committee of the Hungarian Academy of Sciences). Vol. 1. Budapest: MTA Történelmi Bizottsága.

Pach, Zsigmond Pál. 1970. *The Role of East-Central Europe in International Trade, 16th and 17th Centuries*. Budapest: Akadémiai Kiadó.

Pais, Dezső. 1912. "Az aszó elhomályosult összetételei" (Obscure compounds of *aszó*). *Magyar Nyelv* 8: 391–401.

Pais, Dezső. 1933. "Hetevény." *Magyar Nyelv* 29: 37–42.

Paládi Kovács, Attila. 1981. "'Keleti hozadék' a magyar pásztorkultúrában" (Eastern heritage in Hungarian animal husbandry). In *Emlékkönyv a Túrkevei Múzeum fennállásának harmincadik évfordulójára* (Celebrating the thirtieth anniversary of the foundation of the Museum of Túrkeve), ed. Imre Dankó, 109–125. Túrkeve: Finta Múzeum.

Pálffy, Géza. 1995. "A főkapitányi hadiipari műhely kiépülése Kassán és nyersanyagellátó forrásai" (The formation of the principal ordinance workshop in Kassa and its sources of supply). In *Végvár és környezet* (Castle and environment), ed. Tivadar Petercsák and Ernő Pető, 183–219. Eger: Heves Megyei Múzeumi Szervezet.

Pálóczi Horváth, András. 1987. "Kutak Szentkirályon" (Wells in Szentkirály). *Múzeumi Kutatások Bács-Kiskun Megyében*: 86–89.

Pálóczi Horváth, András. 1995. "L'archéologie de l'environnement écologique et les recherches des villages désertés médiévaux en Hongrie." *Ruralia* 1: 262–268.

Pálóczi Horváth, András and Mátyás Szőke. 1995–1997. "A visegrádi királyi palota középkori kertjeinek kutatási programja" (Research programme of the medieval gardens of the Royal Palace in Visegrád). *Magyar Mezőgazdasági Múzeum Közleményei*: 43–59.

Pálóczi Horváth, András. 1996. "Élet egy középkori faluban: Szentkirály régészeti kutatásának eredményei" (Life in a medieval village: Results of the archaeological research in Szentkirály). In *Élet egy középkori faluban* (Life in a medieval village), 7–28. Budapest: Magyar Mezőgazdasági Múzeum.

Pálóczi Horváth, András. 1997. "Recherches archéologiques sur l'environnement écologique du jardin médiéval du palais royal à Visegrád." In *Environment and Subsistence in Medieval Europe. Papers of the 'Medieval Europe Brugge 1997' Conference*, ed. Guy De Boe and Frans Verhaeghe, 181–186. Zellik: IAP.

Pálóczi Horváth, András. 1998. "Multidisciplinary Archaeological Research of the Medieval Royal Palace Garden at Visegrád." In *Archaeometrical Research in Hungary*, vol. 2, ed. László Költő and László Bartosiewicz, 177–182. Budapest, Kaposvár, and Veszprém: MNM.

Pálóczi Horváth, András and Andrea Torma. 1999. "Environmental Archaeological Research at Visegrád in the Medieval Garden of the Royal Palace." In *Archaeology of the Bronze and Iron Age: Experimental Archaeology, Environmental Archaeology, Archaeological Parks*, ed. Erzsébet Jerem and Ildikó Poroszlai, 343–350. Budapest: Archaeolingua.

Pálóczi Horváth, András. 2000a. "Lakóház és telek rekonstrukciója Szentkirályon, egy alföldi késő középkori faluban. I." (Reconstruction of a house and a plot in Szentkirály, a late medieval village in the Great Plain: part 1). In *A középkori magyar agrárium* (Hungarian agriculture in the Middle Ages), ed. Lívia Bende and Gábor Lőrinczy, 121–149. Ópusztaszer: Ópusztaszeri Nemzeti Történeti Emlékpark and Csongrád Megyei Múzeumok Igazgatósága.

Pálóczi Horváth, András. 2000b. "A visegrádi palota kertje mint környezettörténeti forrás" (The garden of the royal palace at Visegrád as an environmental historical source). In *A táj változásai a Kárpát-medencében a történelmi események hatására* (Anthropogenetic landscape changes in the Carpathian Basin), ed. György Füleky, 96–100. Budapest and Gödöllő: Kulturális Örökség Igazgatósága.

Pálóczi Horváth, András. 2001. "A késő középkori népi építészet régészeti kutatásának újabb eredményei" (New results in the research on late medieval vernacular architecture). In *Népi építészet a Kárpát-medencében a honfoglalástól a 18. századig* (Vernacular architecture in the Carpathian Basin from the Conquest to the eighteenth century), 221–260. Szentendre and Szolnok: Szabadtéri Néprajzi Múzeum and Jász-Nagykun-Szolnok Megyei Múzeumok Igazgatósága.

Patay, Pál. 1969. "A Csörsz árka" (The Csörsz ditch). *Természet Világa* 100: 409–411.

Pécsi, Márton, ed. 1989. *Magyarország nemzeti atlasza* (National Atlas of Hungary). Budapest: Cartographia.

Penyigey, Dénes. 1980. *Debrecen város erdőgazdálkodása a XVIII. században és a XIX. század első felében* (Woodland management of the town of Debrecen in the eighteenth century and in the first half of the nineteenth century). Budapest: Akadémiai Kiadó.

Pesty, Frigyes. 1880. *Az eltűnt régi vármegyék* (The disappeared old counties). 2 vols. Budapest: MTA.

Pesty, Frigyes. 1882. *Krassó vármegye története* (A history of Co. Krassó). Vol. 3. *Oklevéltár* (Charters). Budapest: Krassó-Szörény vármegye közönsége.

Peterken, George F. 1981. *Woodland Conservation and Management*. London: Chapman and Hall.

Peterken, George F. 1996. *Natural Woodland: Ecology and Conservation in Northern Temperate Regions*. Cambridge: Cambridge University Press.

Pickett, S. T. A., S. L. Collins, and J. J. Armesto. 1987. "Models, Mechanisms and Pathways of Succession." *Botanical Review* 53: 335 371.

Pilcher, Jon R. 1983. "Radiocarbon Calibration and Dendrochronology: An Introduction." In *Archaeology, Dendrochronology, and the Radiocarbon Calibration Curve*, ed. Barbara S. Ottaway, 5–14. Edinburgh: University of Edinburgh, Department of Archaeology, Occasional Paper no. 9.

Pilcher, Jon R and Seán Mac an tSaoir, ed. 1995. *Woods, Trees and Forests in Ireland*. Dublin: Royal Irish Academy.

Pintér, László. 2001. "Egy leégett tetőszerkezetű Árpád-kori veremház Mosonszentmiklós, Egyéni földek lelőhelyről" (An Árpádian period semi-subterranean house with a burnt down roof from Mosonszentmiklós, Egyéni földek). In *Népi építészet a Kárpát-medencében a honfoglalástól a 18. századig* (Vernacular architecture in the Carpathian Basin from the Conquest to the eighteenth century), 461–466.

Szentendre and Szolnok: Szabadtéri Néprajzi Múzeum and Jász-Nagykun-Szolnok Megyei Múzeumok Igazgatósága.

Pollard, Ernest, Max D. Hooper, and Norman W. Moore. 1974. *Hedges*. London: Collins.

Poole, Austin Lane. 1958. *From Domesday Book to Magna Charta 1087–1216*. Oxford: Clarendon Press.

Postan, M. M. 1975. *Medieval Economy and Society: An Economic History of Britain in the Middle Ages*. Harmondsworth: Penguin Books; Reprint, 1984.

Priszter, Szaniszló. 1986. *Növényneveink* (Hungarian plant names). Budapest: Mezőgazdasági Könyvkiadó.

Průsa, Eduard. 1985. *Die böhmischen und mährischen Urwälder – ihre Struktur und Ökologie*. Prague: Nakladatelství Československé Akademie věd.

Pyne, Stephen J. 1997. *Vestal Fire: An Environmental History, Told through Fire, of Europe and Europe's Encounter with the World*. Seattle and London: University of Washington Press.

Rab, János. 1993. "Az etnogeobotanika – mint történeti ökológiai segédtudomány" (Ethnogeobotany – as an auxiliary science of historical ecology). In *Európa híres kertje: Történeti ökológiai tanulmányok Magyarországról* (The famous garden of Europe: Historical-ecological essays on Hungary), ed. Ágnes R. Várkonyi and László Kósa, 223–257. Budapest: Orpheusz Könyvkiadó.

Rackham, Oliver. 1972. "Grundle House: On the Quantities of Timber in Certain East Anglian Buildings in Relation to Local Supplies." *Vernacular Architecture* 3: 3–8.

Rackham, Oliver. 1975. *Hayley Wood: Its History and Ecology*. Cambridge: Cambs & Isle of Ely Naturalists' Trust.

Rackham, Oliver. 1976. *Trees and Woodland in the British Landscape*. London: Dent.

Rackham, Oliver. 1978. "Archaeology and Land-use History." in *Epping Forest – The Natural Aspect?* ed. David Corke, *Essex Naturalist*, n.s., 2: 16–57.

Rackham, Oliver. 1979. "Neolithic Woodland Management in the Somerset Levels: Sweet Track I." *Somerset Levels Papers* 5: 59–61.

Rackham, Oliver. 1980. *Ancient Woodland: Its History, Vegetation and Uses in England*. London: Edward Arnold.

Rackham, Oliver. 1986. *The History of the Countryside*. London: Dent.

Rackham, Oliver. 1989. *The Last Forest: The Story of Hatfield Forest*. London: Dent.

Rackham, Oliver. 1991. "Introduction to Pollards." In *Pollard and Veteran Tree Management*, ed. Helen J. Read, 6–10. London: Corporation of London.

Rackham, Oliver. 1996. *Trees and Woodland in the British Landscape*. London: Phoenix, Revised edition.

Rackham, Oliver. 1998. "Savanna in Europe." In *The Ecological History of European Forests*, ed. Keith J. Kirby and Charles Watkins, 1–24. Wallingford and New York: CABI Publishing.

Rácz, Lajos. 1993. "Éghajlati változások a középkori és kora-újkori Európában" (Climatic changes in medieval and early modern Europe). In *Európa híres kertje: Történeti ökológiai tanulmányok Magyarországról* (The famous garden of Europe: Historical-ecological essays on Hungary), ed. Ágnes R. Várkonyi and László Kósa, 67–86. Budapest: Orpheusz Könyvkiadó.

Radkau, Joachim. 1996. "Wood and Forestry in German History: In Quest of an Environmental Approach." *Environment and History* 2: 63–76.

Rady, Martin. 1999. "The Filial Quarter and Female Inheritance in Medieval Hungarian Law." In *The Man of Many Devices, Who Wandered Full Many Ways: Festschrift in Honor of János M. Bak*, ed. Balázs Nagy and Marcell Sebők, 422–431. Budapest: CEU Press.

Rapaics, Raymund. 1918. "Az Alföld növényföldrajzi jelleme" (The plant-geographical characteristics of the Alföld). *Erdészeti Kisérletek* 20: 1–97, 183–247.

Rasmussen, Peter. 1990a. "Leaf Foddering in the Earliest Neolithic Agriculture. Evidence from Switzerland and Denmark." *Acta Archaeologica* 60: 71–86.

Rasmussen, Peter. 1990b. "Pollarding of Trees in the Neolithic: Often Presumed – Difficult to Prove." In *Experimentation and Reconstruction in Environmental Archaeology*, ed. D. E. Robinson, 77–99. Oxford: Oxbow Books.

Redon, Odile. 1996. "L'arbre et la forêt dans la Toscane méridionale aux XIIIe et XIVe siècles." In *L'homme et la nature au Moyen Âge: Paléoenvironnement des sociétés occidentales*, 133–137. Paris: Editions Errance.

Reuter, Camillo. 1963. "Gyümölcsény." *Magyar Nyelvőr* 87: 368–370.

Reuter, Camillo. 1965. "Tölgy és haraszt" (Oak and *haraszt*). *Magyar Nyelv* 61: 80–89.

Reuter, Camillo. 1975. "Adatok a régi magyar fa- és erdőnevek ismeretéhez" (Regarding the ancient Hungarian names of trees and woods). In *Az erdőgazdálkodás története Magyarországon. (Tanulmányok)* (The history of forestry in Hungary (Essays)), ed. Szabolcsné Kolossváry, 80–87. Budapest: Akadémiai Kiadó.

Roberts, Neil. 1989. *The Holocene: An Environmental History*. Oxford: Blackwell.

Rómer, Flóris. 1860a. "Magyarország földirati és terményi állapotáról a középkorban" (On the geographical and agricultural status of Hungary in the Middle Ages). *Magyar Akadémiai Értesítő*: 226–385.

Rómer, Flóris. 1860b. *A Bakony, természetrajzi és régészeti vázlata* (Bakony, its geographical and archaeological description). Győr: Sauervein.

Romhányi, Beatrix. 1993–1994. "The Role of the Cistercians in Medieval Hungary: Political Activity or Internal Colonisation?" *Annual of Medieval Studies at the CEU* 1: 180–204.

Romhányi, Beatrix. 1996. "Monasteriologia Hungarica Nova: Monasteries, Friaries, Provostries and Collegiate Churches in Medieval Hungary." PhD diss., Eötvös Loránd University, Budapest.

Romhányi, Beatrix. 2000. *Kolostorok és társaskáptalanok a középkori Magyarországon* (Monasteries and chapters in medieval Hungary). Budapest: Pytheas.

Rubner, Heinrich. 1964. "Vom römischen Saltus zum fränkischen Forst." *Historisches Jahrbuch* 83: 271–277.

Rubner, Heinrich. 1965. *Untersuchungen zur Forstverfassung des mittelalterlichen Frankreiches*. Wiesbaden: Steiner.

Rus, Vasile. 1999. "Aestimatio, aestimare, aestimator: Observații pe marginea unor articole de glosar" (Aestimatio, aestimare, aestimator: Remarks upon some glossary articles). *Medievalia Transilvanica* 3: 125–154.

Saád, Andor. 1972. "Adatok a kaptárkövek eredetének, korának és rendeltetésének meghatározásához" (Regarding the identification of the origins, age, and function of *kaptárkövek*). *Hermann Ottó Múzeum Évkönyve* 17–18: 33–86.

Sabján, Tibor. 2000. "Lakóház és telek rekonstrukciója Szentkirályon, egy alföldi késő középkori faluban. II." (Reconstruction of a house and a plot in Szentkirály, a late medieval village in the Great Plain: part 2). In *A középkori magyar agrárium* (Hungarian agriculture in the Middle Ages), ed. Lívia Bende and Gábor Lőrinczy, 151–182. Ópusztaszer: Ópusztaszeri Nemzeti Történeti Emlékpark and Csongrád Megyei Múzeumok Igazgatósága.

Salvestrini, F. 2000. "Law, Forest Resources and Management of Territory in the Later Middle Ages: Woodlands in Tuscan Municipal Statutes." In *Forest History: International Studies on Socio-Economic and Forest Ecosystem Change*, ed. Mauro Agnoletti and Steven Anderson, 279–286. Wallingford: CABI Publishing.

Schenk, Winfried. 1995. "Forest Development Types in Central Germany in Pre-Industrial Times: A Contribution by Historical Geography to the Solution of a Forest History Research Argument about the 'Wood Scarcity' in the 18th Century." In *L'uomo e la foresta secc. XIII–XVIII*, ed. Simonetta Cavaciocchi, 201–225. Prato: Instituto Internazionale di Storia Economica.

Schenk, Winfried. 2000. "Pre-Industrial Forests in Central Europe as Objects of Historico-Geographical Research." In *Methods and Approaches in Forest History*, ed. Mauro Agnoletti and Steven Anderson, 129–138. Wallingford: CABI Publishing.

Schweingruber, Fritz Hans. 1983. *Der Jahrring: Standort, Methodik, Zeit und Klima in der Dendrochronologie*. Bern and Stuttgart: Paul Haupt.

Sedlák, Vincent, ed. 1980–1987. *Regesta diplomatica nec non epistolaria Slovaciae*. 2 vols. Bratislava: Academia Scientiarum Slovaca.

Semmler, Josef, ed. 1991. *Der Wald in Mittelalter und Renaissance*. Düsseldorf: Droste Verlag.

Sheail, John. 1980. *Historical Ecology: The Documentary Evidence*. Cambridge: Institute of Terrestrial Ecology.

Simonyi, Zsigmond. 1881. "A magyar gyakorító és mozzanatos igék képzése" (Forming iterative and momentaneous verbs in Hungarian). *Nyelvtudományi Közlemények* 16: 237–269.

Smout, Chris, ed. 1997. *Scottish Woodland History*. Edinburgh: Scottish Cultural Press.

Smout, Chris, ed. 2003. *People and Woods in Scotland: A History*. Edinburgh: Edinburgh University Press.

Soltész, Elisabeth, ed. 1982. *Das Corvinus Graduale*. Budapest: Magyar Helikon and Corvina.

Soltész, Zoltánné, ed. 1983. *Flamand kalendárium* (Flemish book of hours). N.p.: Helikon.

Solymos, Rezső. 2000. *Erdőfelújítás és -nevelés a természetközeli erdőgazdálkodásban* (Woodland renewal and cultivation in naturalistic forestry). Budapest: Mezőgazdasági Szaktudás Kiadó.

Sólymos, Szilveszter. 2001. "Az első bencés szerzetesek hazánkban" (The first Benedictine monks in Hungary). In *Paradisum plantavit: Bencés monostorok a középkori Magyarországon* (Paradisum plantavit: Benedictine monasteries in medieval Hungary), ed. Imre Takács, 48–60. Pannonhalma: Bencés Főapátság.

Solymosi, László. 1984. "Veszprém megye 1488. évi adólajstroma és az Ernuszt-féle megyei adószámadások" (The 1488 taxation list of Co. Veszprém and the Ernuszt taxation county account books). In *Tanulmányok Veszprém megye múltjából* (Essays from the past of Co. Veszprém), ed. László Kredics, 121–233. Veszprém: Veszprém megyei Levéltár.

Solymosi, László. 1996. "Albeus mester összeírása és a pannonhalmi apátság tatárjárás előtti birtokállománya" (The Albeus conscription and the estates of the abbey of Pannonhalma before the Mongol invasion). In *Mons Sacer 996–1996. Pannonhalma 1000 éve* (Mons Sacer 996–1996: One thousand years of Pannonhalma), ed. Imre Takács, 515–526. Pannonhalma: n.p.

Solymosi, László. 1998. *A földesúri járadékok új rendszere a 13. századi Magyarországon* (The new system of seigneurial allowances in the thirteenth century in Hungary). N.p.: Argumentum Kiadó.

Somogyi, Sándor. 1987. "Történeti földrajzi bevezető" (Historical geographical introduction). In *Magyarország története: Előzmények és magyar történet 1242-ig* (A history of Hungary: Antecedents and Hungarian history until 1242), ed. Antal Bartha. Budapest: Akadémiai Kiadó.

Sonnlechner, Christoph. 2000. "Landschaft und Tradition: Aspekte einer Umweltgeschichte des Mittelalters." In *Text – Schrift – Codex: Quellenkundliche Arbeiten aus dem Institut für Österreichische Geschichtsforschung*, ed. Christoph Egger and Herwig Weigl, 123–223. Vienna and Munich: R. Oldenbourg Verlag.

Soó, Rezső. 1926. "Die Enstehung der ungarischen Puszta." *Ungarische Jahrbücher* 6: 258–276.

Soó, Rezső. 1931. "A magyar puszta fejlődéstörténetének problémája" (Problems of the development history of the Hungarian *puszta*). *Földrajzi Közlemények* 59: 1–17.

Soó, Rezső. 1964–1985. *A magyar flóra és vegetáció rendszertani-növényföldrajzi kézikönyve* (A taxonomic – plant-geographic handbook of the Hungarian flora and vegetation). 7 vols. Budapest: Akadémiai Kiadó.

Soó, Rezső. 1965. *Növényföldrajz* (Plant-geography). Budapest: Tankönyvkiadó.

Soproni, Sándor. 1969. "Limes Sarmatiae." *Archaeologiai Értesítő* 96: 43–52.

Southern, Richard William. 1970. *Western Society and the Church in the Middle Ages*. Harmondsworth: Penguin Books; Reprint 1990.

Sörös, Pongrác. 1903. "Bakonybéli regeszták a XV. század első feléből" (Regesta from Bakonybél from the first half of the fifteenth century). *Történelmi Tár*: 354–372.

Stamper, Paul. 1988. "Woods and Parks." In *The Countryside of Medieval England*, ed. Grenville Astill and Annie Grant, 128–148. Oxford: Basil Blackwell.

Steane, John. 1984. *The Archaeology of Medieval England and Wales*. Athens: University of Georgia Press.

Steane, John. 1993. *The Archaeology of the Medieval English Monarchy*. London: Batsford.

Steensberg, Axel. 1957. "Some Recent Danish Experiments in Neolithic Agriculture." *Agricultural History Review* 5.2: 66–73.

Stefanovits, Pál. 1963. *Magyarország talajai* (The soil types of Hungary). Budapest: Akadémiai Kiadó.

Stefanovits, Pál. 1981. *Talajtan* (Pedology). Budapest: Mezőgazdasági Kiadó.

Stokes, Marvin A. and Terah L. Smiley. 1968. *An Introduction to Tree-Ring Dating*. Chicago: Chicago University Press.

Sugár, István. 1995. "Az egri vár építőanyagainak beszerzési helyei 1548–1564" (The sources of supply of building material of the castle of Eger 1548–1564). In *Végvár és környezet* (Castle and environment), ed. Tivadar Petercsák and Ernő Pető, 175–180. Eger: Heves Megyei Múzeumi Szervezet.

Sümegi, Pál and Róbert Kertész. 1998. "A Kárpát-medence őskörnyezeti sajátosságai" (Palaeoenvironmental features of the Carpathian Basin). *Jászkunság* 44: 144–157.

Sümegi, Pál, Ede Hertelendi, Enikő Magyar, and Mihály Molnár. 1998. "Evolution of the Environment in the Carpathian Basin during the Last 30, 000 BP Years and its Effects on the Ancient Habits of the Different Cultures." In *Archaeometrical Research in Hungary II*, ed. László Költő and László Bartosiewicz, 183–197. Budapest, Kaposvár, and Veszprém: Magyar Nemzeti Múzeum and Somogy Megyei Múzeumok Igazgatósága.

Sümegi, Pál. 1998a. "Ember és környezet kapcsolata a Kárpát-medencében az elmúlt 15 000 év során" (The relationship between people and the environment in the Carpathian Basin during the last 15,000 years). *Panniculus*, ser. B, 3: 367–380.

Sümegi, Pál. 1998b. "Reconstruction of Flora, Soil and Landscape Evolution, and Human Impact on the Bereg Plain from Late Glacial up to the Present, Based on Palaeoecological Analysis." In *Upper Tisza Valley*, ed. József Hamar, 173–204. Szeged and Szolnok: Tisza Klub Târgu Mureş.

Sümegi, Pál, Enikő Magyari, Péter Dániel, Ede Hertelendi, and Edina Rudner. 1999. "A kardoskúti Fehér-tó negyedidőszaki fejlődéstörténetének rekonstrukciója" (A reconstruction of the Quaternary geohistory of Lake Fehér at Kardoskút). *Földtani Közlöny* 129: 480–518.

Sümegi, Pál and Elvira Bodor. 2000. "Sedimentological, Pollen and Geoarchaeological Analysis of Core Sequence at Tököl." In *Százhalombatta Archaeological Expedition, SAX, Annual Report 1*, ed. Ildikó Poroszlai and Magdolna Vicze, 83–96. Százhalombatta: Matrica Múzeum.

Sümegi, Pál, Attila Molnár, and Gábor Szilágyi. 2000. "Szikesedés a Hortobágyon" (Alkalization in Hortobágy). *Természet Világa* 131: 213–216.

Szabadfalvi, József. 1970. *Az extenzív állattenyésztés Magyarországon* (Extensive animal husbandry in Hungary). Debrecen: KLTE.

Szabó, Dénes. 1936. "A dömösi prépostság adománylevele" (The donation charter of the provostry of Dömös). *Magyar Nyelv* 32: 54–57, 130–135, 203–206.

Szabó, István. 1937. *Ugocsa megye* (Co. Ugocsa). Budapest: MTA.

Szabó, István. 1966. *A falurendszer kialakulása Magyarországon* (The formation of the settlement system in Hungary). Budapest: Akadémiai Kiadó.

Szabó, István. 1969. *A középkori magyar falu* (The medieval Hungarian village). Budapest: Akadémiai Kiadó.

Szabó, István. 1975. *A magyar mezőgazdaság története a XIV. századtól az 1530-as évekig* (A history of Hungarian agriculture from the fourteenth century to the 1530s). Budapest: Akadémiai Kiadó.

Szabó, László. 1975. "Néprajzi párhuzam Árpád-kori falvaink árkainak rendeltetéséhez" (Ethnographic parallel to the functioning of the ditches of Hungarian Árpádian period villages). *Archaeologiai Értesítő* 102: 84–87.

Szabó, Péter. 2001. "'There is Hope for a Tree': Pollarding in Hungary." *Medium Aevum Quotidianum* 44: 41–60.

Szabó, Péter. 2002. "Medieval Trees and Modern Ecology: How to Handle Written Sources." *Medium Aevum Quotidianum* 46 (2002): 7–25.

Szabó, Péter. 2003. "Sources for the Historian of Medieval Woodland." In *People and Nature in Historical Perspective*, ed. József Laszlovszky and Péter Szabó, 265–287. Budapest: CEU Medieval Studies and Archaeolingua.

Szabó, T. Attila, ed. 1975–2002. *Erdélyi magyar szótörténeti tár* (Transylvanian Hungarian etymological dictionary). 11 vols. Bucharest and Budapest: Kriterion and Akadémiai Kiadó.

Szamota, István. 1891. *Régi utazások Magyarországon és a Balkán félszigeten* (Ancient travels in Hungary and on the Balkan peninsula). Budapest: Franklin Társulat.

Szamota, István. 1895. "A tihanyi apátság 1055-iki alapítólevele" (The 1055 foundation charter of the abbey of Tihany). *Nyelvtudományi Közlemények* 25: 129–167.

Szamota, István and Gyula Zolnai, ed. 1902–1906. *Magyar oklevél-szótár* (Hungarian charter dictionary). Budapest: Hornyánszky Viktor.

Székely, György, ed. 1987. *Magyarország története* (A history of Hungary). Vol. 2. Budapest: Akadémiai Kiadó.

Szentgyörgyi, Viktor. 2000. "Az elveszett tető" (The lost roof). In *A középkori magyar agrárium* (Hungarian agriculture in the Middle Ages), ed. Lívia Bende and Gábor Lőrinczy, 183–204. Ópusztaszer: Ópusztaszeri Nemzeti Történeti Emlékpark and Csongrád Megyei Múzeumok Igazgatósága.

Szentkirály: Élet egy középkori faluban (Szentkirály: Life in a medieval village). 1996. Budapest: Magyar Mezőgazdasági Múzeum.

Szentpétery, Imre and Iván Borsa, ed. 1923–1987. *Regesta Regum Stirpis Arpadianae Critico-Diplomatica*. 2 vols. Budapest: Akadémiai Kiadó.

Szentpétery, Imre. 1930. *Magyar oklevéltan* (Hungarian diplomatics). Budapest: Magyar Történelmi Társulat.

Szentpétery, Imre, ed. 1937–1938. *Scriptores Rerum Hungaricarum*. 2 vols. Budapest: MTA.

Szentpétery, Imre. 1938. "Szent István király oklevelei" (The charters of King St. Stephen). In *Emlékkönyv Szent István király halálának kilencszázadik évfordulóján* (A collection of essays in memory of King St. Stephen on the nine hundredth anniversary of his death), ed. Jusztinián Serédi, 133–202. Budapest: MTA.

Szily, Kálmán. 1919. "A -mán és -ván képző történetéhez" (On the history of the suffixes *-mán* and *-ván*). *Magyar Nyelv* 15: 92–95.

Szinnyei, József. 1879. "Az aj- alapszó és családja" (The basic word *aj-* and its family). *Magyar Nyelvőr* 8: 97–102.

Szőke, Mátyás. 1986. *Visegrád: Ispánsági központ* (Visegrád: county centre). N. p.: TKM Egyesület.

Szűcs, Jenő. 1984. "Az 1267. évi dekrétum és háttere: Szempontok a köznemesség kialakulásához" (The decretum of 1267 and its background: The formation of Hungarian common nobility). In *Mályusz Elemér emlékkönyv* (Studies in honour of Elemér Mályusz), ed. Éva H. Balázs, Erik Fügedi, and Ferenc Maksay, 341–394. Budapest: Akadémiai Kiadó.

Szűcs, Jenő. 1985. *Les trois Europes*. Paris: Editions L'Harmattan.

Szűcs, Jenő. 1993a. "Sárospatak kezdetei és a pataki erdőuralom" (The beginnings of Sárospatak and the Forest of Patak). *Történelmi Szemle* 35: 1–57.

Szűcs, Jenő. 1993b. *Az utolsó Árpádok* (The last Árpáds). Budapest: História and MTA Történettudományi Intézete.

Tagányi, Károly, ed. 1896. *Magyar Erdészeti Oklevéltár* (Charters of Hungarian forestry). 3 vols. Budapest: Országos Erdészeti Egyesület.

Tagányi, Károly. 1950. *A földközösség története Magyarországon* (The history of common land in Hungary). Budapest: Athenaeum, Reprint.

Takács, Imre. 1992. "Egy 13. századi kút töredékei a pilisi ciszterci monostorból" (Fragments of a thirteenth-century well from the Pilis Cistercian monastery). *Művészettörténeti Értesítő* 41: 1–19.

Takács, Imre. 1994a. "Gertrudis királyné síremléke" (The shrine of queen Gertrudis). In *Pannonia Regia: Művészet a Dunántúlon 1000–1541* (Pannonia Regia: Art in Transdanubia 1000–1541), ed. Árpád Mikó and Imre Takács, 248–251. Budapest: Magyar Nemzeti Galéria.

Takács, Imre. 1994b. "A pilisi ciszteri apátság" (The Pilis Cistercian monastery). In *Pannonia Regia: Művészet a Dunántúlon 1000–1541* (Pannonia Regia: Art in Transdanubia 1000–1541), ed. Árpád Mikó and Imre Takács, 236–238. Budapest: Magyar Nemzeti Galéria.

Takács, Imre, ed. 2001. *Paradisum plantavit: Bencés monostorok a középkori Magyarországon* (Paradisum plantavit: Benedictine monasteries in medieval Hungary). Pannonhalma: Bencés Főapátság.

Takács, Károly. 2000a. "Árpád-kori csatornarendszerek kutatása a Rábaközben és a Kárpát-medence egyéb területein" (Research of Árpádian period channels in Rábaköz and in other areas of the Carpathian Basin). *Korall* 1: 27–61.

Takács, Károly. 2000b. "Árpád-kori csatornarendszerek kutatásáról" (On the research of Árpádian period channels). In *Táj és történelem* (Landscape and history), ed. Ágnes R. Várkonyi, 73–106. Budapest: Osiris.

Takács, Károly. 2000c. "Fokgazdálkodás vagy valami más?" (*Fok* management or something else?) *Korall* 2: 149–150.

Takács, Károly. 2003. "Medieval Hydraulic Systems in Hungary: Written Sources, Archaeology and Interpretation." In *People and Nature in Historical Perspective*, ed. József Laszlovszky and Péter Szabó, 289–311. Budapest: CEU Medieval Studies and Archaeolingua.

Takács, Lajos. 1969. "A régi gazdálkodás emlékei földrajzi neveinkben" (Remnants of ancient agriculture in Hungarian geographical names). *Magyar Nyelvőr* 93: 120–123.

Takács, Lajos. 1976. *Egy irtásfalu földművelése* (Agriculture in a cleared village). Budapest: Akadémiai Kiadó.

Takács, Lajos. 1980. *Irtásgazdálkodásunk emlékei* (Remnants of Hungarian clearance-based agriculture). Budapest: Akadémiai Kiadó.

Takács, Lajos. 1987. *Határjelek, határjárás a feudális kor végén Magyarországon* (Boundary signs, perambulations at the end of the feudal era in Hungary). Budapest: Akadémiai Kiadó.

Takács, Miklós. 2001. "Az Árpád-kori köznépi lakóház kutatása, különös tekintettel az 1990-es évekre" (Research on vernacular architecture in the Árpádian period, with special emphasis on the 1990s). In *Népi építészet a Kárpát-medencében a honfoglalástól a 18. századig* (Vernacular architecture in the Carpathian Basin from the Conquest to the eighteenth century), 7–54. Szentendre and Szolnok: Szabadtéri Néprajzi Múzeum and Jász-Nagykun-Szolnok Megyei Múzeumok Igazgatósága.

Tálasi, István. 1939. "A bakonyi pásztorkodás" (Herding in Bakony). *Ethnographia* 50: 9–37.

Tari, Edit. 2001. "Faépületek az Árpád-kori népi építészetben" (Wooden buildings in the vernacular architecture of the Árpádian period). In *Népi építészet a Kárpát-medencében a honfoglalástól a 18. századig* (Vernacular architecture in the Carpathian Basin from the Conquest to the eighteenth century), 157–186. Szentendre and Szolnok: Szabadtéri Néprajzi Múzeum and Jász-Nagykun-Szolnok Megyei Múzeumok Igazgatósága.

Taylor, Robert E., Austin Long, and Renee S. Kra. 1992. *Radiocarbon after Four Decades*. New York and London: Springer Verlag.

Torma, Andrea. 1996. "Archaeobotanikai maradványok a középkorból" (Archaeobotanical finds from the Middle Ages). *Agrártörténeti Szemle* 38: 317–342.

Torma, István. 1981. "Mittelalterliche Ackerfeld-Spuren im Wald von Tamási (Komitat Tolna)." *Acta Archaeologica Academiae Scientiarum Hungaricae* 33: 245–256.

Torma, István. 1993–1994. "Veszprém megye az újkőkorban és a rézkorban" (Co. Veszprém in the Neolithic and in the Copper Age). *Veszprém Megyei Múzeumok Közleményei* 19/20: 63–68.

Tóth, Melinda, ed. 1975–1978. *Documenta Artis Paulinorum*. 3 vols. Budapest: MTA Művészettörténeti Kutatócsoport.

Török, József and László Legeza. 2001. *Kartauziak* (Carthusians). Budapest: Mikes Kiadó.

Tringli, István. 1999. "Pest megye története 1301–1526" (A history of Co. Pest 1301–1526). PhD diss., Eötvös Loránd University, Budapest.

Vadas, Jenő. 1896. *A selmeczbányai magyar királyi Erdőakadémia története és ismertetője* (The history and description of the Selmecbánya Hungarian royal Academy of Forestry). Budapest: n.p.

Vaday, Andrea, ed. 1999. *Pannonia and Beyond: Studies in Honour of László Barkóczi*. Budapest: Archaeological Institute of the Hungarian Academy of Sciences.

Vaday, Andrea. 2003. "Történeti áttekintés" (Historical overview [of the non-romanized parts of Hungary in the Roman Period]). In *Magyar régészet az ezredfordulón* (Hungarian archaeology at the turn of the millennium), ed. Zsolt Visy, 265–267. Budapest: Nemzeti Kulturális Örökség Minisztériuma and Teleki László Alapítvány.

Vajda, Ernő. 1969. *Öreg fák* (Ancient trees). Budapest: Natura.

Vajkai, Aurél. 1959. *A Bakony néprajza* (The ethnography of Bakony). Budapest: Gondolat.

Végh, András. 2001. "Bakonybél Árpád-kori padlótéglái" (Árpádian period floortiles from Bakonybél). In *Paradisum plantavit: Bencés monostorok a középkori Magyarországon* (Paradisum plantavit: Benedictine monasteries in medieval Hungary), ed. Imre Takács, 124–128. Pannonhalma: Bencés Főapátság.

Vera, Frans W. M. 2000. *Grazing Ecology and Forest History*. Wallingford: CABI Publishing.

Virágos, Gábor. 1997. "Noble Residences in the Middle Ages: Pomáz and its Owners." MA thesis, CEU, Budapest.

Virágos, Gábor. 1999. "The history of the Cyko family in Pomáz." In *The Man of Many Devices, Who Wandered Full Many Ways: Festschrift in Honour of János M. Bak*, ed. Balázs Nagy and Marcell Sebők, 539–549. Budapest: CEU Press.

Visy, Zsolt, ed. 2003a. *Magyar régészet az ezredfordulón* (Hungarian archaeology at the turn of the millennium). Budapest: Nemzeti Kulturális Örökség Minisztériuma and Teleki László Alapítvány.

Visy, Zsolt. 2003b. "A légirégészet Magyarországon" (Aerial archaeology in Hungary). In *Magyar régészet az ezredfordulón* (Hungarian archaeology at the turn of the millennium), ed. Zsolt Visy, 25–28. Budapest: Nemzeti Kulturális Örökség Minisztériuma and Teleki László Alapítvány.

Volf, György, ed. 1876. *Érdy Codex*. Magyar Nyelvemlékek 4–5. Budapest: MTA.

Vörös, István. 2000. "Adatok az Árpád-kori állattartás történetéhez" (Regarding husbandry in the Árpádian period). In *A középkori magyar agrárium* (Hungarian agriculture in the Middle Ages), ed. Lívia Bende and Gábor Lőrinczy, 71–120. Ópusztaszer: Ópusztaszeri Nemzeti Történeti Emlékpark.

Waitz, Georg, ed. 1878. *Annales Hildesheimienses*. Hannover: Hahn.

Wallner, Ernő. 1937. "A Bakony." *Pannónia* 3: 46–147.

Wallner, Ernő. 1941. "A Bakony erdőtakarójának átalakulása a XVIII. század végéig" (Changes in the woodland of Bakony until the end of the eighteenth century). *Földrajzi Közlemények* 69: 1–29.

Wallner, Ernő. 1942. "A Bakony erdőtakarójának pusztulása a XIX. században" (Destruction of woodland in Bakony in the nineteenth century). *Földrajzi Közlemények* 70: 32–42.

Wallner, Ernő. 1943. "A Bakony erdőtakarójának jelen képe" (Current woodland in Bakony). *Földrajzi Közlemények* 71: 260–277.

Warde, Paul. 2000. "The Ecology of Wood Use in Early Modern Württemberg ca. 1450–1650." PhD diss., University of Cambridge.

Wefer, Gerold, Wolfgang H. Berger, Karl-Ernst Behre, and Eystein Jansen. 2002. *Climate Development and the History of the North Atlantic Realm*. Heidelberg and Berlin: Springer.

Wenzel, Gusztáv, ed. 1860–1874. *Árpádkori Új Okmánytár / Codex Diplomaticus Arpadianus Continuatus*. 12 vols. Pest and Budapest: n.p.

Werbőczy, István. 1897. *Tripartitum opus juris consuetudinarii inclyti regni Hungariae per magistrum Stephanum de Werbewcz personalis praesentiae regiae majestatis locum tenentem accuratissime editum*. Budapest: Franklin Társulat. Reprint 1989.

Wertner, Mór. 1897. "Az Árpádkori megyei tisztviselők" (The county-officials of the Árpádian period). *Történelmi Tár*: 439–86, 652–79.

Wertner, Mór. 1902. "A Losonczi Bánfi család őse" (The ancestor of the Losonczi Bánfi kindred). *Erdélyi Múzeum* 19: 261–263.

Wickham, Chris. 1994. "European Forests in the Early Middle Ages: Landscape and Land Clearance." In *Land and Power: Studies in Italian and European Social History, 400–1200*, 155–199. London: British School at Rome.

Willis, K. J., M. Braun, P. Sümegi, and A. Tóth. 1997. "Does Soil Change Cause Vegetation Change or Vice Versa? A Temporal Perspective from Hungary." *Ecology* 78: 740–750.

Willis, K. J., P. Sümegi, M. Braun, K. D. Bennett, and A. Tóth. 1998. "Prehistoric Land Degradation in Hungary: Who, How and Why?" *Antiquity* 72: 101–113.

Winiwarter, Verena. 1999. "Landscape Elements in the Late Medieval Village: Can Information on Land-Use Be Derived from Normative Sources?" *Medium Aevum Quotidianum* 41: 22–42.

Young, Charles R. 1979. *The Royal Forests of Medieval England*. Leicester: Leicester University Press.

Zadora-Rio, Elizabeth. 1986. "Au carrefour de l'imaginaire et de l'économie: la forêt médiévale." In *Archéologie de la France rurale de la préhistoire aux temps modernes*, 120–125. Paris: Belin.

Zatykó, Csilla. 1997. "Morphological Study on a Fifteenth-century Village, Csepely." *Acta Archaologica Academiae Scientiarum Hungaricae* 49: 167–193.

Zatykó, Csilla. 2003. "Medieval Villages and Their Landscape: Methods of Reconstruction." In *People and Nature in Historical Perspective*, ed. József Laszlovszky and Péter Szabó, 343–375. Budapest: CEU Medieval Studies and Archaeolingua.

Zolnay, László. 1961. "Esztergomi érsekek palotái" (The palaces of the archbishops of Esztergom). *Művészettörténeti Tanulmányok*: 201–228.

Zolnay, László. 1971. *Vadászatok a régi Magyarországon* (Hunting in old Hungary). Budapest: Natura.

Zolnay, László. 1983. *A középkori Esztergom* (Medieval Esztergom). Budapest: Gondolat.

Zólyomi, Bálint. 1936. "Tízezer év története virágporszemekben" (The history of ten thousand years in pollens). *Természettudományi Közlöny* 68: 504–516.

Zólyomi, Bálint. 1952. "Magyarország növénytakarójának fejlődéstörténete az utolsó jégkorszaktól" (The history of the vegetation of Hungary since the last glacial). *MTA Biológiai Osztályának Közleményei* 1: 491–527.

Zólyomi, Bálint. 1957. "Der Tatarenahorn-Eichen-Lösswald der zonalen Waldsteppe (Acereto tatarici-Quercetum)." *Acta Botanica Academiae Scientiarum Hungaricae* 3: 401–424.

Zólyomi, Bálint. 1969. "Földvárak, sáncok, határmezsgyék és a természetvédelem" (Earthworks, ditches, boundaries and nature protection). *Természet Világa* 100: 550–553.

Zólyomi, Bálint, Imre Máthé, István Précsényi, and Zoltán Szőcs. 1972. "A vegetáció produktivitásának vizsgálata az újszentmargitai IBP mintaterületen" (Research on the productivity of vegetation in the IBP sample area in Újszentmargita). *MTA Biológiai Osztályának Közleményei* 15: 31–43.

Zólyomi, Bálint and István Précsényi. 1985. "Pollenstatistische Analyse der Teichablagerungen des Mittelalterliches Klosters bei Pilisszentkereszt." *Acta Archaeologica Academiae Scientiarum Hungaricae* 37: 153–158.

Zsoldos, Attila. 1998. "Visegrád megye és utódai" (Co. Visegrád and its successors). *Történelmi Szemle* 40: 1–32.

Zsoldos, Attila. 2000a. "A magyar királynék és Veszprém az Árpád-korban" (Hungarian queens and Veszprém in the Árpádian period). In *Válaszúton: pogányság – kereszténység, kelet – nyugat: konferencia a X–XI. század kérdéseiről* (At the crossroads: Paganism – Christianity, East – West: a conference on the questions of the tenth-eleventh centuries), ed. László Kredics, 177–184. Veszprém: MTA Veab.

Zsoldos, Attila. 2000b. "Confinium és marchia: Az Árpád-kori határvédelem néhány intézményéről" (Confinium and marchia: On some institutions for border protection in the Árpádian period). *Századok* 134: 99–116.

Zsoldos, Attila. 2001. "Szent István vármegyéi" (St Stephen's counties). In *Államalapítás, társadalom, művelődés* (Foundation of the kingdom, society, culture), ed. Gyula Kristó, 43–54. Budapest: MTA Történettudományi Intézete.

Zsoldos, Attila. 2002. "Somogy és Visegrád megyék korai története, valamint a 'várelemek spontán expanziója'" (The early history of Counties Somogy and Visegrád, and 'the spontaneous expansion of castle-people'). *Századok* 136: 679–685.

www.ingramcontent.com/pod-product-compliance
Lightning Source LLC
LaVergne TN
LVHW071211210926
848482LV00015B/780